Lecture Notes in Computer Science 1029

Edited by G. Goos, J. Hartmanis and J. van Leeuwen

Advisory Board: W. Brauer   D. Gries   J. Stoer

Springer
Berlin
Heidelberg
New York
Barcelona
Budapest
Hong Kong
London
Milan
Paris
Tokyo

Ed Dawson   Jovan Golić (Eds.)

Cryptography:
Policy and Algorithms

International Conference
Brisbane, Queensland, Australia, July 3-5, 1995
Proceedings

Springer

Series Editors

Gerhard Goos, Karlsruhe University, Germany

Juris Hartmanis, Cornell University, NY, USA

Jan van Leeuwen, Utrecht University, The Netherlands

Volume Editors

Ed Dawson
Jovan Golić
Queensland University of Technology, Information Security Research Centre
GPO Box 2434, Q 4001 Brisbane, Australia

Cataloging-in-Publication data applied for

Die Deutsche Bibliothek - CIP-Einheitsaufnahme

Cryptography : policy and algorithms ; international
conference, Brisbane, Queensland, Australia, July 3 - 5, 1995 ;
proceedings / Ed Dawson ; Jovan Golić (ed.). - Berlin ;
Heidelberg ; New York ; Barcelona ; Budapest ; Hong Kong ;
London ; Milan ; Paris ; Santa Clara ; Singapore ; Tokyo :
Springer, 1996
  (Lecture notes in computer science ; 1029)
  ISBN 3-540-60759-5
NE: Dawson, Ed [Hrsg.]; GT

CR Subject Classification (1991): E.3-4, G.2.1, D.4.6,F.2.1-2, C.2, J.2, K.4-5

1991 Mathematics Subject Classification: 94A60, 11T71, 11Yxx, 68P20,
68Q20, 68Q25

ISBN 3-540-60759-5 Springer-Verlag Berlin Heidelberg New York

© Springer-Verlag Berlin Heidelberg 1996
Printed in Germany

Typesetting: Camera-ready by author
SPIN 10512392   06/3142 – 5 4 3 2 1 0   Printed on acid-free paper

PREFACE

The Cryptography: Policy and Algorithms Conference was held at Queensland University of Technology (QUT) from July 3-5, 1995, in Brisbane, Queensland, Australia. The conference was sponsored by the Information Security Research Centre (ISRC) at QUT in association with the International Association for Cryptologic Research (IACR) and the Distributed Systems Technology Centre (DSTC), an Australian Government Co-operative Research Centre. A total of 91 delegates from 10 countries attended the conference.

We wish to thank all members of the organising committee for their assistance in running a smooth and enjoyable conference. The organising committee consisted of Paul Ashley, Andrew Clark, Ed Dawson, Mark Looi, Lauren Nielsen, Rick Nucifora and Christine Orme.

The aim of the conference was to discuss issues on both policy matters and the related theory and applications of cryptographic algorithms. Over the past few years issues relating to crypto policy have made headline news, particularly those concerned with the rights to privacy of the individual, who may choose to use cryptographic systems to maintain confidentiality, against the needs of legal authorities to conduct wiretapping to help combat crime. This has lead to the development of "key escrow" based cryptography such as the United States' "Clipper Chip", export restrictions on both hardware and software containing cryptographic algorithms and interfaces, and laws making it an illegal act to use cryptography on public telecommunications channels unless authorised.

At the same time there has been a rapid development in new cryptographic algorithms over the past twenty years. These techniques have been widely publicised in the open literature and efficient implementations of many of these algorithms are readily available (at least in software). However, there is a continuing need to design new cipher algorithms. The advent of high speed computers has led to the development of efficient methods of attacking encryption algorithms. At the same time there has been a revolution in the development of computer networks. There is movement towards a "global village" based on electronic commerce. Such networks require secure systems, otherwise many of the basic institutions which are the lifeblood of modern society are at risk. All of these different issues were discussed in depth at the conference.

The conference consisted of both invited talks and refereed papers. All together there were seven invited papers, 34 refereed papers and one poster paper. The invited speakers to the conference were Willis Ware (keynote speaker, USA), Bill Caelli, Mark Ames, Ross Anderson, Steve Orlowski, Gilles Brassard and Dieter Gollmann. The first five of these speakers were in the area of crypto policy. In addition, all five of these speakers were panel members in an open forum on crypto policy chaired by Bill Caelli entitled "Cryptography: Personal Freedom and Law Enforcement - Is it possible to get agreement?". Another member of this panel was Whit Diffie. All the refereed papers were concerned with either the theory or the application of cryptographic algorithms.

Following the conference all authors were invited to resubmit their papers to be refereed for publication in this volume of the Springer-Verlag series "Lecture Notes in Computer Science". These papers were sent to each member of the program committee to be refereed. From these papers 26 have been selected for this volume consisting of six of the invited papers and 20 of the remaining papers. In addition, Bill Caelli has provided an overview paper of the open forum.

It should be noted that three of the papers presented at the conference were not published in this volume since they have been accepted for publication elsewhere, namely:

(i)   G. Brassard, *"Quantum Computing: The End of Classical Cryptography?"* to appear in Vol. 1000 Lecture Notes in Computer Science.

(ii) E. Dawson and L. Nielsen, *"Automated Reproduction of Plaintext Strings From Their XOR Combination"*, to appear in Cryptologia.

(iii) J. Seberry, X-M. Zhang and Y. Zhang, *"Relating Nonlinearity to Propagation Characteristics"*, Journal of Universal Computer Science, Vol.1, Number 2, pp.136-150, 1995.

We would like to thank all members of the program committee for their help in selecting the papers for this volume. The program committee consisted of Mark Ames, Ross Anderson, Bill Caelli, Ed Dawson, Jovan Golić, Dieter Gollmann, Luke O'Connor and Josef Pieprzyk.

Brisbane, Queensland, Australia   Ed Dawson
November 1995   Jovan Golić

CRYPTOGRAPHY: POLICY AND ALGORITHMS CONFERENCE '95

Queensland University of Technology (QUT)
Brisbane, Australia, July 3-5, 1995

Sponsored by
Information Security Research Centre
in association with
International Association for Cryptologic Research
and
Distributed Systems Technology Centre, Brisbane

Conference Chair:
Ed Dawson (QUT)

Organising Committee:
Paul Ashley (QUT)
Andrew Clark (QUT)
Ed Dawson (QUT)
Mark Looi (QUT)
Lauren Nielsen (QUT)
Rick Nucifora (QUT)
Christine Orme (QUT)

Program Committee:
Ed Dawson (QUT)
Mark Ames (Telstra, Australia)
Ross Anderson (Cambridge University, UK)
Bill Caelli (QUT)
Jovan Golić (QUT)
Dieter Gollmann (Royal Holloway, UK)
Luke O'Connor (DSTC, Australia)
Josef Pieprzyk (U of Wollongong, Australia)

CONTENTS

Cryptographic Policy

Key Escrow and Secret Sharing

Block and Stream Ciphers

Authentication Techniques, Public Key Protocols and Hashing Techniques

Smart Cards and Cryptography

Applications of Cryptography

Open Forum - Cryptography:
Personal Freedom and Law Enforcement
Is it Possible to Get Agreement?

Chaired and Summarized by:
Professor William J Caelli

Summary of Panellist Presentations and
Question / Answer Session

Panellists:

Professor William J Caelli   School of Data Communications
Queensland University of Technology
Brisbane, Australia

Mr Mark Ames   Telstra Corporation
Australia

Dr Ross Anderson   Cambridge University
UK

Dr Willis Ware   RAND Corporation
Santa Monica, USA

Mr Steve Orlowski   Attorney-General's Department
Australia

Dr Whitfield Diffie   Sun Microsystems
USA

1 Panellist Presentations

Prof Bill Caelli:

The key theme is the approach which must be taken by information technology (IT) and telecommunications professionals as this particular topic comes to the forefront politically, technically and in business. To illustrate the problem, I pose the question:

"Will it become in the 1990s, and possibly into the 21st century, an illicit act for a computer professional to incorporate cryptographic technology into, say, some software system or application for either private use or for the use of his or her employer to maintain privacy of information flows ?"

This is a realistic question. It is important to IT professionals, given that throughout the world, the various professional associations such as the Australian Computer Society here in Australia, the British Computer Society, the A.C.M. and the IEEE Computer Society in the U.S.A. and other groups all have codes of ethics and conduct that talk quite specifically about professional responsibility.

This responsibility relates to privacy. It relates to the incorporation of confidentiality mechanisms and services, usually via cryptographic technologies, specifically into increasingly integrated computer and telecommunications systems. In other words I am asking "Is it likely that we should see, enacted, appropriate legislation that covers such incorporation?" Personally, I feel that the answer is "No !". I believe there will be agreement or at least a compromise. But will there be a "meeting of the minds", as the Asian term goes. In other words, will cooperation between IT professionals and law enforcement agencies in this area be one of complete accord. That, I think, is most dubious.

There are other related questions as follows :

"Will the supply and sale of computer programs over the counter, that include cryptography for privacy services, likewise become illegal - possibly a civil or even a criminal offence or, at least, be in some way subject to controls by state, federal or local government?"

Well, in some ways, that already exists here in Australia. Indeed imported products subject to an export licence from a foreign government might indeed be subject to at least registration in this country.

"Will the manufacture and sale of cryptographic equipment and systems be illegal, with only government agencies or approved organisations authorised to supply such requirements ?"

We have seen that trend occurring in some countries.

"Will the importation and export of cryptographic systems, hardware and software, be banned in Australia or the nations of the O.E.C.D., if not already?"

These questions all fall into the realm of meeting an agreement. Having met an agreement, however, precisely how will that agreement be implemented?

Will the possession of cryptographic hardware or software systems not supplied or approved by government be an illegal act? Will academic and industrial research into cryptography be a banned, or become at least an officially discouraged, activity, with only approved government organisations encouraged and promoted or even financed to work in this particular area?

These are all questions that flow from the major question "Is it possible to get agreement?"

Mr Mark Ames:

Tomorrow marks the 219th anniversary of the signing of the United States Declaration of Independence. That was a seminal event, and I believe it started with words like : "We recognise certain inalienable rights to life, liberty and the pursuit of happiness."

I believe there is some mention that "men" are entitled to this, and certainly at that time, I think that that was the point of view. Some who interpret the U.S. constitution would say that it wasn't all "men" who were referred to but rather that it was "men in positions of power". Now the U.S. Declaration of Independence and the Constitution stand out as critical documents of the "Age of Enlightenment", a period of social change about two hundred years ago. I believe that what we are talking about today goes back at least that far.

To say that people have an inalienable right to life, liberty and the pursuit of happiness is very noble. The United States today, of course, is one of the few developed industrial nations that executes prisoners; it has the highest incarceration rate in the world, I believe, and the pursuit of happiness tends to be "whatever turns you on". This is not an anti-American comment. There are certainly lots of concerns in that regard from people within the United States.

The real fight that the U.S. Declaration of Independence led to was "Who controls the flow of resources in terms of taxation?" "Taxation without representation is tyranny" is one of the themes of the time. In the United States, it came to the point where the British could no longer exercise their control over taxation.

Thus, what we are talking about is the problem of who can exercise control over our ability to hold our information private and to ourselves, as we see fit. Now what was happening then, near the end of the Age of Enlightenment, when the U.S. Declaration of Independence was signed, was a major change in concepts of social relationships and the structure of power and, indeed, in the views of what were deemed to be human rights. What did they mean?

Now I believe that we are going through another similar paradigm shift at the moment; one that is not yet fully realised. So, we need to bear in mind the question of just how much can legislation restrict the use of cryptography? How much can legislation say "You do not have the right to the privacy of your communications." Certainly we know that anyone who wants to can have privacy in their communications through the use of publicly available software. It really comes down to having the information; not military resources, not political power, but having the information and knowing how to apply the technology.

By and large, most of the regulations that are being imposed have an impact on so-called "law-abiding" citizens. This may be an "Orwellian big-brother" move; perhaps not intentional. But we have to bear in mind that any terrorist, drug pedlar, paedophile, enemy of the state, or anyone who wants to use cryptography, can and will. Corporations and individuals who want to abide by the laws are subject to having their communications listened into and eavesdropped. Now there is often an argument: "Trust us. We are the government. We have laws in place. We will operate via due process." There has been too much history since 1766 where due process has not occurred. This is a major concern.

There are two points, then, that I would like to make.

One is: "Can governments really legislate against cryptographically, effectively?" And, secondly, if they do legislate, who are they going to impact? Are they going to impact people who are up to "no good"? I believe that people who are up to "no good" can get up to "no good" regardless.

Do we need to have a different view of what rights people have? Do we need to have a different view of what rights the state has to legislate?

These are the major issues.

Dr Ross Anderson:

I simply don't believe that there is much relationship between cryptography, privacy and law enforcement or between any pair of these three topics. Cryptography is almost always fielded nowadays for the purpose of securing transaction processing. Whether it is the token you put in your electricity meter, or the pin that you put into a cash machine, or the card that you put into a GSM telephone, its primary purpose is to prevent fraud rather than to secure privacy. As for privacy, the main threats that we see have got nothing whatsoever to do with crypto because they don't concern external opponents which crypto can shut out; they concern internal opponents.

One example is the new national health service network in the U.K. At present most doctors in Britain have got a PC on which they keep a record of what is wrong with you; what they have prescribed for you; what your address is and so on. There is also some billing information so that they can get paid by the health service. The health service is now putting in a computer network in order to co-ordinate all this billing information, and they have got the idea they can sell this to doctors by saying "well this will help you to give better patient care ?". If somebody turns up in your surgery in Bournemouth and their normal doctor is in Cambridge and, well, they complain of abdominal pain, which could be any of hundreds of different things, then you simply go into the computer and you fetch the records from Cambridge. You say "Aha, the irritable bowel syndrome three years ago responded to Valium, or whatever, so we will try that again." The problem is that once you give many people access to so much information, you create a target for privacy violations.

Paper based files, for example, aren't particularly secure because I can walk into a hospital while wearing a white coat and I can take a file out of the patient notes trolley. But so long as there are only twenty files in the note trolley, nobody is going to try to do that very often.

Once, however, I can go into a computer system and I have got the choice of the files on fifty or sixty million people then that becomes an irresistible target. And once you have got fifty thousand or a hundred thousand doctors in a position to do this, it is inevitable that some doctors will sell records for money. If they don't, then there will be nurses or receptionists who will sell them and suddenly there is no privacy any more. This is the real threat that we are trying to fight in the U.K. nowadays. As I have said, it has got little to do with crypto. Crypto may be part of the solution especially if the government rules out all other alternatives.

As for law enforcement, I don't believe that law enforcement agencies need cryptography to be controlled. The amount of information which police forces gain from wiretaps is, in many cases, very very low. We all get a strong feeling, when we consider the arguments for control by law enforcement agencies that these are really arguments that are being put forward by national signals intelligence agencies. They are worried that their ability to tap third- world military and diplomatic communications will be eroded once there are more and more cryptography based products on the market and more and more skilled people who have got some means of putting together cryptographic protection schemes or even of performing attacks.

This, I believe, lies at the heart of the whole debate. When it is looked at in these terms, what we actually see is that the national signals intelligence agencies are being massively incompetent. Why? Because they should have kept quiet, kept their heads down and done nothing.

The U.S.A.'s "Clipper" initiative has sensitised tens, perhaps hundreds of thousands of Americans and other people around the world to the fact that crypto exists; to the fact that you can secure messages, and to the fact that bad guys can listen in on messages. All of a sudden there are tens and possibly hundreds of thousands of people using PGP. Without the Clipper initiative that wouldn't have happened. We see the same sort of thing in other countries that attempt to control cryptography. Attempts to interfere with research, backfire !

So I would say that, to the question of whether we can get agreement between the various goals of cryptography, privacy and law enforcement, I would say that there is probably no agreement possible but it is not necessary because these are entirely different goals to start off with. In relation to the "Clipper" initiative, and similar attempts to control crypto, which is what this debate is at its heart about, I would say that governments are basically "shooting themselves in the foot". If they wish to retain their capability to eavesdrop on third-world governments for as long as possible, then the best thing that they could reasonably do is to basically back off and let the market take its course.

Dr Willis Ware:

Well, first of all, you should know that I am a pragmatist, and I have learned by now that important questions do not go around unanswered. So my practical answer to the question that has been posed is "of course an agreement will be reached", but the significant issue is how it gets reached. There are things that we must take as given, I believe. One is that technology is going to keep "charging right on". All of the things that we do daily in our professional life we will continue to do. All of the business pressures that want more and more information-based systems will continue. New applications will emerge. Politicians, or slightly differently, federal and local bureaucrats will continue to think of new things that they would like to do with information about people. All of those forces are irresistible and they will continue. There may or may not be legislative action and that issue will depend upon the individual country, but for all practical purposes a decision will be reached.

Now whether it is a decision that we like as a society is an entirely different matter. At best the decision might come out as a "decision not-to-decide". Optimally it would come out a well-considered one supported by public dialogue. If we do nothing or if we attempt to ignore the issue, then we are going to get an end point which is the result of a "drifting" sort of environment. It may or may not be that agreement will evolve or emerge from an orderly dialogue. That is what one has to hope for if the interests of society are to be properly adjudicated vis--vis law enforcement. If we allow the situation to "drift", the end point may very well be one that reflects the preferred position of one of the players and not a considered compromise among the interests of all the players.

Thus, from my point of view, any kind of "well, let's see how it comes out" attitude is much less desirable than a thoroughly orchestrated and carefully conducted public dialogue in the open, visibly, and with participation by any group of "players". Now, why that point of view? From my perspective, that point of view is based on the observations of a "jury" adjudicating in a conflict, and that is really what is going on. There is a conflict between the interests of law enforcement and the citizen. This implies an adjustment in the balance point between society and the law enforcement authorities of a country. Now that balance point has emerged over decades upon decades of experience. It has emerged from endless debates and legislative bodies as to what laws should be on the books. Now we want to modify the balance point, or at least we are talking about modifying it. That is a treacherous concept to consider if it is done blindly and if it is done without thorough public visibility.

So my view is that casually changing the balance point between the society of any country that you care to pick and the law enforcement community of that country, must be supported by careful and thorough public dialogue. A country has to do it slowly, efficiently, carefully and thoroughly with as long as it takes to decide where that compromise should evolve.

Mr Steve Orlowski:

I happen to agree with Willis that the question is not one of is it possible to get agreement, but how we can get agreement ? We are reaching a stage where the world is going through a great change in technology. There is an investment of billions of dollars going into that change and there are billions of people being involved. Up until now the debate has tended to focus on government and corporate needs. It has now come to the stage where we have got to look at user needs and what the community is expecting. That is the role of government; to try and get this community expectation and to try and find an appropriate balance between what the community expects in terms of law enforcement and what the community expects in terms of their privacy when they are using information infrastructures. This has got to be done both at a national and an international level.

The Australian Government and the O.E.C.D. have both been working very heavily on issues relating to the security and privacy and, in fact, intellectual property rights, in global information infrastructures. We are trying to find a compromise. It is the only way it can be done.

You have got to look at the way in which you can compromise in order to meet the requirements of all elements of the society. With that in mind, for example, I mentioned earlier in this conference "differential key escrow" where an encryption scheme could be provided at a lower level which would meet community privacy requirements, but would not cause a problem for law enforcement agencies should they require to get information under the appropriate warrant arrangements. Higher encryption standards, which would be more difficult to break, would be subject to key escrow and that sort of encryption would be mainly reserved for the corporate sector like banks and also for government sectors.

But basically it has all got to come down to a way of finding out how we are going to be able to do it, and that is going to involve compromise between the various parties. It is going to involve talks between the various parties, and that is where the role of government and fora like this and the O.E.C.D. are going to come into it. It is basically a question of "how are we going to get to that agreement ?".

Dr Whitfield Diffie:

In a continuing approach of exploring a "small part of the elephant", I am going to look at two things. In the first place, we hear talk about the need, somehow, to control cryptography, and about two societal interests, without much differentiation in most of the debate. Those two interests are law enforcement and national intelligence. Now, I think they really function quite differently. Suppose there were nothing here except a law enforcement issue. Suppose it were true (and I don't know whether it is true, I am not sure how to determine it and I don't find the accumulations of opinions on both sides particularly persuasive), that cryptography presented a clear and present danger to the effective enforcement of law, and that if cryptography were allowed to proliferate, as an individual prerogative, that there would be serious damage to society.

Now, there is general agreement in that direction about various things; most conspicuously nuclear weapons. Essentially nobody holds that nuclear weapons are an individual prerogative. They are at best a societal prerogative, perhaps restricted to certain societies. So suppose that that fact were to be concluded about cryptography - it is a national prerogative; countries get to set the rules for how cryptography will be used within the countries; countries get to negotiate in bilateral treaties about how cryptography can be used between countries. I believe all of that could happen.

Whether it represents an outcome I like or not, I don't think there is any logical inconsistency in that. There have been treaties and international agreements of various kinds about police cooperation for quite some time, and exactly the same thing could happen in the form that is envisioned by some elaborate plans called International Key Escrow and things of that kind.

But, the trouble is, that if you actually inquire into what information there is about the importance of communications intercepts and wiretapping, in particular to law enforcement agencies, you find that the prima facie argument that this is not a very important aspect of law enforcement, is substantial. In the United States there is an official wiretap report put out by the Administrative Office of the Federal Courts, and that report summarises all the legal wiretaps. The year before last there were about 950 of them. Last year it took a big jump. I think it reached an all-time high since this report has been issued. It went up to 1130. Every year in the U.S.A. there are a quarter of a million federal prosecutions. There are some comparable number of state prosecutions. The prima facie argument says "wiretaps are a drop in this hat".

Now we may ask "where is the driving force behind any interest in controlling cryptography ?". Why should the police be wasting their effort on this? And once again the answer, the prima facie answer, comes up rather quickly. Well, who might really be profiting from this? And the likely answer is national intelligence. There seems very little question that national intelligence organisations make extensive use of communications intelligence. Now there really is a big difference between law enforcement and other activities, where essentially it is implicit in the structure of a nation to feel that law enforcement has some rights of control over its own citizens. So nations have a generally unified objective in that direction.

On the other hand, by its very nature, intelligence is a competition among nations. So it is not at all clear how you could get any sort of international agreement about control of cryptography that would serve the interests of the intelligence communities of all nations. I am rather inclined to agree with Ross that what the U.S. has done by trying to steal a march on people with this rather optimistic notion that it could put over the "Clipper chip", is to have stirred up exactly the kind of paranoia in the world that leads people to do security evaluations of their communications systems, which typically leads to somebody else losing sources of intelligence.

So the only closing remark I have is that I am not at all convinced that within individual societies honest people can't be kept from using cryptography. But, if you are really thinking of "playing" in the international arena, you are "swimming upstream" against the flow of technology.

Incidentally, communications intelligence shouldn't be sold short. People have been saying since the end of the second world war that the triumphs of communications intelligence in that war would not be repeated. We know how to build secure cryptographic systems. Soon lots of people will know how to do it. We are just not going to repeat the successes that the western alliance had against the Germans and the Japanese. But communications intelligence has steadily improved over the intervening fifty years by a very simple and fundamental mechanism; namely more information has migrated into vulnerable channels than can be protected. That might continue for quite some time. The availability of cryptography may or may not at any time in near future stem that tide.

But the critical point is, in terms of the feasibility of applying cryptography to communications and whether somebody else likes it or not, the fundamental costs are dropping and the feasibility of doing it now appears fairly clear. Even that doesn't mean somebody won't be successful in "swimming upstream". Look at the attempt to protect copyright in light of the impact of everything from xerox machines to data written over networks or on disks. It is pretty clear that the feasibility of protecting copyright a century ago, and even much more recently, was deeply rooted in the controllability of printing presses. The fact is that the mass production of printed materials dropped the costs dramatically. Well all that has changed, and yet a climate of copyright protection has emerged, and that has succeeded to a remarkable degree.

So let me close by saying I don't know the answer to this question. Will people reach agreement? I actually suspect "probably not". But I think, probably, in the absence of an agreement, one situation or another - availability of cryptography or a stifling of cryptography - will come about and some people will like it and some people won't.

2 Questions and Answers (Summaries)

1. Wiretaps and their Usage by Law Enforcement.

Are reasons for wiretaps given in the U.S.A. wiretap summary
reports ?
How are they used ?

Diffie:
They publish a summary of reasons. That is a little hard to determine from looking at the wiretap report. What you would like to know is how many of them have to do with drug investigations, how many of them have to do with bank robberies, or whatever. I think that information is implicit in the report but it is a bit hard to extract.

Law enforcement seems overall, in presentation of public arguments in this area, to be surprisingly evasive if one assumes that there is a solid argument to be made. Louis Freeh, who is the director of the FBI, testified to the Senate about fifteen months ago and his written testimony lists about 170 convictions based on wiretap information. However, he is very careful always to say "electronic surveillance" which is what the law says. At least one was conspicuously something that had been a room bug, not a wiretap.

The general argument is that wiretaps are used in only a very small number of cases, but these are very very important high-profile drug cases, terrorism cases, organised crime cases. On the other hand, almost no wiretap evidence appears in court in these cases. For example, the highest profile organised crime case in the last several years in the United States was a prosecution in which electronic surveillance figured very heavily. But it was all "bugs". Wiretap evidence didn't figure in the case. Of these approximately thousand legal electronic surveillances per year, about two-thirds are wiretaps. But the other third seem to produce all the evidence that actually emerges in court.

I have been talking to police for the last year or more about how wiretaps are used. I have been working with somebody who is trying to analyse the information that is in the wiretap reports. So far this has been quite a difficult process because, in so far as wiretaps may be used for background investigative purposes, that information, since things don't actually come out in court, is very difficult to unearth. They don't need to release information that, by and large, doesn't actually appear as testimony in court. And so that makes, so to speak, the societal oversight process of discovering whether the police view of these things is an accurate one or not, a difficult problem to undertake.

Ware:

I just wanted to observe that the document that we are talking about is, in the United States, a public document, so it is yours for the asking. Write to the Department of Justice in Washington 20505 and ask for it.

Diffie:

It actually comes from the Administrative Office of the Federal Court.

Orlowski:

I'd also like to make a point that that sort of information is also available in Australia where there is an annual report on telecommunications interception. There has also been a recent review into the long-term cost effectiveness of telecommunications interception. You can get a copy of it from the Attorney-General's Department (Australian Government).

Anderson:

I don't think that the onus is on the civil liberties side to show that wiretaps are not necessary. I think that the onus is on the government side to show that they are. In Britain when you prosecute someone, the police have got to hand over all the information that they have got to the defence, and if they conducted a wiretap under a warrant and they don't tell the defence about it, that is an offence and they can go to jail. It's as simple as that.

2. Privacy, Confidentiality and Health Care.

In relation to health care, there is a problem in that information needs to be disclosed for health care purposes but at the same time such information needs to be kept confidential. Cryptography can be used but when information is shared it must also become available, if needed.

Ames:

I refer to concepts of human rights and human privilege, and that comes down to the social issue which you mentioned. What information do I need to disclose for the benefit of society? That is an ethical and judgemental issue.

In the case of infectious diseases and information to an employer it can be very clear cut. I think the real issue is that someone who is engaged in criminal activity will encrypt their information on their computer disk and we won't be able to prosecute.

There are rules. If an investigation, whether it is a wiretap or not, suggests that you are in possession of pornography or information on drug deals, and that encrypted information is found in your possession, if you refuse to give up that information in clear text you can be held in contempt of court. So it mainly comes down, I believe, not to the use of cryptography but to what the law is entitled to know, and if you are not prepared to file that information there are sanctions against that.

Ware:

I don't know Australia's legal foundation well enough to judge but I do know that in the United States there is no right to privacy built into the constitutional structure. There is a derived right that is based on interpretations of the first, fourth and fourteenth amendments which collectively imply that there is such a right. It goes back to a man named Justice Brandeis and some things that he wrote earlier in this century. The whole privacy action in the United States is very much a derived one that has grown up as a result of amounts of information held primarily by the Federal Government and used for all sorts of reasons.

Anderson:

Well, the sort of problems that are concerning doctors in Britain at the moment are driven, to a large extent, by the situation in the United States of America where doctors, in order to get paid, supply your medical file to insurance companies who are not bound by the Hippocratic Oath and who, for the most part, sell it on to whoever will pay the money for it. Now over half of the Fortune 500 companies in the United States use employee medical records for purposes such as screening of potential employees, decisions on promotion, and so on. Among people who have experienced this in the United States, it is universally felt to be an extremely bad thing.

It is also the case, for example, that in the United States of America it is now impossible to get health insurance if you have ever had psychotherapy. So if you are ever in a plane crash and you feel nervous about it, you had better go to a "quack doctor" rather than to a proper psychiatrist or you will find yourself without medical cover for the rest of your life. There are many many "horror stories" which come up.

3. Cryptography and the Internet.

Do you believe that the spread of the Internet and the increasing use of encryption in e-mail changes the ways in which crimes are planned and the nature of crimes committed and thus crypto will become a more important concern for law enforcement?

Orlowski:

In terms of law enforcement, the majority of telecommunications interception is more an enforcement thing than a preventative thing. It usually occurs after someone has come to notice. So things like planning an offence by e-mail with encryption would probably not make any difference, because the person hasn't come to police attention anyway and is unlikely to have been subject to interception. Thus the encryption in question is not relevant.

Ware:

I think there is an important point to be made about the Internet. It is a strange environment because it is remarkably good at self-policing. Now why it works, I don't know, but for its whole time of existence it's done an outstanding job of policing itself and, generally speaking, taken care of people who misbehave.

Caelli:

The problem is that maybe we are facing a discontinuity now in the use of the Internet for the first time. It is becoming essentially a commercial vehicle and with the advent of EFTPOS "on the Net", "heaven help us". If via ATM (Automatic Teller Machine) transactions, banking transactions and other financial transactions, the Internet moves to what we would normally call a financial transactions network, we are qualitatively and quantitatively shifting what was the original Internet into something that competes, essentially, with banking and commercial business networks.

Now, if that shift can become self-regulatory, I think it would be very interesting. Let's wait and see. I think that over the next six months we are going to face the problem of electronic commerce on the Net, most likely in a totally uncoordinated, uncontrolled and quite disparate way as we have various groups fighting for different standards. We may have that discontinuity problem at the present moment.

Ames:

I'd like to make two comments. I'll first pick you up on the banking matter. When I worked in banking, systems were designed so that they could run over any network. It is not a problem to put an EFTPOS terminal on the end of an Internet server and run it through to a bank. Now if that comes down to an application on a PC with an integrated "smart card" the Internet is just a transmission medium.

This leads me to the second point. When computers became much more widespread, primarily in the eighties, there was a lot of talk about computer crime. Most of it was just ordinary crime using computers. It was fraud. It was invoice fraud and standard things that people do on paper. Computers made it possible to do it quicker and make it, perhaps, harder to detect.

Orlowski:

In terms of law enforcement we are starting to get to the stage where we are moving from "RoboCop" to "CyberCop". We have been starting to talk to law enforcement agencies about how we can use the technology to detect crime on the technology itself, rather than using traditional police methods. We are starting to look at how we are going to use the technology to police things like intellectual property violations and fraud on the network once it is up and running on a large scale.

Caelli:

Interesting. Perhaps we are all about to launch "intelligent agents" out onto the Net.

4. Smart Cards and Traceable Transactions.

> Smart cards offer a means of privacy in commercial transactions through anonymous payment technologies, and other means. They also offer a government traceability for every transaction that is originated using the card. There seems to be two sides to the picture.

Anderson:

There is awful lot of rubbish talked about this. People have said you can't possibly allow anonymous electronic cash because it will facilitate kidnapping, extortion and money laundering. Now if the people who said this had bothered to sit down and read the papers on digital cash they would have seen that current systems provide anonymity for the payer but not for the payee. So kidnapping, extortion, money laundering and so on are unaffected by this technology. They are still as traceable.

Ames:

I would say that, in the banking world at least, there is a different mindset. When people pay cash, out on the street or at the marketplace, the bank has no interest or control and makes no money out of it. However, if a payment is routed through a bank, for whatever purpose, then the bank makes money out of it. A bank is probably not that concerned with tracing the movements of its customers. Lately, however, the concept of "information mining" including topics such as "what is the cash flow", "what are they spending money on", "what products can we target towards them", and so on, has become more of an issue. It is quite possible to trace the movement of someone through the use of their plastic card, whether it's an EFTPOS terminal or an ATM machine, as they move around the country along with the details of the purchases that they make.

Orlowski:

By limiting the amount of money on that smart card you can, in fact, restrict money laundering because people would then have to carry out many transactions over a lengthy period of time .

5. Privacy Legislation.

What advice would the panel give a (State) premier with regard to developing necessary legislation for privacy in the (State) government sector?

Caelli:

This question is particularly pertinent as moves are made to further data integration across Government department systems.

Ames:

You can implement a series of access controls and encrypt files but the critical parameters are the people in the system, procedural controls and recognition of the value of information and what people are prepared to pay for it.

Anderson:

I think you also have to create a culture in government whereby you think about the data protection implications of each new move before you take it. You really have to look before you leap; before you enact programs that, however you administer them, have some kind of highly unpleasant side-effects.

Orlowski:

From my point of view, I have no problem with governments aggregating databases provided that the data is returned only to the people that are entitled to use that data. In other words there has to be no "crossroad", for example, health data does not come into a central database, get aggregated and then go out to, say, the police.

The various fields of data for a particular individual must be clearly identified so that the department that originates the data gets that data back and nothing else. So you have to look at, not so much the aggregation, even though there is a certain minimal weakness there, but the controls over the environment in which it is kept. The main concern is that the data that is collected only goes back to the people that are entitled to have it. You keep the elements separate.

Caelli:

As an aside, Queensland and Western Australia are "fellow" States in size. Just one point I would like to make also concerns "mechanisms". As you move to more integration of information bases and networks, you are "upping the price-tag", you are "upping the ante" in the sense that integrated data is far more valuable and subject to attempts at illicit access.

A major question, then, that may be posed is the use of "commodity-level" software for management of such integrated collections of data. By integration of cross-enterprise data collections the value of the data, in a new cross-referenced form, is markedly increased. My advice to the Premier would be to look at exactly the level of mechanism required to protect the integrity, confidentiality and availability of that integrated system and to set up appropriate legislation regarding reasonable safeguards.

The problem we have here in Australia is that we are dependent, very much, on imported commodity-level software and hardware products that are now being used in such important, integrated Governmental information system environments. This may be totally unsatisfactroy against reasonable risk assessment and management procedures. In the case of commodity systems, in particular, I take this opportunity to state that it is now urgent that we initiate full security testing of such products. After all, we do that with cars !

Ware:

I would like to "tag" onto that point. Australia may have an opportunity to do the privacy issue better than the US has been able to, primarily because you are doing it now instead of 1975. When the US cranked up its laws in 1975 it addressed the kinds of issues like credit reporting agencies, in which the request for information was properly answered by the entire record. So the operational habit has become "give them the entire record, no matter what you are asked". And so we (USA) have got a proliferation of data all over the place that is totally unnecessary, but it is hard to fix because we have got a whole capital city full of ancient hardware and equally ancient software. You (Australia) are probably much more current and contemporary in terms of the hardware and software base.

Caelli:

There is also this other problem, one of "cost pressure". The question may become one of the ability to incorporate our own security technologies into imported software and hardware systems to meet any Government level requirements or legislation covering security of information systems. I have maintained recently that maybe we have to start looking at penalties on those importers who wish to bring a product into this country but then don't give us the "hooks" onto which we can attach our own cryptographic and associated security system. We have to work towards the provision for doing that.

This question cropped up in September last year at a conference in Washington. The question was whether or not the ban on export (from the USA) of details and implementations of software "hooks" for cryptographic systems could be taken out of the current Department of State "ITAR" system, would be possible. This was very strongly rejected by the Department of State.

Ware:

I can give you the current history on that. Netscape built in hooks to allow users to attach their own cryptography and they also had a version with the cryptography already built in. Netscape was quietly advised that if it wished to have the product sold overseas, they would do well to remove the hooks. And they did.

Caelli:

So that at the present moment in Australia, if we wish to incorporate our own cryptographic systems into imported systems from the USA, we have a genuine problem on our hands. And that problem does not appear to be going to go away in a short period of time. Maybe that is an international and national policy challenge for us all.

Ames:

Can I comment on that problem just briefly? A major organisation started rolling out "Lotus Notes" from about 1992. I discussed "Notes" with Ron Rivest's group, who had developed the cryptography, about providing hooks for cryptography or about providing the US domestic version. They stated that there was no way that either of these could be done. As it is, "Notes" comes to Australia with a fairly easy-to-break crypto system.

What this major organisation had to do, when they finally realised what level of information was being passed around using this commodity software, including sensitive strategic planning and contractual information. Now, the whole "Lotus Notes" network in this enterprise is encrypted using line encryptors, encrypting modems; all of this at great cost. Now, certainly, that organisation would rather have the domestic version and save themselves lots of money. It took them a long time to bite the bullet and pay the extra dollars to protect that network, especially when the software vendor said "Yes, there is security. We have got encryption. We have got public key distribution." But, it wasn't quite what management thought it was.

6. "Outsourcing" of Government Information Systems.

In relation to government "outsourcing" to private enterprise to run the government information systems function, the trend is becoming a very popular thing. Any comments from the panel ?

Ware:

The US Privacy Law covers any contractor that is acting on behalf of a function that an agency could do, if it chose to. So, contractors supporting an agency, which I think is generally known as "outsourcing" would come under Federal Law (in the USA).

Orlowski:

The (Australian) Privacy Commissioner has, in fact, issued guidelines for Commonwealth Government agencies' outsourcing in respect to privacy. For the State Governments, it would depend on the State having its own privacy legislation. Without legislation, the Commonwealth hasn't got the power to deal with it.

Ames:

One of the organisations I used to work for was outsourcing a great deal of information on individuals including credit-worthiness and realted data. We had a discussion with the Privacy Commissioner on this. We had bound the outsourcing bureau by non-disclosure agreements and so on. However, there were no legal penalties applicable.

Diffie:

I was just going to remark that, in the US, there seems to be a view that, in certain circumstances, you can get "better" privacy by outsourcing. However,avoid linking the subject to the "Freedom of Information Act".

Anderson:

In the UK there is some concern about the fact that most Health Service data processing has been contracted out to private organisations. There are growing problems that we have detected from the fact that, in private organisations, the auditors' main concern is to be reappointed as auditors ! You can't reasonably expect to get the same independence of mind that you could in the old days from the internal auditors supplied by the Treasury or people like that.

3 Closing Comments

Ames:

One thing that has become quite clear from our discussion tonight and from some of the papers today and, I am sure from others that we will hear over the next couple of days, is that the whole issue of privacy is the key and there is a link, in terms of controlling cryptography or the use of it, to the privacy issue. However, I see these two as quite separate. I think that this fact has come out this evening because in a number of the areas where we see privacy abuses there is very little that cryptography could do to actually assist in controlling the situation. Privacy is very much a social perception. Cryptography is only going to be a tool, an aid, to assist that.

Anderson:

I tend to be slightly pessimistic. Perhaps with an engineering background I tend to assume that things will break, and to look at how one can stop that from occuring. But in this business I think that pessimism is borne out with experience. I think that what is likely to happen is a "lose, lose, lose" situation in that firstly privacy will continue to be eroded by market pressure; by a failure of resoluteness among governments to enact sufficiently tough data protection legislation, at least in most countries.

I think we are also going to see a proliferation of crypto that will in the immediate term seriously erode the SIGINT capability of various government intelligence organisations. I think it is more feasible for governments to stop crypto than it was for them in the 1880s and the 1890s to force a chap with a red flag to walk in front of a motor car.

As for the question of whether or not there will be harmony eventually between the various players, I think that this is unlikely to happen. There will continue to be people who are worried about privacy and resolve this in crypto terms. There will continue to be secret policeman who worry about SIGINT and who resolve their concerns also in terms of access to crypto. So, we are probably going to see a rather bad-tempered debate "fizzling" on for the next ten or twenty years.

Ware:

I am slightly "more optimistic" than Ross. I am not optimistic; I am just less pessimistic. It seems to me in this privacy issue there is a given that has to be accepted. Like it or not, we, the world, have collectively built a very complex society. We enjoy it every day. We like it. And that complex society needs enormous amounts of information about people to function.

It not only needs enormous amounts of information; it collects enormous amounts of information. So, I submit that it is a fact of life that this will continue, "period, amen" ! The name of the game in the privacy issue is, in spite of this, how do we erect a framework that provides preventative measures so that the individual is not worse off. That is a tough question.

Orlowski:

In the fifties and sixties (information) technology really raced away from us and we spent the next thirty or forty years trying to catch up, particularly with security. At this stage we are identifying these issues at an early stage of the development of the next level of technology, and I think that is helpful. We are getting into debates between individuals and between countries on what is happening. I think that is an optimistic sign and so I am more optimistic about what the future will hold. We are addressing the issues and we can find solutions.

Diffie:

I think actually that Ross is rather optimistic from the point of view of those of us who are making a living conducting this acrimonious debate. But let me just make one remark to put things, as I see it, in perspective.

Whatever impact cryptography has on any notion of national security or national sovereignty today seems to me small by comparison with the impact that the rise of international telecommunications had on those notions in the nineteenth century. You notice that nation states survived the rise of international telecommunications which, for the first time, provided channels by which news, propaganda, information and doctrine could cross national borders without having to confront the border police or the customs authorities. And I think that nation states are very likely to survive the onslaught of cryptography.

Caelli:

I will finish off this evening with a little quote from a speech that I found recently. This speech starts off by posing a question, as follows :

"What belongs to the citizen alone. What belongs to society. Those, at bottom, are the questions we face - timeless questions of the nature and place and destiny of man. All of us together must seek the answers, answers which make the force of technology, the wealth of great repositories of information, the servant of the individual, to run his cities, to cure his diseases, to lighten his darkness with education and, above all, to confirm and strengthen him in freedom. Moreover, I believe we in industry must offer to share every bit of specialised knowledge we have with the users of our machines, the men who set their purposes in a determination to help secure progress and privacy, both together. For the problem of privacy, in the end, is nothing more and nothing less than the root problem of the relation of each one of us to our fellow men."

That quotation, you might be interested, is from a speech by Thomas J. Watson, Jnr. of IBM, on the fifth of April, 1968, almost thirty years ago.

Privacy and Security Policy Choices in an NII Environment

Willis Ware*

RAND Corporation
Santa Monica, USA

INTRODUCTION[1]

The United States is actively promoting a concept nicknamed the National Information Infrastructure (NII). While the NII does not yet have an overall design and many conceptual details have yet to be resolved, it approximates what Singapore indicates that it is implementing, except on a vastly larger scale. If one imagines that all the information and entertainment services that exist throughout U.S. society and its institutions were to be coalesced into one overall entity, the result would equate to what the NII is intended to be. Other countries, including Australia, undoubtedly are also planning such a "national level information entity"; but for expedience in this discussion, the acronym NII will be used because it is widely known and generally understood, both as a specific undertaking in the United States and as a categoric term for similar actions elsewhere.

This paper will discuss the author's views about the probable architecture of a particular segment of the NII, and the policies for security and privacy that are likely to be required. The segment in question is that which deals with computer-based services and data.

In the United States, the NII[2] is anticipated to have other parts, notably entertain-

*This invited paper was presented at the Policy and Algorithms Cryptography Conference, Queensland University of Technology, Brisbane, Australia; July 4-6, 1995. It is to be published in the proceedings of the conference by Springer-Verlag.

[1] This paper addresses public policy issues which are far from being resolved. Following its presentation in July 1995, other events have occurred which are not included in the discussion. For example, the United States proposed a new cryptographic policy position, nicknamed by some as Clipper-II or Son-of-Clipper, whose details and implications have yet to play out as of the date on which this final manuscript was submitted for publication.

[2] The NII - National Information Infrastructure - as it is called in the United States is the culmination of several events. In the beginning the ARPAnet was created to interconnect the computer systems of the research community of the United States, notably the academic and near-academic organizations that were

ment. This part is to include many video services such as television, on-line shopping, probably audio, video/movies on demand, and possibly even telephony. This paper will exclude discussion of all such delivery systems, in part because it is not of interest to the present conference and in part because the industries involved in the respective parts have so far been disjoint and have approached many problems, including technical ones, quite differently. Moreover, their policy needs will probably be somewhat different, especially in the short term[3]. It is not necessary to examine the NII in great depth in order to give a first order discussion of relevant security and privacy aspects; and hence, related policy issues. Indeed, the discussion need not address other components of the information infrastructure for the present limited purpose.

The discussion also will include views about cryptography and its role in the data aspects of the NII and views about privacy. Again, neither will be in depth but intended only to introduce the issues and make an initial pass at structuring their policy content.

TERMINOLOGY

Of direct interest to the present discussion are three terms: confidentiality, security, and privacy. Integrity is also of relevance as an adjunct to security considerations[4].

> CONFIDENTIALITY (of data) – the assertion about a body, or category, of data that it is sensitive and must be protected.

"Protection" means not only safeguarding it against destruction or unauthorized change, but also limiting access only to authorized consumers or users. "Authorized users" may be established by law, by regulation, by professional custom, by organizational policy, simply by established historical uses, by operational needs or by agree-

---

deeply involved in computer science and the use of computers for scientific and engineering research. Its success led to the creation of the Computer Science Net which was intended to serve the specialized academic community. Eventually the two blended together and as connectivity among the communities grew internationally, the whole ensemble came to be called Internet. Roughly concurrently, the U.S. government also supported a High Performance Computing initiative which was intended to link the computer-intensive users of the country, in particular the organizations engaged in innovative exploitation of the largest available computers; e.g., weather prediction, theoretical chemistry. nuclear physics, bio-genetics.

The election of a new administration by the United States in 1992 led to a call for "reinventing government" which implied that federal agencies were to review and reconsider how business had been conducted, and at the same time, intensively embrace contemporary computer and communication technology, especially as reflected in the Internet achievements. There was a concomitant drive to "informatize" the federal government, and the phrase National Information Infrastructure came into existence as an advocacy focus. Along with it came such new media phrases as "information super-highway".

[3] As of the date of this paper, various mergers between corporations have begun to occur or are under discussion. In general, such mergers combine a carrier (e.g., television network, one or more cable-TV companies) and a content source (e.g., book publishers, media organizations, video/movie producers). Such actions will partially negate the statement made here, namely, that the content industry and the carrier industry are disjoint. Also, there are regulatory decisions that will open the local telephone-service market to competition with the result that there may be corporate combinations involving both the traditional telephone companies and cable operators.

[4] This section is based on the author's paper "A Brief Essay on the Terminology of Security", May 4 1995; it was circulated privately but has not been published.

ment among the members of some organized community; for example, insurance companies in the case of the Medical Insurance Bureau[5].

Sometimes a specified body of data is intended (e.g., the accounting records of a specified corporation); but sometimes, especially in law, it is expedient to reference a general category (e.g., tax records). Occasionally the term is applied to other than data (e.g., confidential software) but while it still connotes protection, it suggests the notion of secrecy. Other terms such as "proprietary" can also be used for data, software, processing, even physical location of facilities.

Protection is usually considered to be in the context of exchange of data between two communicants, but avoiding a third-party intrusion. In the case of information systems, the "two parties" in question generally imply a person requesting data from a system; the person is said to be "authorized" to have the data. In a network context, however, the scope must be extended to include other systems, software, or software processes which are authorized to access data.

> SECURITY (information system): [A] the protection of a computer system and/or network together with its data and human resources against a prescribed threat, plus assurance that information from it is delivered only to authorized recipients. [B] The totality of safeguards that give assurance (1) a defined system (2) operating in a defined operational environment, (3) serving a defined set of users, (4) containing prescribed data and operational programs, and (5) having defined network connections will be satisfactorily protected (6) against a defined threat and (7) against defined risks.

The last is a very precise definition that is often framed more simply as:

> SECURITY (information system): The totality of safeguards that protect a system, its resources, and its data against a defined threat and defined risks, assure the confidentiality and integrity of the data, and control access to the system and to its assets.

In general the protective safeguards include appropriate ones in hardware, software, personnel control, object control, physical arrangements, procedures, administration, and management oversight. Their collective intent is to give high assurance that the system, and its resources and information are protected against harm, and that the information and resources are properly accessed and used by authorized users.

It can also be defined in terms of data only, and in particular, to its relevance to privacy.

> SECURITY (data) – the collection of safeguards that assure the confidentiality of data, protect it, control access to it; and hence, enforce the rules of privacy.

It can also be defined in terms of telecommunications.

---

[5] An organization that supports the life insurance industry by collecting data about policy holders and their health status.

SECURITY (communications): The set of protective safeguards that give assurance against some prescribed threat that traffic in transit through a communications system will not be delivered to and/or readable by unauthorized or unintended recipients.

Turning to privacy:

PRIVACY (as a societal concept) – The view that an individual (and by extension, a group of individuals or an institution or all of society) must be able to determine to what extent information about oneself is communicated to or used by other individuals or organizations[6].

The term can also be defined in terms of its implementation.

PRIVACY – the rules that govern how data about people (1) will be collected, (2) will be allowed to be used and by whom, and (3) will be available for review and correction by the data subject.

The authority establishing such rules can derive from law, from regulation, from organizational policy, from tacit agreement between society and holders of personal data, or from combinations of such things. The general intent of United States privacy law is to protect individuals against harm, unwarranted intrusion, or possibly serious damage[7]. If personal information is used for a purpose other than for which initially collected, it is often said to be a "secondary use".

Within government confidentiality usually is a matter of law – a legal concern; and law always reflects implicitly or explicitly a policy statement by some level of government. In the private sector, it can reflect law, sound business practices, or organizational policy.

Privacy is a matter of policy, law, regulation, common agreement, organizational ethics or societal attitudes. It is an information-use issue.

Security is partly technical, partly procedural and partly management arrangements; hence, it must be an issue of concern for any organization that collects, holds, and utilizes data about people.

INTEGRITY – The property that a given body of data, a system, an individual, a network, a message in transit through a network, computer equipment and/or software, etc. or any collection of these entities meets an a priori expectation of quality that is satisfactory and adequate in the circumstance.

The set of attributes that determine "quality" in any circumstance can be general in nature and implied by the context of the discussion; or specific and in terms of some intended usage or application.

Early 1965-definitions of the term were usually cast only in terms of data. For example:

---

[6]This is a definition of the privacy principle, and is limited to the context of data about people. It is the meaning of the word as it is used in such phrases as the Privacy Act of 1974, privacy law, Privacy Protection Study Commission. There are other dimensions of privacy which are not of interest here; e.g., physical privacy or psychological privacy, the general right "to be left alone".

[7]European privacy law often also protects legal persons such as corporations.

Data has integrity if it is free from surprise; i.e., it is what it is expected to be, it has the quality expected of it.

Such a formulation also conveyed the notion of an a priori expectation of quality defined in some way. Such characterizations are limited in application and not broad enough for system-level or policy-level considerations, but they will easily fit under the broad definition by attaching an appropriate characterization of quality in each case.

Note that the definition as given here does not require absolute accuracy, freedom from errors, or complete specification of the entity in question. It only requires that whatever something was thought to be before the fact is indeed what it proves to be after examination. In some limited contexts, integrity is taken to mean assurance against unauthorized change.

Since security safeguards control access, there is a clear interplay between the interests of security and those of integrity. Some safeguards, possibly all in some circumstances, will contribute to both end goals. Technical individuals, attempting to formulate a strict technical foundation for security policy, introduced the term "integrity policy" in the 1970s; but the term was not well defined, is currently in little use, and has not been well understood as to intent.

OVERALL ARCHITECTURE

As a basis for considering policy aspects, consider first what the architecture of an (inter)national network might be. The only extant such network is the Internet whose general structure is shown in Figure 1.

At the core is a telecommunications arrangement built upon and from the various telecommunications industries; e.g., telephone industry, satellite communications industry, independent communication carriers. Around the core and built upon it is a value-added layer that provides packet-switched connectivity. Finally at the outer rim, subscriber systems are attached, although some may have direct connection to the core as a route into the packet layer. The latter communicate with one another via the packet switched connections that are actually accomplished through the telecommunications core. Any one may be a provider of services or information to the overall ensemble or may be a consumer of services or information from others, or in general will be both.

It is argued that the architecture of an NII data network will generally replicate that of the Internet.

There are several reasons to support such a position.

First, we know how to do it. Based on Internet experience in both design and operation, we know how to implement such as architecture;

Second, such an architecture is growable and expandable; and

Importantly, it can be built incrementally by accretion.

INTERNET DATA ARCHITECTURE

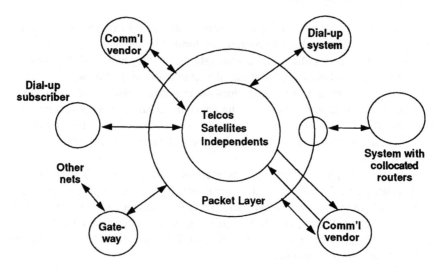

Figure 1

POLICY CONCERNS

Such an Internet-like approach distributes policy issues manageably, as will develop in later discussion. In the context of this paper, the policy concerns of interest are the related ones of:

- Confidentiality – of sensitive data;
- Privacy – of personal data; and
- Security – of all data.

Of these three, the first two are predominantly a provider issue because it will be they who traffic in personal and other information; they who will have to protect such information; they who will have to control its dissemination according to laws, regulations, social custom, industry agreements, business practice and business considerations.

It is conceivable that privacy can become an issue for individual end-users depending upon what use-limitations might exist for such material in their hands.

Security, of course, will be everybody's concern.

- Telecommunications must at minimum protect its facilities and its operational integrity;

NII DATA ARCHITECTURE

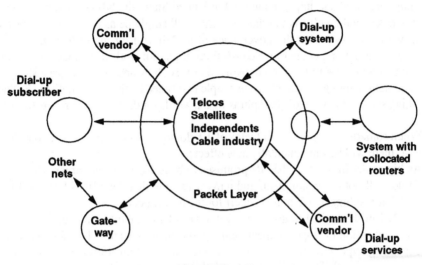

Figure 2

- Valuable data, wherever it resides or whenever it is in transit, must be protected, which implies that access to it must be controlled; and

- Security controls, as that phrase is known in the context of computer security and network security, will be a prerequisite to accommodating data sensitivity obligations

NII ARCHITECTURE

Thus, a probable architecture for the data part of NII networks is suggested in Figure 2.

- There will be a telecommunications core with service providers as suggested;

- Surrounding it will be a packet-switching value-added layer; and

- At the outer rim will be the subscribers.

It is crucial to observe that, as with the Internet, a subscriber to the NII can be a consumer of services or data supplied by it, or a provider of services and data to it, or a subscriber can be both.

In reality, the situation will probably not be as simplistic as sketched. For example, telecommunication carriers might offer packet services, packet routers might be collocated with host subscribers, or dial-up connections might connect directly to the core, or alternatively, to one of the subscriber systems.

NII VS. INTERNET

As being discussed and implemented in the United State, the NII is obviously a huge effort. Corresponding efforts in other countries will surely be as complex. At the moment, we don't know what the governance of an NII will, or might have to, be; but the Internet model of a cooperative consortium might not work. It is not clear whether federal governments will need to be involved or if so, how and on what aspects.

As a caution, though, remember a principle first stated by the mathematician Richard Hamming some 30 years ago[8]. Paraphrasing his words for the present purpose, he noted that:

When something changes by a factor of 10, do not expect a simple extrapolation of the past; there will be fundamentally new effects.

As big as the Internet is, the NII promises to be at least 10-fold larger. Hence, we would do well not to be simplistic and naively assume that the NII will equate to the Internet, but just bigger. We must be alert for the new aspects.

The NII in the United States is not yet a well-defined entity. In some ways, it as largely a collection of hopes, expectations, good intentions and advocacy. However, there is strong impetus to make it go and there is a government effort to examine related policy issues. Nonetheless, policy issues relating to security and privacy have not, as of this writing, been completely thought through or resolved.

NETWORK SECURITY

There is a lot of security technology presently available, both technical safeguards and procedural and administrative ones. There are software and hardware security features; there are locks, fire protections, security procedures, management oversight mechanisms. It is also true, regrettably, that system security is not being attended as thoroughly as it should be; organizations seemingly have not taken the matter seriously. System and network security as operational goals have not diffused through the information infrastructure of today.

There is one place for which some technology development is required; namely, in the network context, especially network with a capital "N" as would be implied by the Internet with all of its worldwide links.

Since arbitrary connectivity among systems, among subscribers, among purveyors and consumers is the way of the future, means are needed for inter-system identification. Inter-system data flow controls must be innovated; the operational integrity of the whole ensemble must be addressed. In spite of such gaps in security technology, at the moment the dominant and controlling aspect is the lack of policy guidance.

Thus, with ample technology available at least for a starter, the security matter becomes one of motivation and policy. Hence, the emphasis in this discussion is on the policy dimension of security.

---

[8]"Impact of Computers", R.W. Hamming, American Mathematical Monthly, vol. 72, no. 2, Part 2, pp. 1-7, February 1965.

SECURITY POLICY

The first order of business would seem to be a statement of an overall security policy architecture.

We would expect it to do many things:

a. To offset the threat;

b. To assure controlled and proper distribution and dissemination of data;

c. To assure dependable operation of the ensemble;

d. To protect a payments mechanism for use of intellectual property; and

e. To support, possibly help to implement, social policies.

This list is by no means complete; a complete top-level security policy statement would undoubtedly be much longer.

RESPONSIBILITY

Just as it has been suggested that the data segment of the NII will replicate the general architectural image of the Internet, so it is also argued that the responsibility for daily operation and overall governance of the NII will replicate the distributed nature of such obligations in the Internet.

There many reasons for such a position. Doing so,

a. Will facilitate our getting to the end-goal;

b. Will help accommodate, even take advantage of, present institutions, their mechanisms, and societal behaviors and attitudes;

c. Will facilitate drafting of whatever law and/or regulations prove to be necessary; and

d. Will facilitate international discussions which will be essential.

Distributing responsibility can cut the pie into slices that we have a chance of dealing with individually, instead of all at once.

If responsibility is going to be distributed, there will need to be boundaries between parts of the overall NII, so to speak, the dividing lines between the slices of pie. The importance of this view is that such boundaries are convenient places at which to:

a. Assign legal and operational obligations;

b. Assign oversight responsibilities;

c. Structure the interface policies;

d. Establish technical standards for connection.

Telephone Service

The point can be illustrated with a useful simile, the traditional residential telephone service as it has evolved in the United States.

At one time, the service boundary stopped at the handset; the operating telephone organizations were responsible for everything. Then, triggered by the well know Carterphone court decision, it changed. The consumer could connect anything he wished, subject to certain rules; henceforth, the service would stop at the plug on the wall.

Subsequently it was changed again, and now service stops at the end of the service drop to the building. The subscriber can do whatever he pleases inside the residence – again subject to rules. Whatever he does, it is his responsibility to maintain it, to repair it, to upgrade it, or to modify it.

Thus, in today's telephony world in the United States, the telephone companies and their subscribers have well-defined and well-understood obligations and responsibilities on the respective sides of the system boundary. The tariff structures reflect and support such a division.

NII BOUNDARIES

With respect to the NII, one obvious place to put a boundary is at the edge of the telecommunications core. If we do so, what are some of the security policy considerations likely to be for telecommunications?

The answer would seem to be quite straightforward. They will generally be the ones that are well understood by the telecommunications industry today. Among them will be an obligation to:

a. Protect physical facilities and personnel;

b. Assure continuity of service;

c. Protect against crackers and phreakers;

d. Protect the privacy aspect of customer records; and

e. Provide en route protection against illicit eavesdropping.

The technical solution to the last is of course encryption, perhaps only on the main communication links, but perhaps also end-to-end from the calling subscriber to the called subscriber.

Telecommunications Security vs. Law Enforcement Interests

It is well know that the security considerations of an NII, or for that matter, any information network, is in head-on collision with law enforcement interests. Law enforcement, when authorized by a court to do so, wants to be able to intercept telephony and data traffic when investigating terrorist, drug, organized crime, or other criminal activity. Encryption, if implemented by traditional technology, is squarely in opposition to such goals.

A SECOND BOUNDARY AND ITS SECURITY CONCERNS

Another boundary could be established at the point of connection to the packetized layer. It would be in the spirit of the way telephone service is now structured and delivered. What might be some of the security policy issues for subscriber systems and networks?

In general, it will be what today is called computer or network security, topics understood in some places by some people but by no means universally understood or implemented.

There will be a few new wrinkles. For example, some security aspects will straddle a boundary and thus, be the joint responsibility of different players in the overall NII. For example, packet routers often reside on the premises of subscriber systems. Thus the security of the packet-switching function will depend in part on the security of the subscriber installation, especially its physical, personnel and procedural safeguards; and, in turn, impact the security and integrity of the entire connectivity function.

SOCIAL POLICY

The United States government and various social rights organizations are insisting that the NII honor a number of social policies. Among them are:

a. equal access for all;

b. non-discriminatory access;

c. proper treatment of intellectual property, including payment for its use.

Hence, security controls, notably those which protect information and control its dissemination, will be obligated to support and implement such positions. Social decisions will require corresponding security policy that leads to the installation of proper technical safeguards, procedures, management oversight, etc.

SOVEREIGN VULNERABILITY

Governments have seemingly not yet thought about the real impact of stitching together all the information assets of a country. When they do, national interests may dictate safeguards for the protection of critical parts of the national information infrastructure. For example, a government might mandate protection of all in-government or government-related sensitive information.

While such interventions would be a proper obligation of a government acting in behalf of the common good, such policy positions will imply both security policy and safeguard consequences.

CRYPTOGRAPHY

With a conceptual foundation now in place, consider two specific issues, each of which will be different for the NII; namely, cryptography and privacy.

Basically, cryptography will support and/or provide such security services as might be required by the threat, by law or regulation, by operational integrity, or by legal obligations levied on business conducted through the NII.

A shopping list of cryptographically related security services that might be required in and for an NII can be divided into three groups.

Comsec Services

The first are the classical comsec services; namely, to protect against eavesdropping on telecommunication links only or alternatively end-to-end, and to protect against traffic analysis – with whom is any one subscriber communicating and on what schedule[9].

Data Protection Services

The second set is protection of data and/or software, a safeguard frequently talked about. Its principle intent is clear. It functions as a superior form of access control, and limits access to protected material to only those individuals, software processes, or other systems that are authorized access and have the necessary cryptographic keys. Hence, it can protect against intrusion for both data and software. For example, it could protect against unauthorized changes by aberrant software that misbehaves and accesses the wrong data.

Business Related Services

The third set and rest of the services are of a different nature. They provide features of various kinds that are important in the conduct of business affairs, and they also bolster the overall security of the system. Collectively, they might be thought of as security functionality; their roles are as follows:

  a. Key distribution provides a trusted mechanism for moving encryption keys from place to place.

  b. Non-repudiation protects against impersonation of individuals, systems, software or business processes.

  c. Digital signatures protect against forgery, usually for an individual but in principle for other things as well; e.g., a system, a process, a host.

  d. Integrity locks protect against unknown accidental or unauthorized modifications of data or programs while in storage or in transit;

---

[9]End-to-end implies that encryption protection begins with the traffic originator and (ideally) at the workstation or terminal in use, and is uninterrupted through and including the corresponding place for the recipient. "Traffic analysis" is an attempt to elicit useful information by examining the origin, destination, schedule, length, frequency, etc. of transmissions but without having access to the substantive content of them.

e. Authentication protects against spoofing or accidental events that might compromise the integrity of processes, of data, or of the system.

f. Secure control protects operational integrity. An interloper can not be allowed to seize an information system and manipulate it to his desire by capturing its control mechanism; for example, down-loading bogus software throughout the system.

g. Selective dissemination restricts a requestor to getting only what is he/she/it is authorized to receive. It is phrased in that unusual way because a "requestor" must be thought of as an individual, some other system, a software process, or combinations of them.

h. The digital notary feature provides trusted time-stamping and archiving of the action.

Cryptography is essential to implement some parts of this functionality; it is one way to implement others.

Should it become important in some application to assure that certain software processes interact only with defined data sets, or to assure that certain processes interact only with one another, or to assure that operational personnel have access to only certain actions from the console, an internal digital signature approach would be one way to meet such a requirement.

THIRD PARTY KEY ACCESS

For this topic, there is a terminology issue to first be addressed; it is captured in the question: "What is escrow?"

The term "escrow", in the context of cryptography, was introduced by the so-called Clipper Initiative in the United States, April 1993. Key escrow is a technical approach whereby a special and usually permanent key (often called the chip key) associated with a hardware-based encryption process has been split into two or more components which are deposited in a secure manner with two or more highly secure escrow agents. With proper court authorization, law enforcement authorities can retrieve such components from the escrow agents, combine them to recreate the chip key which functions as a master key to allow recovery of the session key[10] currently in use; and hence allows decryption of whatever digital traffic has been been protected by the encryption process. The intent of escrowed-keys was to assure that encryption would not thwart authorized actions by law enforcement to conduct wiretaps.[11]

---

[10]"Session" is used in the usual computer-oriented sense; namely, a time period during which a user is connected to and using a computer, or a time period in which two computer systems are interacting. A session key is a cryptographic key which typically prevails for the duration of the session only; commonly the session key changes each time although sometimes it might transcend several sessions and be operative for a specified period of calendar/clock time. In the latter case, the specified time interval is usually called the "crypto period".

[11]In the United States, such wiretaps are authorized by a law called the Wiretap Act of 1968. The procedures for getting a court authorization to conduct a wiretap and the procedures stipulating how it must be conducted are also set forth by the law in detail.

As initially used, escrow implies [1] a chip- or master-key used only for law enforcement access, [2] splitting of the master key into parts held by [3] extra-organization trusted third parties, [4] automatic inclusion of the master key in the message header with every use of the encryption but [5] protected with yet another secret key embedded in the encryption chip which is itself [6] protected against reverse engineering or external access to its details. The concept was oriented to communications intercept and hence, [7] a "law enforcement access field" would be a required part of every transmission. The LEAF contains chip identity which in turn allows law enforcement to solicit components of the master key from escrow agents.

Subsequently the concept of mimicking the hardware-based Clipper approach in software arose, and it has come to be called "commercial key escrow." Some of the details are the same as Clipper; in particular, something is to be held by trusted third parties which, in some proposals, are called Data Recovery Centers. Variations inevitably arose. Some proposals plan to replicate Clipper by depositing split master keys with trusted third parties; law enforcement with proper authorization can acquire the parts of the master key, deduce the actual key used for a given transmission (i.e., the session key) and so decrypt the traffic. Other proposals depart from Clipper and plan to store split actual session keys with the trusted agents.

Inevitably, the concept was extended further to include the possibility that corporations could retain the keys internally in a specially trusted part of the organization. Legal entities, such as corporations, would be subject to law enforcement court order or subpoenas to provide the keys under authorized circumstances. As to be expected, such intra- organizational protective retention of keys came also to be called escrow, sometimes self-escrow. However, the departure from Clipper was even more extensive because there might not be need for splitting such keys, there probably would be no need for a second-tier master-key approach, and there might not need to be a "law enforcement access field" in externally transmitted messages.

At the present writing, the terminology is not clearly established; but in this discussion, the term escrow will be confined to situations in which some key, split or not, master or session, will be stored with extra-organizational trusted parties. For the intra-organizational situation, the term "key backup" might be used in the image of conventional backups of data bases and files; but the term "archive" or "key archive" has been chosen instead, hopefully to avoid confusion.

Turning now to the cryptographically based security functional features, there is an important observation to be made because some governments could conceivably try to insist on some form of key escrow in connection with them. Namely,

> For such features, there must be no third party access to keys or to the processes which are involved.

To put it more strongly, the concept of government-escrowed-key cryptography is inappropriate, and even a privately-escrowed-key approach involving trusted third parties is questionable for these features. Yet, at the same time, organizations will need protection against lost or misused keys.

What is this point all about? Features such as non-repudiation, digital signatures, integrity locks and the digital notary in reality implement and support long standing legal precepts that are essential to the conduct of electronic commerce. The legal validity of any of them must never come into question. From a legal point of view:

- Without unforgeable digital signatures, the concept of a binding contract is seriously weakened;

- Without trusted digitally notarized documents, questions of time precedence might not be legally resolvable;

- Without unbreakable integrity locks, the notion of certified copies of digital documents is empty; and

- Without strong authentication and unquestionable non-repudiation, the analog of registered delivery in postal systems is open to suspicion.

From an overall system security point of view:

- Without digital signatures, there cannot be high assurance that two individuals, two processes, or two systems are communicating with mutually known identified parties;

- Without unbreakable integrity locks, the quality of data cannot be guaranteed with high assurance; and

- Without strong authentication and unquestionable non-repudiation, the identity of the many entities that must communicate with one another in an NII environment cannot be verified with high confidence.

If one permits extra-organizational escrowing of keys for the cryptographically based business-related security services, or if there is any other third party access to such keys:

> There can be a de facto subversion of the intended safeguards that could bring the validity of the associated legal protection into question and/or weaken essential security safeguards in the system.

From the organization's point of view, the protection it expects against the consequences of breaking certain laws or assuring legal conduct of business will have been compromised and potentially weakened. At minimum, there would be a questioning of the validity – integrity – of the protective safeguards, and there might be grounds for legal challenge.

Key Archival

On the other hand, an organization does need to protect itself against lost keys, especially since some documents, and therefore digital signatures, will have lifetimes of decades. In principle, a signature on an electronic document may need to be verified, or the signer identified, 50-60-70-100 years after the event.

Well managed organizations know how to implement backup procedures for data files and software. One could argue that "backing up of encryption keys" is similar in nature, but must be done with much more rigid control and temporal assurance. Hence, "key archival" better connotes the process required.

Offhand, it would seem that any archival protection or supplemental access scheme for the cryptographic keys involved with business-related security services must come under the purview and control of the organization deriving or expecting the concomitant legal and operational protections. On the other hand, one could also argue that an organization can exhibit rogue behavior, or people within it commit criminal acts. Such a possibility suggests the best legal posture might be to preserve the keys in some form of external storage, possibly ultra-secure escrow. But again, there is always the vulnerability associated with major damage or catastrophic destruction of a commercial escrow facility. Indeed, destruction or heavy damage to a public-key certificate organization is sometimes identified as a central point of vulnerability for such systems.

There might be enterprise-unique reasons for external escrow; for example, required by a public auditing firm as a preferred business practice, required by law, seen as providing a "reasonable man behavior" in legal proceedings. On balance, it is not at all clear whether the preferred choice will be escrow or archival, but it is clear that it will be a business decision.

There is a corollary statement of this point; namely, that for some markets and some uses of cryptography, there is no compelling argument for promoting or forcing government escrow-key cryptography on the private sector. While so-called "commercial key escrow" which is software based and uses private fee-for-service third party escrow organizations is certainly applicable for the keys protecting data files and software, it may or may not be relevant to other encryption-based functionality. There are serious legal questions involved with some uses of cryptography in commercial systems.

Rogue Encryption

A related point is that of rogue encryption, the circumstance in which (say) an employee uses either non-standard encryption or a non- authorized key to (say) hold organizational data hostage. In general, an archival or escrow scheme cannot offset such a risk.

On the contrary, it is a management obligation to guarantee with appropriate technical and procedural controls that only valid software is functional on the organization's machines, that there is protection against the introduction of rogue software into the system, and that there is assurance that unauthorized encryption keys cannot be selected arbitrarily. Moreover, there are technical options that management can implement to offset the rogue threat. For example, the system can be designed to maintain shadow data-files that are in lock step with the operational files, but thoroughly isolated from normal work station access and accessible only by special multi-party actions. If there were extremely high certainty that software processes could not be circumvented, it might be feasible to incorporate some form of assured automatic second-tier master-keying for any encryption or key that had not been provided by the hardware/software system itself.

PRIVACY

The other issue which will have different considerations in an NII context is privacy. Remember that in the present context, privacy is the issue of who may be allowed to use personal information and for what purpose(s). As we argued before, privacy is a purveyor and possibly a consumer issue because it is they who will traffic in personal information and use it.

Since it is an information-use issue, security mechanisms for controlling access to and dissemination of data will be paramount. Therefore, appropriate computer and network security is a necessary prerequisite to the assurance of privacy, but the question of "who can do what" with personal information is a policy issue. It might be framed for an organization, for a government, for a consortium of organizations, for a group of countries, as a de facto policy derived from socially accepted custom, or derived from internationally accepted agreements.

Getting such policy statements into place will be tedious and difficult. It is hard enough to do within one country but much harder internationally because of different cultural attitudes, different ethnic customs, different social views, even different national needs for the use of such information.

So far, the United States has chosen to address the privacy issue piecemeal[12]. The country does not have a broad umbrella law nor is there a single federal authority responsible for the issue, no analogue of the federal-level data commissioner prevalent in many European countries.

Unless there is a sharp turn onto a different direction, about all the United States can do is consider privacy concerns by information category and incorporate appropriate privacy restrictions into laws as they come along. Such a posture is not satisfactory in some ways, but the climate for strong umbrella privacy law in the United States continues to be unfavorable.

An obvious and immediate example is the privacy of medical records. Presumably, all law that restructures the health-care delivery system will also incorporate proper confidentiality and privacy safeguards. But, there are other trying examples, such as the aggregation of public records into extensive personal dossiers; or the commingling of public records data with private data bases[13].

---

[12]The best known United States law is the Privacy Act of 1974. Others also deal with privacy but less obviously. Among them are: the Fair Credit Reporting Act, Fair Credit Billing Act, Equal Credit Opportunity Act, Fair Debt Collection Practices Act.

[13]Public records data is that information collected by governments in the United States that has been deemed to be the property of the citizenry and hence, freely accessible to them; e.g., birth records, court records, criminal history records, property ownership records. When such things were kept manually on paper, there was little risk of cross-correlation. Most public-record systems are now computerized and there is some exchange of data among them but extensive correlation by commercial information purveyors with the result that in combination with other databases (e.g., census data, auto ownership data, property tax records, credit-reporting data, sometimes medical records) an extensive dossier can be created about individuals. At the moment, only the original laws identifying such records as public exist; there are no limitations on use except in a few cases; e.g., automobile and driver license records are dissemination-controlled because there have been instances of malicious use.

A New Privacy Principle

There is a new dimension of privacy that needs attention, one brought on largely by networking of systems. In such circumstances, the expectation is that there will be much trafficking and sharing of data among subscribers to the network – an NII.

The governing philosophy in United States privacy law has been the long standing Code of Fair Information Practices[14]. It stipulates (among other things) that the collection of data shall be minimized for the purpose intended. Moreover, the process of collection has always implied "getting the data from the individual." Because of this point of view, there has been, in effect, an unstated assumption that the total content of the record is pertinent to the purpose for which the data was collected. Hence, a request for access to data about an individual has been, and is often still is, commonly serviced by divulging the complete record.

Such a view dates to the early 1970s when the Fair Code was formulated, an era of mainframe computer systems often running in the old fashioned batch mode. Moreover, such systems generally replicated the processes and data structures of the manual recordkeeping systems which they replaced. This particular aspect of privacy has never been reconsidered and adapted to a technology base which has become one of workstations, servers, and extensive networks.

The combining and sharing of databases by federal agencies, together with older hardware/software systems that cannot do fine grained dissemination control, have together gradually weakened the principle of "collect only what is required for the intended purpose". The interconnection of systems and databases has diluted the concept of "collect the data from the individual".

Thus, collection has increasingly included acquisition or transfer of data from other systems and/or databases. In fact, "collection" has come to have a secondary connotation which often reflects sources that are more desirable than the individual; e.g., other data bases or data systems that include additional information, that have been verified for accuracy, and can yield information in a more timely fashion.

The Fair Code of Information Practices, particularly in regard to divulgence of the record and data collection, needs to be extended to match the contemporary data-sharing environment and the network context. It would seem that access to and divulgence of the record should now provide in addition to its original scope:

That the minimum data for the purpose intended shall be provided to either standing or on-demand requests, and that this be the policy whether the requestor be an individual, a system or a software-based process.

---

[14]The Code was first described in the report of a federal study committee established in 1970 to examine the consequences of widespread use of personal data held by United States federal agencies. SEE: "Records, Computers, and the Rights of Citizens," Report of the Secretary's Advisory Committee on Automated Personal Data Systems, U.S. Department of Health, Education and Welfare, July 1973. DHEW Publication (OS)73-94. Contains a good bibliography.

The Code was subsequently incorporated into the Privacy Act of 1974, the law which is commonly thought of as "the privacy legislation" of the U.S. This act created the Privacy Protection Study Commission which examined recordkeeping practices of the private sector over a period of 30 months. SEE: [2] "Personal Privacy in an Information Society," the Report of the Privacy Protection Study Commission, July, 1977. There are also five appendices on specialized topics, including a discussion of how the 1974 Privacy Act had been working.

To put it another way: just because a requestor asks for data about individual X, it does not follow that the entire record on X should be provided. On the contrary, the requestor should get only what is necessary for the intended purpose that stimulated the request, and this principle should pertain whether the request is directly to an individual or to another database/system. To remind the reader, a "requestor" might be an individual, a system, a process in the same or different system. There is an even more subtle aspect; namely, a process might be operating in behalf of an individual (e.g., an on-line inquiry) or it might be operating in behalf of another process or in behalf of another system (e.g., a scheduled automatic update of one data base by another, an inquiry from a remote system for data access in behalf of an on-line inquiry).

An example will indicate what can be ahead in a commercialized environment of the NII.

With regard to the on-demand electronic delivery of video services, a service anticipated to be a prominent feature of the entertainment segment of the NII:

- The purveyors of the video content – typically entertainment producers – need only know how many consumers have looked at each product. They need no personal identifying consumer information although they might legitimately have demographics for market analysis.

- On the other hand, the operators of the system – cable or telephone company or satellite – need only know the price of each item, how long or how often it was played, perhaps when it started, and billing identification. They certainly have no need for knowing the content consumption of each consumer, but they might legitimately have access to aggregated usage data to assist in programming decisions.

In this example, the social privacy policy would acknowledge the difference between content and carriage. The corresponding security policy and technical design problem would be to provide systems with selective dissemination controls coupled to databases with fine-grain structure.

SUMMARY

For national information structures – the "NII" of a country – that are being or will be built, the country will need an overall security policy architecture that:

a. Honors the social policies expected from or levied on the NII;

b. Protects against vulnerabilities of the NII;

c. Guards against perceived threats to the NII;

d. Supports privacy obligations wherever they exist;

e. Assures continuity of NII operation and reliability of its performance;

f. Enforces dissemination control;

g. Enforces a payments mechanism for use of intellectual property; and

h. Enforces access control to NII system assets, data, and other services.

There are probably several additions that will emerge as discussion of these top level policy issues continue.

Such a security policy architecture cannot ignore present societal or institutional mechanisms and established protections – ones already in law and regulation, and already in place and operational.

Quite the opposite, to avoid having to "start all over at square zero", the policy would do well to capitalize on the foundation of established mechanisms, institutions, and protections to facilitate the emergence of national-level information infrastructures.

For similar reasons, a country would do well to implement its NII security policy through various existing mechanisms, organizations, and industrial or societal groups – again to avoid "having to reinvent the wheel".

If an NII becomes what present discussions portend, then every country will sooner or later realize the necessity to protect its sovereign and national interests. This must obviously be done at the government level, and would be a proper intervention.

Such national considerations are ones that could lead to such things as:

a. Mandatory cryptographic protection of specified traffic;

b. Mandatory redundancy in critical parts of infrastructure;

c. A commonly agreed-on statement of threat to guide the security considerations of all subscribers to the NII; and

d. Perhaps, even legal controls on some things such as access restrictions for minors.

Such actions by sovereign powers would certainly impact both the internal security policy of any one country as well as international social and security policy.

The people who have brought the Internet to its present level of maturity may find the future to be quite unlike the past.

EPILOGUE

The vision of a National Information Infrastructure (NII) internal to the many countries of the world and a Global Information Infrastructure (GII) among them is one of the future. Some of the policy aspects raised in this paper are not yet of high or any importance because the corresponding issues have yet to arise. Some of the security safeguards based on cryptography are not yet significant because the corresponding threats, risks or operational requirements have not appeared; but on the other hand, much of the security issue is already pertinent to present levels of networked environments.

While the picture portrayed in this paper is incomplete and hazy in places, so are the details of NIIs and so are those of an international GII. At the topmost federal levels of countries, the security and privacy policy issues have yet to be addressed thoroughly. Both will need careful attention.

Commercial Key Escrow: An Australian Perspective

Professor William J Caelli, FACS, FTICA, MIEEE
Head, School of Data Communications
Faculty of Information Technology
Queensland University of Technology
GPO Box 2434, Brisbane QLD 4001
Australia

Abstract

"Commercial Key Escrow (CKE)", and an earlier "Software Key Escrow (SKE) scheme, have been proposed by Trusted Information Systems Inc. (TIS) in the USA as a possible compromise scheme to meet the demands of commerce and industry for new levels of information security, particularly transaction and message confidentiality in an international and national networked environment, while meeting law enforcement demands for continued effectiveness of telecommunications line-tapping ability. These latter requirements relate to the perceived need by law enforcement agencies to make use of legitimate authorised line-tapping capabilities for the gathering of appropriate intelligence and/or evidence for the purpose of fulfilling perceived roles in the protection of society from criminal activity against the potential case where such line-taps produce intercepts that are encrypted. CKE, involving the incorporation of software based cryptography in computer and network systems with associated key recovery data transmitted during data network activity and provision of "Data Recovery Centres (DRC)", is seen as presenting a new solution to the problems encountered in the USA with the "Clipper" initiative in that country announced in 1993.

This paper examines the CKE/SKE proposals in an Australian and international context and sets the proposal against the more general debate on cryptography, its technology and usage, and public policy. A likely scenario is suggested for Australia involving the incorporation of backup and recovery and network directory services into the encryption scheme and the use of *Australia Post*, and indeed any national post office structure, as an ideal candidate for trials of both the technology and public/business acceptance of this overall structure. More basic principles of "freedom-of-speech" are also raised in conjunction with this overall analysis of a concrete proposal.

" This room is bugged. This telephone is bugged. This is all part of the reports on us. My residence, and John's residence, are hot. So we are making it public. We have already talked to some press. We are going to talk to more press. If nothing happens - nothing happens - thank God.

> *Russ Smith - Commander, Texas Constitutional Militia"*
>> "Background Briefing", Australian Broadcasting Corporation (ABC)
>> Sunday 11 June, 1995

Dictionary Definitions of "Escrow"

United Kingdom:

escrow   b. A deposit held in trust or as security; *in escrow*, phr. used of money, etc, so held

The Oxford English Dictionary, 2nd Edition, Volume V, Dvandva-Follis [OXFO-89]

United States of America:

escrow   1: a deed or bond, money, or a piece of property delivered into the keeping of a third party by one party to a contract or sometimes taken from one party to a contract and put in trust to be returned only upon the performance or fulfilment of some condition of the contract or to insure such performance or fulfilment by some other disposition

escrowee   the one holding an escrow in trust; the depository of an escrow.

Webster's Third New International Dictionary of the English Language - Unabridged Vol. 1 - A to G [WEBS-86]

Australia:

escrow   1.

a contract, deed, bond, or other written agreement deposited with a third person, by whom it is be delivered to the grantee or promisee on the fulfilment of some condition.

The Macquarie Dictionary, 2nd Revised Edition [MACQ-90]

1.   Introduction - Crypto and Public Policy

1.1   The Magnitude of the Problem - "Genie Out of the Bottle"

The "Clipper" announcement of the United States Government in 1993 focussed worldwide attention on public policy related to the science, technology and usage of cryptography. Hoffman [HOFF-95] went further recently and stated that the announcement *"... set off a frenzy of discussions about cryptography policy in the technological community. ..."*. He went on to summarise the situation as follows:

> *"... We still don't have good answers for some of the questions that have been raised. As the Global Information Infrastructure is being built, we are writing portions of the Constitution of Cyberspace..."*

That same attention, interestingly, has not been so focussed for some twenty years. In a United States context a peak of discussion arose in 1975, with the proposal for an encryption standard that would be open and public. This became the famous "DES" cipher and it was intended for non-military usage for sensitive Government information, but soon also became the backbone of security in the banking and finance industry. The 1993 proposal essentially set out the proposition that should encryption be required to provide confidentiality in information and telecommunications systems, then government should be able to decrypt any information transfers through the availability of the necessary encryption keys used. No such discussions were evident in the 1970s.

The "Clipper" proposal employs special cryptographic hardware incorporated into systems. This hardware added the feature that encryption keys, split in two, could be held in "escrow" by two "third parties", specified as Government organisations, and later recovered by law enforcement groups should it prove necessary. The main theme here was one of the need for law enforcement to continue to successfully use "line-tapping" facilities to combat organised crime and so on. The information thus intercepted needed to be in "plain-text" or as close as possible to it.

There are some overwhelming technical and political facts that must first be agreed upon in consideration of cryptographic policy development. These are simple but they have tremendous implications.

Technical Facts

A. *Cryptographic mechanisms are the only known technologies capable of providing security services in a distributed computing and integrated information environment of computers, telecommunications networks and information hosts/providers;*

and

B. *Cryptography is a well known and available technology for information technology (IT) professionals and amateurs as well as users of information systems, alike.*

Political Facts

C.   *(Australia) While Australians may not appear to be concerned with cryptographic matters, their ability to rapidly react to perceived threats to privacy and freedom is well known (as evidenced by the "Australia Card" debate of the late 1980s);*

D.   *Information networks are mission critical, support systems for enterprises, public and private, and are interconnected on an international basis across enterprises and, as such, are under attack;*

E.   *Information technology (IT) professionals are those who will be charged with the actual implementation of the appropriate technology and its incorporation into operational information systems and data networks.*

and

F.   *There is evidence, particularly in the United States, of growing distrust of "big Government" to a level that overrides that of personal privacy concerns.*

Interestingly, the two technical principles may not have been so evident in the IT industry some ten to twenty years ago.

For example, in considering the privacy of applications and associated data files in a single time-shared computer systems with attached "dumb" terminals, other *separation technologies* incorporated into the computer system itself may have been capable of providing the necessary security features. This could include such devices as "memory fence registers", etc that aimed at the prevention of overlap of applications operating on the one computer system. This was common in such machines as IBM's System/360 - 370 (1960s-1970s), DEC's VAX (1970s) and so on. In a distributed computing system with computer-telephone integrated services and with processors interlinked by public telecommunications facilities, this form of protection is not possible. The associated "client-server" style applications operate over computer systems and data networks that may be separated by countries, time-zones and distinct legal jurisdictions and traditions. Information flow is beyond the control of the single computer and IT professional. Cryptography is the only known mechanism that can protect information flows in this environment.

Twenty years ago cryptography was still essentially a military technology. Today, this has totally changed. There are numerous national and international conferences on public research into the area as well as many educational courses on the topic at universities worldwide. It is now a *tool of the IT professional* as are programming, information/enterprise analysis, and so on are.

At the same time the demand for such network based distributed computing applications has "exploded". In the banking and finance industry the concept of the "virtual bank", for example, has grown over the last three years or so to the extent that the concept of all forms of "electronic commerce" on data networks now seems normal rather than futuristic. This has been highlighted recently [HEAD-95] as follows:

> " .... *By 1997, pundits predict that only 44 per cent of transactions will take place in a branch - the remainder will be made through automated teller machine and eftpos networks, home banking and telephone service centres. Australia is facing the first wave of the virtual bank phenomenon: technology 1, tellers 0... .....*
>
> *Chris Bird, assistant general manager of information services at the Commonwealth Bank ... cites the emergence of interactive television, the information superhighway and smart phones as the main drivers behind the movement away from the branches.* "

This means that former well controlled information environments and applications, such as banking, and not just the underlying IT systems themselves, have moved to an insecure environment, that of distributed computer systems and telecommunications networks. Moreover, these new distributed applications in banking link disparate enterprises in a common structure, ie merchants, message switches, telecommunications services providers, banks, customers at home and in the office or even in a mobile environment, and so on, into a common set of electronic services. All these services must be protected, ie confidentiality, integrity and availability must be guaranteed.

At the macro level, it may well be argued that the "wealth of a nation", as manifest in its banking and finance industry, is now becoming dependent upon the security of national and international data networks and distributed, small scale computer systems. Cryptography is the only known, suitable protection mechanism and it has been carefully regulated by national governments. The problem is that the very protection of the information base of a nation now depends upon the efficient and timely integration of encryption services into new IT applications by computer professionals worldwide.

The "political facts" above highlight a further phenomenon outlined by the USA's Admiral Inman, a former Director of the National Security Agency in the USA, in a talk reportedly given at the Massachusetts Institute of Technology (MIT) in November 1994 [LETH-95]. Essentially, while privacy may be of concern to the populace, Inman pointed out that *"... the public cares about and opposes Big Government."* Inman seems to be suggesting that even though the public does care about privacy, it cares *more* about the potential excesses of large Government. In addition, information technology may contribute to this perception since it is seen as making such potential excesses easier to perpetrate.

There are grounds to argue that this same spirit came to the fore in the Australia Card debate, an aborted government sponsored scheme in the late 1980s to mandate personal possession of an identity card for access to many government services. People readily give their credit card details over the phone to make payments for goods and services of all types, even on the Internet for international transactions. The benefit is seen as outweighing the privacy problem and the threat from the purveyor is also perceived as minimal.

This is all exemplified by the "Fly Buys."[1] "frequent-shopper" programme that has been hugely successful in Australia, with reportedly over two million members by May 1995 [LOYA-95]. The public is apparently unconcerned that buying patterns may be followed by merchants or that a data base of private and personal information is maintained by private enterprise on a large and nationwide scale, even though the public does not really know the background or details of the companies operating such a scheme. Indeed the "Fly Buys." brochure gives sweeping powers for change to the operators of the scheme in the following wording and references totally unknown companies:

> " ...'FLY BUYS.' means Loyalty Pacific Pty Ltd (ACN 057 931 334) acting as agent of the FLY BUYS. joint venturers from time to time, who initially are Lachlen Pty Ltd (ACN 063 729 153) Tickoth Pty Ltd (ACN 062 340 103) and Relationship Services Pty Ltd (ACN 062 806 893) (or any replacement agent) or its authorised representatives ..... " [LOYA-95]

This all gives credence to Inman's suggestion. This overall distrust of "big government" by the populace has become a fact of the late 20th century. At the political level the idea that government can be relied upon not to misuse trust placed in it, is simply not tenable in the eyes of the populace. The crypto debate, outlined in this article, thus becomes one of extreme importance to governments and regulators everywhere as well as to the computer professionals intimately involved in the creation of the associated information systems.

1.2. Differing Cultures

Many papers and discussions by IT professionals on cryptographic policy, particularly in the United States, have made much of the seeming inconsistencies and illogicality in cryptographic export policy in that country. In particular, the case of "free export" of cryptographic algorithms published in program "source code" form in *printed books and articles* is contrasted with the apparent dichotomy of a ban on the exact same programs being exported when recorded on a computer readable diskette, tape or the like.

To the IT professional, it is simply a matter of different forms of the same

---

[1] "Fly Buys." is an Australian customer loyalty programme whereby those who do their shopping with certain merchants and service providers collect points that may be accumulated and used to obtain free flights on a particular airline in Australia. People are asked to fill in a detailed form giving personal details and authorisation for the scheme promoters to use information collected for various purposes.

thing. Anyone, it is claimed, may simply use an optical scanner to "read" the printed form of the algorithm back into a computer system and thus make use of it. The printed form of the algorithm in a book may even be in a type-style that makes such OCR (optical character recognition) a simple matter. The situation has been around for some time but again took prominence with the publication of a book by Schneier entitled "Applied Cryptography" [SCHN-93]. The latter part of this book contained computer programs that implement popular cryptographic systems in printed form. The author and the publishing company offered these programs in diskette form to purchasers of the book. These diskettes were not granted export approval but the book was. To the IT professional this decision seems totally ludicrous such as to hold the responsible organisation up to apparently well-deserved ridicule. The case represents again a lack of understanding, on both sides, of the various bases for opinion formation and decision making. It highlights a major difference in *"culture"*.

To the IT professional a computer programme represents the basic "algorithm" under consideration. Its form of expression may be immaterial, eg a computer program expressed in a suitable programming language stored as a file on a disk, a "printout" of the same program on a printed page, a flow diagram of operations performed, a mathematical description of the procedures involved in sequence, and so on. All represent the "program" and the "intellectual property" of concern. The idea that an intangible asset, such as a cryptographic algorithm, may be rendered a controlled export item purely on the basis of its form of expression seems illogical, arbitrary and indefensible to the IT professional, on any engineering or scientific grounds.

However, to the law enforcement and intelligence communities it may be that such is the nature of their profession that such *minutiae become vitally important*. In dealing at the criminological, political, social and psychological levels, common in the law enforcement arena, matters of perception, ready availability, opportunity, motivation and so on all become vital aspects of the situation. It may be that on technical grounds the IT professional is correct. But in criminological terms, the situation could be vastly different. There could indeed be a marked difference between a crypto system existing in a book, in printed paper form, and one implemented and ready for use on a diskette. For example, assuming that a criminal wishes to make use of the printed form of the cipher algorithm then he/she must "arrange" for the technology to be translated into a really useful form, adding further risks of exposure in the chain of illicit activity, unless of course they already possess IT expertise at the appropriate level. *This highlights the urgent need for further criminological and intelligence work to understand the usage of IT in the criminal arena.*

For example, a drug dealer may simply use a radio based pager and a few code words or phrases ("steganography") to operate an illegal scheme. This is rendered simple and automatic with GSM digital mobile telephony incorporating pager/short message transfer services into the voice telephony services. Jockeys may use signals to indicate information to punters in horse racing, as recently alleged in Australia [MARR-95].

Opportunity and ready availability may be far more important than technical competence or excellence, eg access to a cipher system detailed in a technical book. These types of arguments need to be examined by computing professionals on the one side while the need of professionals to create trustworthy systems must be appreciated by law enforcement.

1.3   Distributed Systems/Computer-Telephony Integration (CTI)

Whatever legal, administrative and technical solutions or compromises are found for the situation, they must be practical for a rapidly changing technological base. Encryption technology and any administrative systems, such as "key escrow" discussed later, must be applicable to the distributed computing environment (DCE), computer/telephone integration (CTI), multi-media services, interactive television and radio, and so on. The IT environment is rapidly moving beyond simple data files and programs to encompass the whole information "universe" of an enterprise as well as the home and personal user.

1.4   Criminology and Defence Policy

" America is now faced with two extreme methods of defending liberty, and both are dangerous. The one is armed and organised militias, which do not trust the government; the other is the apparatus of an over-intrusive state which does not trust its citizens. Each method is, in truth, an enemy of the liberty it purports to protect. " [ECON-95].

The "militia" phenomenon in the United States came to the fore following the Oklahoma City bombing of April 1995. In an article in "Time" magazine, 1 May 1995, Wilson [WILS-95] pointed out that:

"... Political support for intelligence work swings like a pendulum. This quickly changing congressional environment, while understandable in its own terms, is not helpful to a law-enforcement agency. ... " [WILS-95]

Wilson's article points to two sets of guidelines, *"one secret and one public"*, that govern the intelligence gathering activities of the United States' Federal Bureau of Investigation, the FBI. The secret set of rules apparently govern penetration of groups thought to be associated with a foreign power while the second, and public set, refer to US domestic groups. Activities referenced in such intelligence gathering operations include "..electronic surveillance..". These policy guidelines must be placed into context and, in turn, understood by the IT professionals involved in the cryptography discussion.

1.5   Commercial Key Escrow (CKE) vs "Clipper"

An emerging proposal to handle the cryptography situation and to *"relieve the tension"* between technologists, business and the law enforcement community is the CKE proposal. "Commercial Key Escrow (CKE)" as devised and outlined by Trusted Information Systems, Inc. (TIS) of the United States, and the main subject of this paper, may be separated into two distinct proposals [WALK-95], as follows:

a.   usage of a scheme which is an outgrowth of on earlier "Software Key Escrow (SKE)" concept and technology, itself a TIS proposal, as an essential part of any cryptographic mechanisms used for confidentiality services in an information system, and

b.   employment of commercial enterprise or "independent" and autonomous government funded organisations as key escrow facilities.

In actual fact the use of software key escrow and associated ciphers and protocols are not absolutely required and hardware forms of the same technology could be used. In a strict sense the CKE proposal from TIS incorporates the essentials of the SKE concept.

It is reasonably assumed that computer and IT product manufacturers will not readily add a new electronic component to their designs for incorporation into every mass-produced product manufactured, eg a "Clipper" or equivalent chip added to every Intel 486/Pentium, IBM/Motorola/Apple Power PC-60X, SUN Sparc or like CPU based "motherboard". As a start, the component itself would need to be exported from the United States to the point of manufacture of the board, eg in South-East Asia, Mexico, etc with associated control problems almost equivalent to the ones to be solved. The alternative of compelling every application incorporating encryption for confidentiality services to make use of a special "Smart Card" or "PCMCIA" card is also not realistic at present since most computers used do not have the appropriate reader/writer units incorporated into them *as standard items*, although this may soon change.

Users have shown marked reluctance to add security related hardware to PCs unless it is for specific industry purposes and is "compulsory", eg in the banking and finance industry, etc. Moreover, the cost of "adding-it-later", after delivery of the computer, is markedly higher than if such units as "reader/writers" were incorporated at time of manufacture. Reluctance on the part of consumers to buy software products that include mandatory "anti-piracy" devices, such as "dongles" or some like token that needs to be attached to the PC for the program to function, usually via a serial or printer interface, readily demonstrates the problems involved in hardware based schemes. Indeed, many of these anti-piracy devices have fallen into disuse by software developers for commodity software products.

One main point to be recognised early is that key escrow is not, itself, concerned at exactly how encryption keys are created, transferred between users or managed. Rather, it is only concerned that any "session keys" used to encrypt the flow of data between users or host computers may be reconstructed when necessary by an authorised group. This key management requirement could be a major problem with the Software key escrow situation.

1.5.1 Software-based "Key Escrowed" Cryptographic Technology

The "first step" is the development, on the technical side, of software based key escrow encryption technology and recognition, on the political side, that this offers the only realistic solution (over hardware based solutions). However, it must be pointed out that Denning [DENN-95)] still seems to maintain that hardware based solutions may be the only feasible way ahead.

Technically [BALE-94, NIST-94, WALK-94] application and/or system programs would be "Escrow Enabled (EE)" with the associated software packages incorporating the algorithms and protocols for key escrow services embedded in the binary code. Data stored and/or transmitted with the EE application or sub-system would contain a "Data Recovery Field (DRF)" or "Law Enforcement Access Field (LEAF)" in "Clipper" terminology, which would enable an enterprise and law enforcement agencies to recover the "session" key used to encrypt the data. Technically the TIS group, with its earlier (1994) SKE approach, goes somewhat further than the situation with the "Clipper" technology in that the receiver of information can reconstruct the associated LEAF and thus check that the received LEAF is correct for the received data. Other processes used to guarantee integrity and authenticity of data transmitted and received may include cryptographic sum-checks and digital signatures but these are not relevant for this discussion.

The main theme is that what is being subject to "escrow" are the cryptographic keys and, in principle, the actual encryption algorithms to be used do not matter. Walker et al note that, indeed, with such an approach the use of classified or secret algorithms would be pointless since their own confidentiality could not be guaranteed. This admission, however, could be a problem since, if such a guarantee cannot be made (and it realistically cannot be), then possibly equivalent guarantees cannot be made that the key escrow scheme itself is safe from misuse on a large scale. This is discussed later in this paper.

Thus it has to be that the whole scheme must not only be capable of incorporation into commodity level software products from software vendors but also must be capable of being incorporated into "bespoke" software; enterprise specific programs, created for the particular needs of the organisation by computer professionals. (This presents a problem which is not covered adequately in the Walker et al paper. In practice the EE application would need to either incorporate the whole scheme into its own code, i.e. memory address space, or make use of embedded sub-systems in the operating system or "middle-ware" used, eg a screen "windowing" system, network message handling sub-system, etc.) Moreover, the "impregnation" process would need to be carried out by a "trusted party" and in a totally secure manner. Likewise, these administrative and management problems are not discussed in the TIS papers.

1.5.2 Commercial Key Escrow Scheme

The main theme here is that corporate enterprises themselves, or other commercial enterprises acting on their behalf, will become "key escrow" agents for the enterprise, storing appropriate "encryption session keys", used to encrypt data for storage or transmission. So-called "Data Recovery Centres (DRC)" would be established (a name that sounds less threatening than "Key Escrow" unit). The new scheme is a simplified concept based around ideas developed from the earlier SKE scheme. Essentially the new CKE scheme [WALK-95] involves a simpler process as follows.

The important problem with the original "Clipper" proposal was that it offered no advantage to enterprises that used the technology, i.e. encryption keys were only available to government/law enforcement under appropriate legal orders. The key escrow facilities, which were established and maintained by Government, could not be used by the enterprise itself to recover its own information that may have been encrypted and the keys lost. *The important concept here is that enterprise data and programs belong to the enterprise and not to the employee or contractor.* Thus if enterprise data are encrypted by an employee with a cipher and/or key unknown to the enterprise then the enterprise becomes subservient to the employee. Valuable information could be lost or, at best, be difficult and expensive to recover, eg where low level ciphers are used in such commodity software packages as word processing systems, spread-sheet programs, work-group systems, etc.

An enterprise or commercial key escrow arrangement overcomes this problem and enables the enterprise to recover encrypted data should the need arise, through either deliberate or accidental problems emerging. Law enforcement may similarly use the DRC to decrypt data. This is explained by Walker et al as follows:

> "... *If law enforcement authorities ever need to decrypt a file or message of an employee of a company, they need only approach management with appropriate legal authority, and they, too, can obtain the session key in question. The issue of law enforcement access is thus reduced to an already well-understood legal procedure."* [WALK-95]

The scheme is next extrapolated to the international arena. They claim that already established procedures of "international police agreements" mean that authorities could cooperate to enable one law enforcement agency to obtain session keys from an extra-territorial DRC with the assistance of that country's law enforcement agencies. All of this is attractive since it draws upon *already established practice in law enforcement*, eg "search warrants", etc. However, it does assume that the control of the DRC is vested in "trustworthy" management. This assumption needs to be tested.

For example, the DRC data base itself could be super-encrypted using a proprietary scheme set up by the DRC management. After all, the DRC is just another information system established using software on a commercially available

TIS Commercial Key Escrow (CKE) Proposal

a.   Initialisation/Registration

Users install EE (Escrow Enabled) applications on their hosts systems
and provide the DRC with user authentication information. The DRC
returns a "User Identifier" and its own (DRC) public key along with
identifying information for the DRC itself.

> (Note: Solution to the "man-in-the-middle" attack needs to be
> detailed but is not in the original CKE proposal. In this attack
> a third party "sits between the user and the DRC and
> impersonates each to each other. The DRC sees the
> impersonator's public key as does the user.)

b.   Normal Operation

During normal operation the EE application adds the DRF to each
encrypted transmission, which incorporates the session key used along
with the user's identity encrypted under the DRC's public key. For
example, the application may use triple-DES for confidentiality
purposes. The transmission progresses in the normal manner between
parties in the application. Nothing goes to/from the DRC. Some of the
earlier SKE technology could be relevant here.

c.   Session Key Recovery

If the user or his/her organisation later cannot decrypt a stored or
intercepted message, the message is simply sent again to the DRC and
the DRC is requested to return the session key used, obtained by
decryption of the DRF using the DRC's secret (signing) key from its
public key cryptographic pair.

operating system, eg Windows NT, UNIX, OS/2, etc. At the legal level, is it
compulsory for the owner of the DRC to surrender "plaintext" or key information
under a search warrant or equivalent legal instrument? Moreover, can the DRC
owner be held responsible for guaranteeing that the information (escrowed keys)
divulged is "good"?

But, is it really that simple? Does criminology indicate to us that the main
problems are with "employees" and not the "management" itself? Can the DRC itself
be subverted so as to provide its services to the enterprise and not to legitimate law
enforcement agencies? Could individual employees of the organisation, those "in the

know" or party to criminal activity, be separated from others in the organisation in relation to the DRC, by a rogue management?

On the positive side and contrary to general opinion, cryptographic key escrow in a vaguely related form is a "way of life" in at least one industry, although it does not take the form of the proposal as set out in 1993 in the USA, i.e. where "master" encryption keys are split in two and stored by two separate Government entities for later recovery and usage under court order. In the banking and finance industry, for example, so-called "master encryption keys" are used to encrypt other "session keys" that are in turn used to encrypt data. Possession of the master keys, usually by an "issuing" bank, implies that the session keys can be reconstructed if required. This structure is a familiar and widely used one in the security of Australian electronic funds transfer operations and is set out in the SA 2805 series of standards.

The overall structures of the SKE/CKE and "Clipper" technologies are summarised in the following table (Table 1.5.1).

Structure	"Clipper"	SKE	CKE
Technology	Hardware	Software or Hardware	Software of Hardware
Cipher	SKIPJACK	RSA/DES Suggested	RSA
Initialisation of cipher/key sub-system	In Factory	By Supplier via Special Facility into Package	At first use only, by user
Protection of sub-system itself	Tamper resistant hardware	Careful embedding into package	Careful embedding into package
Performance Penalties	Not observable given cipher hardware performance	Asymmetric encryption needed for each message transmission - performance problem	Asymmetric encryption needed for each transmission - performance problem
CTI Ready	Yes	Maybe	Not really

Table 1.5.1: CTI = Computer - Telephone Integration.

2. Computer Professionals and Cryptography

2.1 Professional Codes of Conduct and Ethics

Before more closely examining the case of cryptographic key escrow schemes it is useful to examine the more general problem of professional obligations and ethics in relation to such technologies.

The dilemma facing an IT professional in the age of connected information systems on a global basis is one of meeting the expectations of the various codes of conduct that have been set out by various professional associations worldwide and by national and international groupings, eg the OECD, etc., while maintaining adherence to differing legal and statutory obligations that may be imposed on a national and state basis, eg export/import/usage restrictions applying to cryptographic systems. The Code of Practice of the British Computer Society serves to demonstrate the problem. It clearly states [BCS-95] under Section 4 ("Privacy, Security and Integrity") that the following practices should be followed:

Code of Practice | British Computer Society

Level One

4 Privacy, Security and Integrity.

4.1 Ascertain and evaluate all potential risks in a particular project with regard to the cost, effectiveness and practicability of proposed levels of security

4.2 Recommend appropriate levels of security, commensurate with the anticipated risks, and appropriate to the needs of the client.

4.3 Apply, monitor and report upon the effectiveness of the agreed levels of security.

4.4 Ensure that all staff are trained to take effective action to protect life, data and equipment (in that order) in the event of a disaster.

4.5 Take all reasonable measures to protect confidential information from inadvertent or deliberate improper access or use.

4.6 Ensure that competent people are assigned to be responsible for the accuracy and integrity of the data in the data file and each part of an organisation's database.

(Source : BCS-95)

At the same time the "Code of Conduct" of the British Computer Society clearly states two principles pertinent to this discussion. These are:

"3. Members shall ensure that within their chosen fields they have knowledge and understanding of relevant legislation, regulations and standards and that they comply with such requirements."

and

"10. Members shall not misrepresent or withhold information on the capabilities of products, systems or services with which they are concerned or take advantage of the lack of knowledge or inexperience of others."

To a professional member of the BCS, then, it is highly likely that a total conflict may occur on a number of levels. The IT professional in the UK may need to incorporate cryptographic mechanisms in a software systems to meet Section 4.5 of the Code of Practice BUT principles set out under Section 3 of the Code of Conduct may mean that that same IT professional may be prevented from incorporating such privacy measures in a system which is destined for or even just possible of export sales. Moreover, it also strongly suggests that IT professionals should be cognisant of the appropriate UK cryptographic export, import and/or usage legislation and/or regulations.

Interestingly Section 10 of the BCS Code of Conduct goes further and could be seen as requiring the professional to report the presence or absence of any cryptographic mechanisms that are pertinent in any product or system under their control where privacy or other related security facilities are required. This could be highlighted in the case of a recently reported US product from a company, "Jones Futurex Inc." [ZEIG-95].

A roughly similar situation applies in relation to the Australian Computer Society (ACS) through its Code of Ethics. Security of information systems is covered in the Code and specific reference is again made to *privacy* as outlined above.

A press report, from the Internet, and attributed to D Zeiger of the "Denver Post" newspaper, USA, discussed two members of that company's range of "SentryLine" encryption products, a computer link-to-link encryptor and an encrypting modem. Mr G Scott, a Vice-President of the company is reported as stating essentially that the encryption scheme used had to be downgraded to receive export sales permission, as distinct to internal USA sales. Essentially Scott claimed that to get such export sales approval the company had to downgrade cryptographic strength so that the US Government had:

".... a comfort level that they can break it. ...."

This apparently led to the creation of a private-key cipher, named "JFX E2", by D Young of the company for incorporation into the product, rather than the more normal DES cipher.

Australian Computer Society Code of Ethics

Values and Ideals Subscribed to by Society Members

The professional person, to uphold and advance the honour, dignity and effectiveness of the profession in the arts and sciences of information processing and in keeping with high standards of competence and ethical conduct, will be honest, forthright and impartial, and will serve with loyalty employers, clients and the public, and will strive to increase the competence and prestige of the profession, and will use special knowledge and skill for the advancement of human welfare.

Code Of Ethics

I will act with professional responsibility and integrity in my dealings with clients, employers, employees, students and the community generally.

Standard of Conduct

1.   Priorities

I will serve the interests of my clients and employers, my employees and students, and the community generally, as matters of no less priority than the interests of myself or my colleagues.

1.1   I will endeavour to preserve continuity of computing services and information flow in my care.

1.2   I will endeavour to preserve the integrity and security of others' information.

1.3   I will respect the proprietary nature of others' information.

1.5   I will advise my client or employer of any potential conflicts of interest between my assignment and legal or other accepted community requirements.

1.6   I will advise my clients and employers as soon as possible of an conflicts of interest or conscientious objections which face me in connection with my work.

....

4.   Social Implications

I will strive to enhance the quality of life of those affected by my work.

4.2   I will consider and respect people's privacy which might be affected by my work.

Where does a system like this place the BCS/ACS professional member? Should such a product be recommended for use in the United Kingdom or Australia? Is it the responsibility of the IT professional to become completely aware of the nature of such a cipher scheme and its implementation and report any findings to management? Both the BCS and ACS Codes of Conduct clearly indicate so. These give an indication of the professional problems that must be faced in relation to the CKE scheme.

2.2 National and International Privacy and Security Guidelines

At the same time IT professionals worldwide have been made aware of two major OECD (Organisation for Economic Cooperation and Development) documents that are pertinent to the CKE scheme. These are the "OECD Guidelines on the Protection of Privacy and Transborder Flows of Personal Data" [OECD-80] and the "OECD Guidelines for the Security of Information Systems" [OECD-92]. These are early attempts to draw together a common body of international agreement on overall information systems security at a time when data networks, and thus information systems, have become global in reach. These likewise must be considered when examining possible responses to the CKE architecture. There appears to be no available study that addresses the problem.

3. Commercial Key Escrow - Solution or Simply Avoiding the Principles and the Problem?

3.1 Personal Freedom and Reality

It has to be proposed that few Governments in the world will enact legislation absolutely prohibiting, under civil or criminal penalties, the possession or usage by its citizens of cryptographic systems, hardware, software or manual processes, in order to create confidential channels of information transfer between consenting people. This would also include steganographic techniques as indicated in a recent horse racing case in Australia [MARR-95], as detailed above. In this case horse jockeys were accused of using a:

> " ... *secret code system whereby jockeys would tip their caps with their riding crops, pass messages on colour-coded paper or give a huge smile to flout racing rules.* "

3.2 Software-based Cryptography

From the viewpoint of non-USA users and creators of IT products and services, there has been a growing perception that the basic operating system "kernels" of popular US developed and supplied operating systems, possibly accounting for over

75% of the world's system software, are increasingly incorporating sub-systems that in the past were separate, identifiable modules. These modules, from the operating system viewpoint, really behave as "applications" and, with appropriate definition of software interfaces to the operating system itself, could be developed by any competent computer scientist or IT professional anywhere in the world. However, this may not be the motivation of the system supplier who wishes to lock further services, and thus sales, into the base system itself.

A simple illustration would be one where a user just wants a stand-alone PC with no network capabilities whatsoever. Is there any operating system today that has all such services, and thus overheads in terms of memory requirements and performance, removed or enables those services to be easily removed?

However, with SKE/CKE, as with other forms of cryptographic systems implemented in software form, the temptation, if not the technical and/or political (export approval, etc.) imperative, is to incorporate such cryptographic sub-systems deeply inside the OS kernel itself. In this way "reverse engineering" of the algorithms and techniques used could possibly be made more difficult to such an extent that such reverse engineering could be seen as impractical. This is an exact parallel with commercial cryptographic algorithms and associated hardware. Design philosophies are set out to make it infeasible for an adversary to "break the system", not to make it totally "tamper-proof", even if that were possible.

If this requirement is combined with operating systems that have worldwide market dominance at the desktop, such as with Microsoft's "Windows 3.1" or "Windows 95" systems or IBM's "OS/2", etc., then the problem of both market dominance and overall control is greatly magnified!

This presents a major problem for nations. Operating systems and associated network "middleware" may be created in the United States with the CKE scheme incorporated into it in an integral fashion with or without the approval or even knowledge of the end-user or their nation. A specific and important example would be Internet related data communications "protocol stacks" incorporated into host operating systems. There would be no feasible way to incorporate "national ciphers" in place of the more usually publicly accepted ones, such as DES and RSA, or proprietary ones of the software manufacturer or even those ciphers set out by the United States government. Moreover, there appears to be no effort at present to incorporate SKE/CKE schemes into Internet related protocols at either the IP (Internet Protocol) or Application Protocol levels. A solution to these problem has to be found while allowing protection of the actual software itself.

Getting around this would mean a change in United States export schemes for cryptographic systems. At present not only are encryption systems for confidentiality restricted exports but so also are all the detailed interfaces to those encryption schemes and their positioning in allied software and/or hardware systems that use the encryption services. Export rules would need to be amended to allow for appropriate cipher libraries to be incorporated into exported software systems by customers in the purchasing country or by some other bilateral/multi-national agreement.

This also presents a problem at the time of extraction of the escrowed keys on an international basis. Not only would the keys be needed to decrypt intercepts but also the details of the actual cipher used. Thus key escrow also needs to be *"cipher escrow*" in an international context. The only other possibility is to have bilateral or multilateral agreements and a set of cipher standards based on published ciphers and a set of pointers that indicate which one or group are being used for particular file storage or message transfer.

4. Services - A Possible Solution?

4.1 Commercial "Data Recovery Facilities (DRF)"

At the commercial level, once the above acknowledgment is made that information resources belong to an enterprise and not its employees, a simple "key escrow" solution comes immediately into focus.

4.2 Post Offices

Australia, and indeed most countries, has a well respected heritage in its postal system. The system is governed by a set of laws and regulations, as well as codes of conduct, that precede the existence of the computer and even the telecommunications industries and the associated national and international services created by these industries. Most settled areas have a local post office or agency. Services are well defined.

A solution to the key escrow dichotomy is one of enabling and encouraging the post office to provide a "Data Recovery Facility (DRF)" service to large and small enterprises nationwide. It is useful to note that most businesses have a post office within a local phone call of the enterprise, including the banking and finance industry. The services provided by an expanded post office operation could be as follows:

A. Backup and Recovery Facilities for Small/Medium Enterprises

A backup and recovery facility at the post office level could be used to automatically provide a service for small to medium business. A "dial-up" call could be made in non-peak times to the business and designated data could be copied to the data recovery facility at the post office. All transfers and data stored could be encrypted using appropriate ciphers and applications that incorporate the necessary CKE encryption processes.

B. Certificate Issuing Facilities

Modern digital signature schemes require that the associated "signing key" for a user be itself guaranteed by a trusted third party. That party could logically be the post office. Keying materials/data could be maintained along with the data recovery facility. This proposal mirrors current proposals and experiments in Sweden as outlined by Lundberg and Persson [LUND-95]. The "Secured Electronic Information Handling in Society (SEIS)" scheme in Sweden is based around an association of interested parties aiming to coordinate a number of related security projects in that country. These include the "Allterminal", a secure PC, and SEDIS (Secure Electronic Documents In Society) projects. In turn the SEDIS project includes Sweden Post and Telia, the former Swedish post and telecommunications organisation to provide for secure information transfer in Sweden. In this work it is planned that Sweden Post will act as the public key certificate issuing authority with Telia supplying the underlying telecommunications network structures.

C.   "Yellow and White Pages" - Directory Services

If the above is possible, then the logical next step is to incorporate the above services into an overall information service, including the required Directory Services needed for electronic commerce of all forms. The post office is the appropriate trusted third party that could be relied upon to securely maintain the overall national directory of businesses and citizens alike.

D.   Key Escrow

Encryption keys used for confidentiality services could be deposited into the post office systems using any appropriate key escrow technology. Law enforcement interests could be maintained by making such key escrow stores available under court order, in the normal manner. Interception of "mail" could be subject to the same rules that apply to the post office at present. This does not preclude any "key sharing" or "master key" only scheme. Moreover, the use of such independent but government "related" organisations as Post Offices may get around the noted concern of law enforcement in suitable control over the actual key escrow centres themselves, the DRCs.

The above philosophy is one of making useful services a major part of any "key escrow" scheme. The third party, "escrow agent" is nominated as "Australia Post", an enterprise with a very high level of acceptance in the community and with a long and well perceived record of security in the handling of the mail. The corporatisation of the enterprise has the added benefit of placing any key escrow operations at a level of distance from direct government control, eg under Defence or Government Department controls.

4.3   Incorporation of SKE/CKE into IT Products and Systems

The proposal as it stands in June 1995 appears to be almost totally aimed at the producers of commodity level software who are of U.S. origin. This is understandable since the "tension" so often referred to in the Walker et al paper [WALK-95] is concerned with the United States perception of the problem. This perception is one of:

a.   growing IT industry opposition to United States Government created and enforced export restrictions covering IT products and systems which incorporate cryptographic software or hardware for confidentiality services, (assuming that such technologies used solely, and demonstrably solely, for authenticity and integrity alone have been more or less approved for U.S. export for some time), and

b.   the need to radically reduce the "tension" between the opposing forces, one emphasising the need for increased privacy in information systems and data networks at all commercial levels; the other, the need by law enforcement agencies to pursue their duties through usage of information gained from authorised "line taps".

Evidence for this U.S.-centric vision is contained in the "international key escrow" solution proposed. It is assumed that DRCs exist in other countries that could be available under bi-lateral/multi-lateral law enforcement agreements. However, the concept that other countries may produce software products and systems that in turn may be imported into the united States is not discussed! This latter scenario means that bi/multi-lateral agreements must exist such that computer professionals in all countries will make their applications and/or system software products "EE" cognisant. This is a radically different scenario to the one considered and draws into the discussion such organisations as national societies of computer professionals, such as the Australian Computer Society, British Computer Society, etc., the IEEE, the International Federation for Information Processing (IFIP) and others.

At a technical level another problem needs to be addressed. Since both SKE and CKE are not concerned directly with key creation and transfer this needs to be done *outside the SKE/CKE sub-system itself.* Keys, once agreed, are then "cemented" into the CKE/SKE scheme implemented in the application or sub-system. However, in practice this is not so simple. For example, the actual process of negotiating a data encrypting session key may itself involve "key-encrypting-keys" or "exponentiation processes". These processes are themselves cryptographic processes that *incorporate confidentiality functions.* A rogue user or application developer could simply use them rather than the CKE/SKE system itself once the time comes for confidential data transfers.

In addition the SKE process envisages the existence of a "Key Escrow Programming Facility (KEPF)" and the CKE process envisages a process of user authentication and software package registration. For the SKE scheme the KEPF, itself an organisation of computer systems and people, must create an appropriate program "family key" for each *type* of program created by a manufacturer or software development facility (in the case of in-house development). Individual instances (copies) of the program that seeks to provide confidential, encrypted data transfer services and make use of key escrowed encryption, is itself provided with what TIS call "program unique identifiers" that need to be hidden in the program binary code for distribution. These values are all used later via the LEAF construction to enable reconstruction of any session key used as needed.

All this envisages an organisation that is simply unrealistic on an international basis. While the United States software package industry may be able to coordinate the activity "on-shore" such facilities would need to exist at other international locations where multiple copies of software for distribution are made. At the "bespoke" level, companies and organisations would need to make the overall process an integral part of transferring applications from "development to production". This latter case may actually be much easier to organise and has definite benefits for an organisation since by using the key escrow production procedures, control over distributed copies of in-house software can be realised.

5.   Conclusions

5.1   Compulsion versus Coercion

On an international basis, the need under the SKE/CKE schemes to make IT applications and systems EE cognisant presents the next major challenge to its proposers. "Building in Big Brother" is a concern of and the province of those computer professionals worldwide (eg through IFIP, IEEE, etc.) who would be required to implement the scheme. At the law enforcement level, is CKE acceptance up to such multinational groupings as UNESCO, OECD, and others or are bi-lateral and existing agreements capable of handling access to other country's DRCs as necessary?

In an Australian context this paper has proposed the use of Australia Post as a commercial enterprise capable of offering "Data Recovery Centres" on a national basis with levels of demonstrated and respected trust that has long been recognised in this country. The main problem with CKE has been demonstrated to be the very basis of its formulation, i.e. the relief of "stress" between the United States IT industry, seeking to meet worldwide demands for improved confidentiality services in products offered, and law enforcement concerns at the ability to use legitimate data/telecommunications interception techniques for combating crime. There are severe questions as to whether or not the scheme is realistic on an international basis, not so much from the point of view of bi-lateral or multi-lateral agreements covering access to DRCs, but the very incorporation of the technology into software products and systems offered for general sale by companies throughout the world or software developed and deployed purely for "in-house" usage.

In the end, there appears to be uniform agreement by most, but not all, that total prohibition of the possession and/or usage of cryptographic systems by the general public or by private companies and enterprises, is unrealistic. The discussion centres upon "mass availability" of IT products and systems that could be readily used for criminal purposes, and not the unsolvable problem of two individuals employing strong encryption technology for private communications between them or to a limited and controlled group. However, only criminological studies will tell whether or not this latter case is the real problem or not and whether or not we are, with CKE and even the original "Clipper" hardware scheme, solving the correct problem.

In summary, this paper has made the following observations:

SKE/CKE Technology

1. At present the SKE/CKE schemes appear aimed at mass produced software products that wish to provide confidentiality services and which are to be marketed worldwide, particularly by United States based software companies. This needs to be extended to "in-house" or "bespoke" software systems as well as to the international IT industry. There are technical problems in doing this, eg incorporation of SKE/CKE ciphers and protocols into operating systems, "middleware" and, in particular, software development tools themselves.
    Verdict: Problems seem surmountable but difficult.

2. It is not clearly demonstrated that the software based SKE/CKE technology could not be bypassed on a mass scale equivalent to publishing "bugs" in system software on the Internet for use by "hackers". For example, it is not clear that session key creation and exchange can be divorced from the SKE/CKE architecture. The case of two, or a small group of, "rogue" users/developers is not covered by the scheme which recognises this problem.
    Verdict: Problems seem surmountable.

CKE and Law Enforcement/Crime

3. There is no evidence that management, controlling the Data Recovery Centre in an enterprise and wishing to hide illicit activity, would be willing to "hand over" keys on request and to guarantee that the "keys" handed over are the correct ones. (It would be easy to claim that there had been technical, operational, management or simple human "failures" in the operation of the DRC itself). In turn this would suggest that the DRCs would need to be operated by *trusted enterprises only.*
    Verdict: Enterprises cannot be their own DRCs. Unless legal and law enforcement problems which seem insurmountable in practice at present, can be resolved.

4. The world has known the mail/post office much longer than telecommunications and the computer. It is proposed that a nation's Post Office, with its already in place international set of agreements and legal obligations, is the ideal controller of DRCs for any country, meeting the needs of both private and public enterprises for recovery of keys if needed for internal purposes, as well as for law enforcement purposes, for maintaining confidence in the integrity of the DRC itself.
    Verdict: Post Offices seem ideally suited to operate DRCs. Trials are suggested.

CKE and IT Professionals

5.   IT professionals worldwide would be the identified group responsible for not only incorporating the CKE technology into general commodity software products for sale but also into in-house software for an enterprise.   These professionals must abide by sets of well established code of ethics and conduct. These must be coordinated with the CKE proposal.

Verdict: National and international IT professional organisations need to be fully aware of the current debate and proposals. Conflict of duty or loyalty (to employer or State) seem likely for an individual IT professional.

Will IT professionals worldwide be compelled to "Build in Big Brother"?

6.   References

BALE-94   Balenson, D. et al, "A New Approach to Software Key Escrow Encryption", (in HOFF-95)

BCS-95   Code of Practice" and "Code of Conduct", British Computer Society (BCS), BCSNet, June 1995, (Source: World-Wide Web Site - *http://www.bcs.org.uk*)

DENN-94   Denning, D., "International Key Escrow Encryption: Proposed Objectives and Options", (in HOFF-95)

ECON-95   "In the name of liberty", Editorial - The Economist, 29 April 1995.

ELLI-95   Ellison, C. M., "Initial Description and Specification of the TIS Commercial Key Escrow System", Trusted Information Systems, Inc., 3060 Washington Road, Glenwood MD. 21738., U.S.A., January 9, 1995, (Available on Internet at *"ftp://ftp.tis.com/pub/crypto/drc/papers/"*)

HEAD-95   Head, B., "Playing the Game", Banking Technology, June 1995, Pg. 24

HOFF-95   Hoffman, L. (Editor), "Building in Big Brother: the Cryptographic Policy Debate", Springer-Verlag New York, Inc.,U.S.A., 1995, ISBN 0-387-944441-9

KEHO-95   Kehoe, L., "IBM in $4.6 bn bid for Lotus", The Australian, 7 June 1995.

LETH-95   Lethin, R., "Admiral Bobby Inman's talk at the Massachusetts Institute of Technology", Internet Security Monthly, February 1995, ISSN 1079-5669

LOYA-95   "Fly Buys.™" Brochure, .Loyalty Pacific Pty Ltd, Australia. 1995. (ACN 057 931 334)

LUND-95   Lundberg, B and Persson, E., "Security on Electronic Highways in Sweden - Worthwhile Inheritance for the Future." in Yngstrom, L. (Editor) "Information Security: The Next Decade/Security on the Electronic Highways in Sweden", Swedish Workshop Addendum to the Proceedings of the Eleventh International Conference on Information Security, IFIP/Sec'95, Capetown, South Africa, May 1995., The Swedish Computer Society, Saltmatargatan 9, 11359 Stockholm, Sweden, 1995.

MACQ-90   The Macquarie Dictionary, 2nd Revised Edition, The Macquarie Library Pty Ltd, 1990, ISBN 0-949757-41-1

MARR-95   Marris, S., "Investigation finds jockeys used secret codes to fix races." The Australian, 8 June 1995.

OECD-80   *"Guidelines on the Protection of Privacy and Transborder Flows of Personal Data"*, Organisation for Economic Co-operation and Development (OECD), Paris, 1980.

OECD-92   *"Recommendation of the Council Concerning Guidelines for the Security of Information Systems, Guidelines for the Security of Information Systems and Explanatory Memorandum to Accompany the Guidelines"*, DSTI/ICCP/AH(90)21/REV6, 18 Sept. 1992., Organisation for Economic Co-operation and Development (OECD), Directorate for Science, Technology and Industry, Committee for Information, Computer and Communications Policy, Ad Hoc Group of Experts on Guidelines for the Security of Information Systems.

OXFO-89   The Oxford English Dictionary, 2nd Edition, Volume V - "Dvandva-Follis", Clarendon Press, Oxford, 1989, ISBN 0-19-861186-2 (Set), ISBN 0-19-861217-6 (Vol V)

WALK-94   Walker, S., "Software Key Escrow: A Better Solution for Law Enforcement's Needs?", (in HOFF-95)

WALK-95   Walker, S et al, "Commercial Key Escrow: Something for Everyone Now and for the Future", TIS Report #541, January 3, 1995, Trusted Information Systems, Inc., 3060 Washington Road, Glenwood MD. 21738., U.S.A., (Available on Internet at *"ftp://ftp.tis.com/pub/crypto/drc/papers/"*)

WEBS-86   Webster's Third New International Dictionary of the English Language - Unabridged, Vol. 1 "A to G", Encyclopedia Britannica, Inc., Chicago, Il. U.S.A. 1986, ISBN 0-87779-201-1

WILS-95   Wilson, J., "The Case for Greater Vigilance", Time, 1 May 1995, Pg.51.

Encryption and the Global Information Infrastructure: An Australian Perspective

Steve Orlowski

Assistant Director
Security Management
Australian Attorney-General's Department

The views in this paper are those of the author and do not necessarily represent the views of the Australian Government.

Abstract. Debate on encryption in Global Information Infrastructures has been complicated by issues relating to law enforcement. This paper looks at a technique for limiting the use of anonymous cash for illicit purposes to an acceptable level. It also argues that the majority of users will not require high level encryption systems to protect their privacy and hence would not require the keys to their encryption schemes to be escrowed. It proposes a scheme of Differential Key Escrow (DKE) where only the keys of high level encryption systems used by government and larger corporations would be held in escrow by the organisation.

In December 1993 the Australian Government established a Broadband Services Expert Group to examine the technical, economic and commercial preconditions for the widespread delivery of broadband services to homes, businesses and schools in Australia. In releasing the Group's Final Report *Networking Australia's Future* [2] the Prime Minister said being linked to the national information infrastructure is a fundamental right for all Australians.

As the Final Report [2] put it:

"In the next decade, large-scale communications investments in Australia will pave the way for many business, government, information and entertainment services. These services could change forever the way business and government operate and how we communicate with our colleagues, families and friends. Over time, even the significance of international borders and the design of towns and cities will change."

Similarly, the OECD in its 1992 Guidelines for the Security of Information Systems [6] said:

"Recent years have witnessed ... growth of computer use to the point that, in many countries, every individual is an actual or potential user of computer and communication networks."

Encryption was for centuries the domain of government, primarily to protect military and diplomatic communications. In the past few decades private enterprise has become an increasingly larger user of cryptography to protect its commercial activities. We have now arrived at the point where individuals are going to become major users of cryptography to protect personal information and finances, and their privacy in general, as they become participants in information infrastructures.

Over the past twelve months, the OECD has embarked on a round of meetings on Global Information Infrastructures. The outcomes of this round are to be provided in a report to the G7 on job creation and the information society. Security, privacy and the protection of intellectual property are some of the issues being addressed as part of this round. Indeed the final meeting will specifically address these issues. In outlining an agenda for this meeting the OECD saw encryption as a pivotal issue in the security of information systems.

The OECD will also be holding a meeting on National Cryptography Policies later this year.

The interest in the Global Information Infrastructure relates not only to the direct impact of the infrastructures on national economies, but also on the economic impact of investment failures if the infrastructure is misused or not used to its expected capacity. User confidence is seen as a key factor in infrastructures reaching their full potential. It is from this position that the OECD is examining issues of security, privacy and the protection of intellectual property.

Turning again to the OECD Guidelines [6], they stated when addressing the question of building confidence:

"Users must have confidence that information systems will operate as intended without unanticipated failures or problems. Otherwise, the systems and their underlying technologies may not be exploited to the extent possible and further growth and innovation may be inhibited."

Obviously if encryption is a pivotal issue in information systems security, confidence in encryption techniques and technology is pivotal to confidence in information infrastructures and therefore to the economic viability of such infrastructures.

At the meeting in Paris last November most of the session on security was taken up with encryption. It was interesting, however, that very little of it was related to security of government or commercial information on systems. The main focus was on verifiable but untraceable transactions on information infrastructures. This highlighted the progression of cryptography towards individual's requirements and their desire for their transactions to be secure but anonymous.

The issue of privacy of an individual's activities in information infrastructures is beginning to receive similar attention in Australia. Individuals are concerned that their activities can be monitored to develop personal profiles such as buying habits. These profiles could then be exploited by organisations such as direct marketing bodies.

The Minister for Justice in a speech to the Australian Share/Guide Conference in March [5] this year identified two areas of concern:

"People want to be assured that information on how they use the network is protected. Usage patterns are of particular interest and value to various groups, for example, direct marketers."

"People also need to be assured that the content of their information is protected both on networked systems and flowing across the network."

Both these concerns can be overcome through the use of cryptography. The first through verifiable but untraceable transactions and the latter through more established message encryption techniques.

The question of verifiable but untraceable transactions has attracted the concern of law enforcement agencies given the potential for the proceeds of crime to be transferred in this way. In Australia the *Financial Transactions Reports Act 1988* requires cash and certain electronic cash transactions above specified limits to be reported. This approach could possibly be extended to put limits on computer cash transactions which can be carried out anonymously. This would allow individuals protection of their privacy on the small transactions which would make up the bulk of their activity but place some obstacles in the way of those who wish to move large volumes of money illicitly. Technology which limits the amount of anonymous cash which can be sent, received or stored per terminal or smart card per day may be able to be developed to overcome the law enforcement concerns.

While such an approach might reduce the problem of cash transactions for illicit purposes, the more vexing problem is that of criminal activities being planned or transacted by telephone or over networks, particularly where encryption is involved. In other words the "key escrow" debate.

In Australia telecommunications interception (TI), both voice and data, is carried out under the provisions of the *Telecommunications (Interception) Act 1979*. In 1993 the Australian Government initiated a *Review of the Long Term Cost Effectiveness of Telecommunications*. The Report [1] stated:

"The evidence suggests that TI is very effective as part of an **integrated** *framework of surveillance by both law enforcement and security agencies."*

A significant finding of the report was:

"Encryption by targets of their communications (both voice and data) is not considered by agencies as a problem for TI at present in Australia, but it is a growing problem in the US and Europe and a potentially significant problem in Australia. It will need to be monitored, particularly with increased availability of cheap voice encryption devices. The issues extend well beyond the scope of the Review."

The report also commented that:

"...Australians have available in the GSM digital mobile services an effective means of encrypting their communications for legitimate privacy and commercial security purposes..."

As a result of the Report, Australia is, among other TI issues, monitoring the impact of encryption in the telecommunications interception area and will re-examine matters in 1997 following the opening of the telecommunications area to full competition.

The average Australian mobile phone user appears to be satisfied with the security offered by the GSM digital mobile services and to date I have not seen a report of instances of communications on that network having been found to be insecure. Individuals and small businesses seem to be the major users of the networks and their requirements for security are relatively low. On that basis there would appear to be a relatively small market for voice encryption devices on mobile phone services. Similarly Australians have, by and large, been comfortable with the standard telephone service and again there has been comparatively little market for voice encryption products, although they have been readily available.

Of course there have been instances of criminals using encryption devices on the existing standard and mobile services, and this will continue. However, most persons involved in this field agree that even if key escrow were introduced, this could be circumvented by determined criminals.

Furthermore we are rapidly moving towards the integration of voice and data services. By the turn of the century, the majority of voice communications is likely to be over data lines. Encryption of both voice and data is therefore likely to be handled by the same products.

Stephen Walker, in his paper *Software Key Escrow A Better Solution for Law Enforcement Needs?* [8] stated:

Since law enforcement's requirements for key escrow appear largely focused, for now at least, on telephone communications, it will probably remain necessary for the government to escrow keys of telephone security devices. (It has been observed that due to the high cost of telephone security devices with or without Clipper, there may never be a significant market for such devices and therefore little reason for an extensive telephone-only based key escrow capability.)

I would therefore argue that the value of key escrow for purely voice encryption would be marginal in the Australian context and probably internationally in the long term. In saying this I am not advocating the abandonment of the conventional field of telecommunications interception, rather I am arguing that resources might be better applied to addressing the longer term problem of the emerging field of interception of telecommunications in the form of data.

That then leaves us with the question of encrypted data communications. Law enforcement concerns have focussed on two aspects; financial transactions associated with criminal activity such as payments for drug deals, and messages such as setting up a drug deal or planning a terrorist attack. These are realistic scenarios which confront law enforcement authorities. Obviously the community expects that law enforcement authorities will take steps to prevent information infrastructures being used for these purposes. Equally users of the infrastructures for legitimate purposes expect that their right to privacy will be respected. The hapless task for governments is to find an acceptable balance between the two.

Firstly I would like to address the question of financial transactions. Earlier I proposed a restriction on anonymous cash transactions which would make it difficult to move large sums of money in this manner. Larger transactions would then have to be moved through traceable transactions. This would mean that records of the transactions and the parties involved would exist in much the same way as they do for financial transactions at present. If the anonymous transaction limit was the same as the cash transactions reporting limit, this would mean that, in Australia or for transactions entering or leaving Australia, the transaction would be reported to the AUSTRAC, the agency which collects and analyses data on cash transactions. Moreover, law enforcement agencies could approach the courts to obtain access to an organisation or individual's records of such transactions.

This leaves the question of messages which may contain evidence of criminal activities. While in some cases, copies of such messages may be recoverable from one of the partys equipment, any serious criminal using these methods would know how to modify or delete all traces of the message. Therefore court orders granting access to the equipment and data held on it would not necessarily provide the evidence sought. This problem would exist whether or not the transmission or the storage media were encrypted.

Before advancing this argument further I would like to make the observation, which I will be expanding on later, that debate to date has focussed on higher level encryption. I feel that the needs of the majority of users of the infrastructure for privacy and smaller financial transactions, can be met by lower level encryption which could withstand a general but not sophisticated attack against it. Law enforcement agencies could develop the capability to mount such sophisticated attacks. Criminals who purchased the higher level encryption products would immediately attract attention to themselves.

Given that a large proportion of the population would not be using the higher level encryption products, application of key escrow for such products is less likely to create the type of adverse reaction seen to date. Government agencies and large financial institutions are more likely to accept the need for key escrow in the type of products which they use.

The *Review of the Long Term Cost Effectiveness of Telecommunications Interception* [1] referred to earlier quoted the following points made by the Australian Federal Police:

· *much valuable TI evidence and intelligence comes from targets talking to people who are not part of a criminal activity and who would not use encryption (arranging hotel, shipping or airline bookings is one obvious example);*

· *call data will not be encrypted and will contain much valuable information about who is involved in an investigation*

The Review did, however, include the following in its findings:

·· *Telecommunications interception is of crucial importance to law enforcement, and*

· *On present indications, it would not be true to say that developments in technology may render telecommunications uninterceptible.*

Given that there is a requirement for telecommunications interception, the question is how is this achieved in the face of changing technology. The answer is to use the new technology to the advantage of law enforcement agencies.

As mentioned earlier, I see encryption being utilised on two levels, a general level being used by the majority of users and a more sophisticated level with much more limited use. Intercepted messages under the first level may be able to be decrypted by the various interception authorities. The second level would probably, however, require more sophisticated techniques in circumstances where the key cannot, for whatever reason, be recovered from escrow. This may be achieved by the establishment of a central decrypting unit which would receive, decrypt and transmit back messages.

Given the standard of equipment and expertise which would be developed at such a centralised unit, it may be more cost effective for that unit to handle all decryption of intercepted messages for all law enforcement agencies within the country. Modern communications technology would facilitate the secure and rapid transmission of messages between the intercepting authority and the central unit. Indeed the "Clipper" proposal, and suggested variations of it, relied on a similar concept for the transmission of escrowed keys to the intercepting authority. This takes the process one step further. It also builds in an additional safeguard to the interception process as the central unit would need to be satisfied of the validity of the interception before it decrypted the messages.

The same concept would apply for the higher level encryption systems where the keys would be escrowed. In this case the central unit would obtain the keys from the escrow agent or data recovery centre.

Regarding the question of data recovery centres, I am attracted by proposals put forward by Stephen Walker in the paper [8] that I referred to earlier, which suggested commercial data recovery centres. Even the term data recovery centre is a positive one of a service rather than the negative image which now surrounds the term key escrow. The concept that I have just outlined could operate for either government escrow agents or commercial data recovery centres.

The suggestion that I have outlined is a rather simplistic version. In practice there would be a number of legal problems to be overcome, especially in a federal structure with a division of law enforcement powers.

In the case of key escrow for corporations, there may already be an implied requirement in corporate affairs legislation which requires records to be held for a statutory period. If the records were encrypted, then the key would need to be available to decrypt them. This could be used as the basis for a formal key escrow requirement.

I put this forward as a starting point for discussion of the concept of differential key escrow.

As mentioned earlier the concept of restricting key escrow to higher level encryption systems would reduce general user concerns about using the GII and provide the confidence which the OECD considered was essential to the economic viability of the infrastructures.

Another area where confidence has to be established is that of content providers. Confidence that provider will receive payment for their intellectual property will be key to the range of material being available on the infrastructure. As the Minister for Justice put it in the speech [5] I referred to earlier:

"*An important aspect of the network will be the quality of the information available on it. the question of intellectual property rights is crucial to the success of the infrastructure.*"

The Government is pursuing the question of intellectual property rights in various international fora. However those rights have to be protected once they have been defined. Encryption will be the key to protecting information to which intellectual property rights attach and to ensuring users pay for what they access. This will involve the more traditional field of data protection as well as access control, user authentication and electronic cash applications.

The Australian Government implements controls on the export of defence and related goods through the *Customs Act 1901* and the *Customs (Prohibited Exports) Regulations*. In March 1994 the Government issued *Australian Controls on the Export of Defence and related Goods - Guidelines for Exporters*. [3] The Guidelines state in part:

"*The Government encourages the export of Australian made defence and related goods where such exports are consistent with Australia's interests including international, strategic, foreign policy and human rights obligations.*"

The controls do allow exporters to apply for permits or licences to export goods.

The controls specifically mention products related to cryptography as follows:

(a) *complete or partially complete cryptographic equipment designed to ensure the secrecy of communications (including data communications and communications through the medium of telegraphy, video, telephony and facsimile) or stored information;*

(b) *software controlling, or computers performing the function of,* cryptographic equipment referred to in paragraph (a);

(c) *parts designed for goods referred to in paragraphs (a) or (b);*

(d) *applications software for cryptographic or cryptanalytic purposes including software used for the design and analysis of cryptologics.*

In November 1994 the Government also issued *Australian Controls on the Export of Technology With Civil and Military Applications - A Guide for Exporters and Importers* [4] defines in more detail equipment, assemblies and components to which the controls apply.

The Strategic Trade Policy and Operations Section, Department of Defence makes recommendations on export applications.

The Government is committed to its policy of encouraging the export of goods where this is not in conflict with the national interest or obligations. To this end it is prepared to cooperate with manufacturers, wherever possible, to advise on products which maight be eligible for export. This is particularly relevant for the type of products which would satisfy the requirements of general users of information infrastructures and thus enhance the development and use of such networks.

Digital signature techniques and public key authentication will play an increasingly significant role as networks expand and the number of users and range of services offered increase.

This is a further area where confidence needs to be engendered to ensure acceptance. There is a need for a mechanism to ensure that techniques are appropriate for the purpose for which they will be used. Similarly there is a need for a structure through which keys can be obtained and digital signatures authenticated.

Within Australia a Government Group has been developing a proposal for a Public Key Authentication Framework. The group's work has been primarily focused on the needs of electronic commerce. In an unpublished paper [7] the group stated: *There needs to be a wide scale informed debate about this issue before any decisions are taken as to choice of technology, the appropriate administrative structure, privacy issues, legal effect, method of implementation and the like. After such a debate the system will need to be introduced in a planned way with appropriate public education, legislation and the like in order that the use of the PKAF system will have the same standing and validity in the eyes of the community as a paper based signature.*

The proposal calls for a management structure to verify various key generation systems, supervise the issue of key pairs and maintain a directory of the public keys.

This proposal has been referred to the Standards Association of Australia which has established a task force to examine the establishment of an Australian Public Key Authentication Facility. The Task Force is required to report by the end of the year.

Australia has also raised in the OECD the need to establish an international framework to ensure the effective use of public keys as a tool for both international electronic commerce and individual use of the global information infrastructure.

While this proposal is driven, primarily, by commercial needs, there is scope for it to be extended to meet the needs of individuals who will also be using the information infrastructure. Any scheme such as this has to be better than the current process of passing credit card information over the network.

The referral of the PKAF proposal to Standards Australia is in keeping with the Australian Government policy of minimal legislative intervention. When commenting on the implementation of the OECD *Guidelines for the Security of Information Systems*, in a speech [5] I referred to earlier, the Minister for Justice outlined the Government's approach as follows: *In implementing the Guidelines, the Government has decided not to use a general legislative approach because of the problems in reaching agreement with State and Territory Governments on legislation where the Commonwealth has no blanket constitutional power.*

Furthermore we recognise that legislation is slow to respond to technological advances, so broad definitions have been used in relevant legislation to allow the courts to consider current technology as cases come up.

This policy extends to electronic commerce and the use of cryptography in general. Any legislation required to support the use of cryptography is likely to be written in broad terms rather than endorsing particular technology or algorithms. It would then be left to groups such as Standards Australia to specify the standards which at that particular point in time would meet the legislative requirement.

By the turn of the century, the major users of the Global Information Infrastructure will be individuals conducting their day to day activities in electronic form. The main concerns of these users will be to authenticate their identity, to conduct their business with privacy and to have a reasonable level of security for the comparatively low level financial transactions they will be performing. To date the cryptography debate has focussed on the higher needs of government and business. There is a need for the debate to be extended to cover the needs of individual users.

For the information superhighway to reach its full potential in terms of both economic viability and social change, cryptographic systems will need to be developed to meet the needs of individual users. These systems will need to be cheap, user friendly, and above all, have public confidence.

For centuries the simple paper wrapper called an envelope has met the needs of the majority of users of the postal service. They come in many forms but most provide an indication of whether they have been tampered with. Also individuals have their own way of opening envelopes no matter what type they are. This basic philosophy needs to be applied to encryption systems for individual users. In other words a simple system which is easy to seal and easy to open and which does not require a wide variety of techniques for either.

Individual users will not be attracted to use services if they each involve different techniques for sending or receiving information. To this end service providers may need to look at providing a number of alternative schemes for distributing material so that they meet the individual's requirements, rather than expecting the user to maintain a number of systems to meet the various providers requirements.

Finally there is the question of public confidence. Users will not use cryptographic systems unless they have confidence in them. Firstly this confidence has to be established. Algorithms and the technology to implement them will need to be tested and the results made public. Once the tests have been completed, endorsement by standards bodies will build public confidence.

There is also an ongoing requirement to continue to test systems to ensure they remain suitable for the purpose for which they are being used. However some caution needs to be exercised in this respect. The main users of encryption systems at this stage are reasonably well equipped to make a considered assessment of the risks involved in using particular systems. This will not be the case initially for most individual users. Messages flowing freely around the network that an algorithm has been broken, even when details of the extent of technology to achieve the result are included, may cause a panic reaction and loss of confidence in the particular system. The resultant lack of confidence could have adverse effects on infrastructure usage. Debate on these issues should be limited to the appropriate parties rather than widely promulgated on the network.

In summary, what I have been saying today is that there is a need for the cryptography debate to be expanded to include the needs of the individual users who will make up the largest percentage of uses of the global information infrastructure.

References

[1] P.Barrett.*Long-term Cost Effectiveness of Telecommunications Interception* Australian Attorney-General's Department, Canberra, March 1994.

[2] Australian Government Publishing Service.*Networking Australia's Future*, The Final Report of the Broadband Services Expert Group December 1994, (also available from **http://www.dca.gov.au** or anonymous FTP from **www.dca.gov.au** (using your e-mail address for password))

[3] DPUBS:7003/93.*Australian Controls on the Export of Defence and Related Goods Guidelines for Exporters* Department of Defence, Canberra, March 1994

[4] Department of Defence.*Australian Controls on the Export of Technology With Civil and Military Applications A Guide for Exporters and Importers*Department of Defence, Canberra, November 1994.

[5] Australian Minister for Justice.*The High Road to Security Policing the Information Superhighway* Speech to the Australian Share/Guide Conference, March 1994, Available from the Australian Attorney-General's Department Canberra

[6] Document Number OCDE/GD(92)190.*Guidelines for the Security of Information Systems Organisation for Economic Co-operation and Development* Paris 1992 Document Number OCDE/GD(92)190.

[7] Unpublished Working Paper,*Public Key Authentication Framework (PKAF)* Australian Attorney-General's Department, Canberra, July 1993.

[8] S.T.Walker,*Software Key Escrow A better Solution for Law Enforcementt's Needs* Trusted Information Systems, Inc, August 1994.

Crypto in Europe — Markets, Law and Policy

Ross J Anderson

Cambridge University Computer Laboratory

Email: rja14@cl.cam.ac.uk

Abstract. The public debate on cryptography policy assumes that the issue is between the state's desire for effective policing and the privacy of the individual. We show that this is misguided.

We start off by examining the state of current and proposed legislation in Europe, most of which is concerned with preserving national intelligence capabilities by restricting the export, and in cases even the domestic use, of cryptography, on the pretext that it may be used to hide information from law officers. We then survey the currently fielded cryptographic applications, and find that very few of them are concerned with secrecy: most of them use crypto to prevent fraud, and are thus actually on the side of law enforcement.

However, there are serious problems when we try to use cryptography in evidence. We describe a number of cases in which such evidence has been excluded or discredited, and with a growing proportion of the world economy based on transactions protected by cryptography, this is likely to be a much more serious problem for law enforcement than occasional use of cryptography by criminals.

1 Introduction

The US Clipper chip initiative has fuelled extensive and acrimonious debate on the privacy versus wiretap issue, and this has spread to other countries too. At this conference, for example, an official from the Australian Attorney General's office has proposed that banks should use escrowed crypto, while ordinary people and businesses should be forced to use weak crypto [Orl95].

We provide an alternative view by looking at the state of play in Europe. We will firstly describe the political situation, then look at what cryptography is actually used for, and finally discuss the real problems of cryptography and law enforcement. Along the way, we will challenge a number of widely held beliefs about cryptology which underpin much research in the subject and condition the public policy debate. These include:

1. the primary role of cryptology is to keep messages secret. So if it is made more widely available, criminals will probably use it to stop the police gathering evidence from wiretaps;

2. its secondary role is to ensure that messages are authentic, and here it provides a useful (if not the only) means of making electronic evidence acceptable to a court. It is thus indispensible to the future development of electronic commerce.

2 Euopean Law and Policy on Cryptography

Some European countries, including Switzerland, Belgium and Germany, used to supply considerable quantities of cryptographic equipment to developing countries. This trade appears to have been tightened up recently as a result of American pressure, and now all European countries appear to enforce export controls on cryptographic hardware. Some even control its use domestically.

The country taking the hardest line is France. There, the "decret 73-364 du 12 mars 1973" put cryptographic equipment in the second most dangerous category of munitions (out of eight); any use required authorization from the Prime Minister, which could not be given to criminals or alcoholics. The "decret 86-250 du 18 fevrier 1986" extended the definition to include software, and specified that all requests be sent to the minister of the PTT with a complete description of the "cryptologic process" and two samples of the equipment. The "loi 90-1170 du 29 decembre 1990" states that export or use must be authorized by the Prime Minister unless used only for authentication [Gai92].

Few people in France seem to be aware of these laws, which are widely ignored. A hard line is still taken by SCSSI, the local signals agency, according to whom the use of PGP even for signatures will never be permitted [Bor95]; but when one looks at the actual text of the Loi No 90-1170 as it appeared in the Journal Officiel on 30th December 1990[1], it is unclear that digital signatures are covered at all.

Germany has no legal restraints on the domestic use of cryptography [Heu95]; indeed, Dirk Henze, the chief of the BSI (the information security agency), recommended that companies which cannot avoid sending data over the Internet should encrypt it, and the interior minister sees encryption as a precondition for the acceptance of electronic communication. However, Henze's predecessor Otto Leibrich took the view that security should rather be provided as a service by network operators in order to stop crypto equipment being available to villains [CZ95]; and a number of politicians, such as Erwin Marschewksi (home affairs spokesman of the CDU), argue for an outright ban [Moe95]. Meanwhile a law has just been passed forcing all telecomms companies to provide wiretap access to government agenices, including various call tracing services [Eis95].

Denmark, Finland, Sweden and Latvia have no domestic restrictions at present [Bor95] and no particular controversy which has come to our attention. But not all northern European countries are so relaxed; the Norwegian government is introducing its own encryption standard called NSK, which will be tightly licensed; Norwegian Telecom will manage the keys of line encryptors which use these chips and will be able to provide access to the intelligence services [Mad94].

Russia seems to be reverting to the policing traditions established under the Tsars and continued under the Soviets; a recent decree by President Yeltsin has

---

[1] Art 28. - On entend par prestations de cryptologie toutes prestations visant a transformer a l'aide de conventions secretes des informations ou signaux clairs en informations ou signaux inintelligibles pour des tiers, ou a realiser l'operation inverse, grace a des moyens, materiels ou logiciels concus a cet effet.

made cryptography illegal without a licence from the local signals agency [Yel95]. At the other extreme, the traditionally liberal Dutch government tried to impose a ban on civilian crypto in 1994 but was forced to back down at once by banks, petrol companies and other business interests.

The UK is mildly liberal at present. Prime Minister John Major stated in a 1994 parliamentary written reply to David Shaw, the member for Dover and Deal, that the government does not intend to legislate on data encryption. However, a spokesman for the opposition Labour party — which appears likely to form the next government — said that encryption should only be allowed if the government could break it [Art95]. This caused a storm on the Internet, and a subsequent policy document backed down on this issue; it did however propose to make warrants for the interception of communications much more easy to get. At present, these are only available to investigate serious arrestable offences; a future labour government would make them available for all offences, for 'racism' and for the 'protection of minors' [Lab95].

Even without a change in government, there is still occasional confusion in government policy. On the one hand, GCHQ permitted the export of over $35m worth of tactical radios to Iraq, which used them against allied forces in the Gulf War; on the other, it has made efforts to suppress academic research in cryptography. Interference with research is also common with the EU in Brussels, whose crypto policy is driven by SOGIS, the Senior Officials' Group (Information Security), which consists of signals intelligence managers. A typical EU project was Sesame, a Kerberos clone supposed to provide authenticity but not secrecy, and to be adopted by European equipment manufacturers. However its many flaws make this unlikely [ano95]: at the insistence of SOGIS, DES was replaced with xor, but the implementers did not even get a 64-bit xor right. Sesame also generates keys by repeated calls to the compiler's random number generator. Another project was RIPE (the RACE integrity primitives project), whose researchers were paid to devise a hash function (since attacked) but forbidden to do work on encryption. Close observers say that defective projects are approved deliberately to provide an excuse to refuse funding for more worthy proposals.

So the overall picture in Europe is one of confusion. Governments, and in particular by their signals intelligence agencies, claim to be concerned that the growth of commercial and academic cryptography might threaten intelligence and law enforcement capabilities. These fears are rarely articulated coherently; in addition to the contradictory behaviour of GCHQ, we would note that the current conference's paper from the Australian attorney general (cited above) says on the one hand that the use of encryption by criminals is not seen as a threat, but on the other hand that controls on crypto should be imposed.

Is there a real case here, or are we just seeing a panicky defensive reaction from bureaucratic establishments for whom the end of the Cold War means the loss of jobs and budgets, and who are looking for something to do? In order to assess the threat to law enforcement operations, we shall have to look first at what cryptography is actually used for.

3   European Applications of Cryptography

Many research papers on cryptography assume that two parties, traditionally called Alice and Bob, are sending valuable messages over an untrusted network. The idea is usually to stop an intruder, Charlie, from finding out the content of these messages. This application, message confidentiality, has historically generated perhaps 85% of research papers in the field.

Confidentiality has indeed been important in the government sector. The available information suggests that the NATO countries' military communications systems have about a million nodes, with the USA accounting for over half of this. This would appear to make governments the main users of cryptology, and they conduct the debate in these terms: for example, a recent report on crypto policy, one of whose authors is Assistant to the Director of the NSA [LKB+94], says *cryptography remains a niche market in which (with the exception of several hundred million dollars a year in governmental sales by a few major corporations) a handful of companies gross only a few tens of millions of dollars annually*.

This assessment is just plain wrong. The great majority of fielded crypto applications are not concerned with message secrecy but with authenticity and integrity; their goal is essentially to assure Bob that Alice is who she says she is, that the message he has received from her is the one she sent, or both. Here Charlie may try to impersonate Alice, or Alice might try to avoid paying for services rendered.

The main commercial cryptographic applications include the following.

Satellite TV decoders: There are tens of millions of these worldwide, with BSkyB having fielded 3.45 million in the UK alone by mid 1994 [Ran94]; they may be the largest single installed base of cryptographic terminal equipment. They are also the one nonmilitary application of cryptography which has attracted sophisticated and sustained technical attacks.

Automatic teller machines: ATMs have been around since 1968, and worldwide there are somewhere between 300,000 and 500,000 of them; over 100,000 are installed in Japan and 70,000 in the USA [AP94]. Many ATMs are networked together, and cryptography is used to manage personal identification numbers (PINs) — in fact this was the first large scale commercial application of cryptography [MM82]. The European ATM population is of the order of 100,000 [CI94].

Electronic funds transfer at point of sale (eftpos) . There is a lot of overlap between ATM, eftpos and credit card systems. In some countries (such as France and Australia) the ATM and eftpos networks are well integrated, with customers using PINs rather than signatures in shops; in others (like Britain), signatures are used to authorise retail transactions, but cryptography is still used to make the cards themselves harder to forge. The installed base of eftpos terminals has overtaken that of ATMs in most countries.

SWIFT: Based in Belgium, this is probably the oldest high security commercial computer network. For the last twenty years, it has transmitted payment instructions between the several thousand banks which own it, and its primary use of cryptography is to calculate a message authentication code (MAC) on each payment message [DP84]. The MAC keys used to be exchanged manually, but are now managed using public key protocols [ISO11166].

Telephone cards: These range from prepaid cards for public telephones to the much more sophisticated 'subscriber identity modules' (SIMs) used in GSM digital mobile phones. The SIMs are smartcards which identify the user of a telephone to the network for billing purposes, manage keys for encrypting the conversation [Rac88], and may even let the subscriber perform banking functions [Rob93] and place bets on horse races [Llo94]. Although only 4 million GSM phones are in use — mostly in Europe — the market is growing at 70% per annum, and 61% of new mobile phone subscribers in the UK now opt for GSM rather than the analogue alternatives [New94]. The market should grow even more quickly once GSM is fielded in countries such as China and India whose land based telephone systems are inadequate.

Utility tokens: The UK has about 1.5 million prepayment electricity meters, using two proprietary cryptographic schemes, and 600,000 gas meters using DES in smartcards. They are mainly issued to bankrupts and welfare claimants. Other European countries have smaller installations; France, for example, has about 20,000. However, prepayment meters are a growth industry in developing countries; technical information on such systems can be found in [AB94].

Computer access tokens: The market leading supplier of software protection dongles, Rainbow Technologies, has sold seven million units since 1984; from this business base, it took over Mykotronx, the manufacturer of the Clipper chip [Rai95]. There are also several vendors of one-time password generators. We have no overall figures for the total European sales of dongles and other access tokens, but they must be in the millions of units.

Building access control tokens: Although many early devices (from metal keys to magnetic cards) do not use cryptography, smartcard vendors are starting to make inroads in this market [Gir93].

Burglar alarms: Under draft CENELEC standards, class 3 and 4 alarm systems must provide protection against attacks on their signaling systems [Ban93], and some manufacturers are already taking steps in this direction. The market leading burglar alarm product in the UK claims to use 'high-level encrypted signalling' [BT93].

Remote locking devices for cars: These are starting to incorporate cryptographic techniques to thwart the 'sniffers' which can intercept and mimic the signals of first generation locking devices [Gor93].

Road toll and parking garage tokens: Some countries may issue these tokens to all their motorists [Sin95]; others may use multipurpose tokens, as with a German scheme to enable road tolls to be paid using the subscriber identity modules of car telephones [SCN94]. As well as a number of pilot schemes there are some fielded systems, including municipal parking garages in Glasgow [Tol93].

Tachographs: The European commission wants the current system for monitoring transport drivers' hours and speed to be replaced with a smartcard system which would be harder to tamper with [Tor94].

Lottery ticket terminals: The UK national lottery uses encryption to ensure that vendors cannot manufacture valid tickets after the draw or otherwise manipulate the system [Haw94]. Similar systems are used in other countries, and remote gambling terminals are becoming popular in the Far East.

Postal franking machines: The latest designs can be replenished remotely, thanks to cryptography: the user can use a credit card to buy a 'magic number' over the phone which lets her machine dispense a certain amount of postage.

Embedded applications: For example, some 40 million users of Novell NetWare use encryption embedded in the authentication protocols with which they log on to the system [Ber94].

These retail applications dwarf the world's military systems. An indication of overall sales figures comes from a French smartcard manufacturer which shipped 53m microprocessor cards last year and estimated that the total world market was 250 - 280m, and set to grow to 600m by 1997 [Rya94]. These microprocessor cards are more expensive than simple memory cards, and are typically used when some kind of crypto protocol needs to be supported. It would therefore seem reasonable to estimate that, whether we measure the size of a secure system by the number of nodes or the number of users, the retail sector is about two orders of magnitude larger than the military sector.

This economic fact is starting to loosen the traditional government control of the technology. On at least two occasions in the past decade — in South Africa in 1986 and the Netherlands in 1994 [Rem94] — a government has tried to ban civilian cryptography, and on each occasion it was forced to back down by pressure from banks, utilities, broadcasters, oil companies and others. So a lot of companies are coming to rely on cryptology, But is it in fact reliable? Does it really do what its advocates claim?

4   The Legal Reliability of Cryptography

In previous articles, we have discussed how cryptographic systems fail. We first looked at automatic teller machines, and the various frauds which have been carried out against them; it turned out that the attacks were not particularly high-tech, but exploited blunders in system design and operation [And93]:

- one bank wrote the encrypted PIN on the card strip. They got away with this for years, until villains found that they could change the account numbers on their own cards to other people's account numbers, and then use their own PINs to steal money from those accounts;

- villains would find out PINs by looking over their victims' shoulders, and then make up cards using the data on discarded tickets. This kind of fraud has been going on for years and is easy to stop, yet some banks still seem vulnerable to it;
- most fraud exploited much simpler blunders, such as insecure card delivery or poorly designed support procedures. For example in August 1993, my wife went into a branch of our bank, and told them that she had forgotten her PIN; they helpfully printed a replacement PIN mailer from a PC behind the counter. This was not the branch at which her account is kept; no-one knew her, and the only 'identification' she produced was her bank card and our cheque book.

We found much the same pattern with prepayment electricity meters. These allow the customer to buy electricity units at a shop and take them home in the form of a coded token, which is inserted into the meter; once the units run out, the supply is interrupted. Here too, most frauds exploited design blunders in simple and opportunistic attacks [AB95]:

- it was possible to insert a knife or a live cable into the card throat of one meter type and destroy the electronics immediately underneath, which had the effect of giving unlimited credit;
- another type of meter could have the tariff code set to a minute amount, so that it would operate almost for ever;
- another would often go to maximum credit in a brownout (a voltage reduction to 160 - 180V). This bug was due to one wrong assembly language instruction in the meter controller; its effect that customers threw chains over the 11KV feeders in order to 'credit' their meters. The manufacturer had to replace and re-ROM over 100,000 units.

A similar failure pattern has been found with satellite TV decoders as well, where, despite using an encryption algorithm which is vulnerable to analysis [And90], the majority of systems have been attacked by manipulating the key management mechanism. We conclude that cryptography does not provide any 'silver bullet' solution for the old problem of software reliability [Bro75]; systems which use it are just as likely to fail in unexpected ways as any other computer system.

This brings us on to the legal reliability of cryptographic systems — after all, as the above list shows, an increasing proportion of GNP is tied up in contracts which are enforced by crypto. If these contracts are broken in some way, the evidence needed for a civil suit or criminal prosecution may depend on crypto. Yet if crypto mechanisms are not reliable, then how can a judge tell whether the system was working at the time of the disputed transaction or not? Of course the system's owner — and his security consultants — will claim that the system is secure, but how is this claim to be tested?

This problem was illustrated by a recent series of court cases about disputed banking transactions. The typical pattern in such cases is that someone has a 'phantom withdrawal' from their account; they go to the bank and complain; the bank says that as its systems are secure, it must be the fault of the account holder, who must have been defrauded by a friend or relative; the victim goes to the police and lays a complaint; and some unlucky person gets arrested.

In years gone by, that was effectively the end of the matter; for example, one Janet Bagwell was accused of stealing money from her father, and was advised to plead guilty as it was her word against the bank's. She did so and then disappeared; much later, the bank manager in charge of the cover-up confessed that it had all been an administrative error. By then, Janet's lawyer could no longer trace her, and we can only speculate at the effects which this incident had on her life.

However, in the last three years, defence lawyers have started to challenge the banks' claims that their systems are secure. In the first such case, charges of theft against an elderly lady in Plymouth were dropped after our enquiries showed that the bank's computer security systems were a shambles, and we demanded full information about their security systems. The same happened in a number of subsequent cases [And94].

The most notorious case so far is that of John Munden. John was a constable at our local police station in Bottisham, Cambridgeshire, with nineteen years' service and a number of commendations. However, his life came apart after a holiday in Greece; he returned to find his bank account empty, and went to the manager to complain.

The manager asked how his holiday in Ireland went; apparently the information he had in front of him indicated that ATM withdrawals had been made in Omagh. When John told him that he had been in Greece, the story changed; the bank claimed that six withdrawals totalling £460 had been made from his home branch just before he had gone on holiday. When he persisted with his complaint, the bank complained to the police that he was trying to defraud them. He was arrested, tried for attempted fraud and — to the surprise of many — convicted.

The description of the bank's systems which came out at the trial was more reminiscent of Laurel and Hardy than of ISO 9000:

- The bank had no security management or quality assurance function. The software development methodology was 'code-and-fix', and the production code was changed as often as twice a week;
- the manager who gave technical evidence claimed that bugs could not cause disputed transactions, as his system was written in assembler, and thus all bugs caused abends;
- he claimed that as ACF2 was used to control access, it was not possible for any systems programmer to get hold of the encryption keys which were embedded in application code;
- he had not investigated the disputed transactions in any detail, but just looked at the mainframe logs and not found anything which seemed wrong (and even this was only done once the trial was underway, under pressure from defence lawyers);

- there were another 150-200 disputed transactions which had not been investigated;
- in the branch itself, the security cameras were conveniently not working, and the branch manager had since left the bank's employment.

An appeal was launched, and a week before it was due to be held, the bank produced a thick expert report from a partner at its auditing firm claiming that its systems were secure. The defence team promptly went to court and asked for the time and the access to prepare their own report as well. The court responded with an order that the defence expert have full access to the bank's 'computer systems, records and operating procedures'.

After five months in which the defence repeatedly demanded this access, and in which the bank refused it, a further application was made, and the judge has now ruled that the prosecution will not be allowed to call expert evidence at the appeal. The date for this has still to be set at the time of writing; it remains to be seen whether the Crown will offer any evidence at all.

This underlines a conclusion which we already drew in [And94]:

> Security systems which are to provide evidence must be designed on the assumption that they will be examined in detail by a hostile expert.

It remains to be seen how the other systems listed above will stand up to the rigours of a trial. One suspects that few if any of them were designed with the above principle in mind; and the lesson does not appear to be getting through.

For example, the Bank of England is building a system called Crest which will be used to register all UK equities. When its security was publicly criticised [Inm95], the Bank's reaction was to keep the design secret; and despite repeated criticism it has evaded the question of how its system will withstand a legal challenge [Boe95].

5   How Realistic is European Public Policy?

Most crypto is about authenticity rather than secrecy, and an increasing proportion of economic activity relies on it to some extent. Thus more and more prosecutions are likely to depend on cryptographic evidence, and law enforcement agencies should be concerned at the difficulty of relying on current systems in court. However, the only official interest so far in liability was in a US Commerce Department study which looked at whether the government could have its cake and eat it too; the idea was that the government could manage everybody's keys without assuming too much liability when things go wrong [Bau94].

The policy debate continues to focus on secrecy, with civil rights groups saying that crypto is important for freedom and privacy in the electronic age, and governments claiming that good crypto would make law enforcement more difficult by making it harder for the police to gather evidence using wiretaps.

This debate misses the point. Quite apart from the liability problem, it is not true that villains will use crypto; it is not true that wiretaps are important to the police; and it is not true that cryptography is important to individual privacy.

1. Clever crooks don't use crypto for secrecy. They are aware that the main problem facing law enforcement is not traffic processing, but traffic selection [LKB+94]: in layman's terms, a ten minute scrambled telephone call from Medellín, Columbia, to 13 Acacia Avenue, Guildford, is an absolute give-away. Instead, a competent villain will try to bury his signals in innocuous traffic. One common modus operandi (in the USA and increasingly the UK) is to use an address agile system — cellular telephones are repeatedly reprogrammed with other phones' identities. In Paris, villains use cordless telephone handsets to make calls from outside unsuspecting subscribers' homes [Kri93]; and in Britain, villains have tapped domestic phone lines in order to make outgoing international calls.

2. The official use of wiretaps varies substantially from one country to another, and even from one local police force to another. In the USA, three states forbid wiretaps completely, and in 1993 there were 29 others that did not use any; 73% of state wiretaps were in the 'Mafia' states of New York, New Jersey and Florida [Han94]. There is similar variation in Europe. Many wiretaps were carried out unlawfully in France by the President's henchmen, and this was one of the scandals which dogged the last years of the Mitterrand administration. In the UK, on the other hand, all legal wiretaps have to be authorised by a minister, and the number reported (both to parliament and by our police informants) is low. The clear conclusion is that wiretaps are not essential, or even very important, for policemen; many admirable police forces function perfectly well without them.

3. Even if crypto were banned, it still does not follow that wiretaps would remain a feasible option for the police. It is very expensive to provide a wiretap capability in a modern digital network; if it is mandated in the USA, phone companies say it could cost $5bn in the first four years alone. Yet US police agencies only spent $51.7 million on wiretaps in 1993 — as close as one can get to an estimate of their value [Han94]. Forcing phone companies to subsidise 96% of the cost of wiretaps makes no more sense than forcing Westland to sell helicopters to the police for the same price as cars. This may become an issue in Europe as well as the USA; senior managers in the European telecomms industry have complained to the author that a similar provision would add to costs, stifle competition and be a disaster for business generally.

4. The real threats to individual privacy have little to do with crypto but are rather concerned with the abuse of authorised access to data.
   - All US police forces have access to the FBI's criminal records system through gateways, and it has proved impossible to impose effective controls on them. As a result, criminal records can be obtained through private detective agencies who in turn buy them from local police officers.

These records have been used on occasion to discredit political opponents and troublemakers [Mad93]. UK criminal records are no different, and the consolidation of European criminal records in the Schengen system will make the problem worse;

- Most of the big UK banks let any teller access any customer's account (one bank even boasted about this when their system was recently upgraded). The effect, as widely reported in the UK press last year, is that private eyes get hold of account information and sell it for £100 or so [LB94];
- a banker on a US state health commission had access to a list of all the patients in his state who had been diagnosed with cancer. He promptly called in their loans [Bar93];
- a study at the University of Illinois found that 40 percent of insurance companies disclose medical information to others, such as lenders, employers and drug salesmen, without the patient's permission; and over half of Fortune 500 companies use medical records in hiring decisions [Con94]. Although the situation is not yet as bad in many European countries as it is in the USA, it is rapidly heading the same way [And95].

The fight against such abuses is political rather than technical. For example, the British Medical Association has recently threatened to boycott a new medical network being installed by the government [Jac95]. Although the doctors want encryption (which the government is resisting), their primary complaint is not about mechanisms but about policy — namely that the system must not repeat the mistakes of the banks and the police; it must limit the number of people who can access any patient's record.

We conclude that the privacy versus police debate is misguided; neither the libertarians nor the policemen have a serious case. Yet this debate continues to wend its weary way across the world; and since about March 1995, there appears to have been a concerted effort by many of the developed world's secret policemen to introduce laws and regulations facilitating wiretapping of digital communications, key escrow, mandatory weak crypto, and various other measures whose ostensible purpose is protecting law enforcement capabilities but whose real purpose may be to hinder the uptake of cryptography by industry and commerce, to preserve employment at signals intelligence agencies, or both.

6 Conclusions

The politics of cryptology is often viewed as a Manichaean struggle between the privacy of the individual and the ability of the police to detect crimes such as money laundering and child pornography. While this perception may drive the actions of legislators, it is at odds with the facts. Villains do not use crypto; wiretaps are almost irrelevant to police work; and there are many much more immediate threats to privacy, such as the wholesale trading of credit records, medical records and other information with the power to do harm.

The real law enforcement problem is that neither prosecutors nor civil litigants can rely on cryptographic evidence, and in an information based society, this kind of evidence is likely to figure in more and more trials. Within the next two to three years, we expect that arguments about whether the crypto was working (and whether the defence experts can examine it) will spread from disputed ATM transactions to investigations of securities fraud and other serious white-collar offences.

This is not inevitable. It can be tackled by insisting that cryptographic systems be built to withstand examination by hostile expert witnesses, just like the alcohol intoximeters and radar cameras used by traffic policemen. If governments are serious about preserving their law enforcement capability, then they should not harrass crypto manufacturers but rather encourage them to get their products up to this standard — whether using government purchasing power in projects such as Crest, product development subsidies such as those currently wasted by the EU in Brussels, or national standards bodies.

But if the real goal is to preserve the payrolls and influence of the secret policemen and their favoured suppliers, then this policy may be painful. The infrastructure built up by GCHQ and its overseas counterparts is of little relevance to commerical crypto. For example, the ITSEC/ITSEM procedure typically takes a year and a million dollars to evaluate a security product, while underwriters' laboratories might do the job in a month for twenty thousand dollars [ESO94]. We can see no reason why military crypto suppliers should be any more able to beat swords into plowshares than the similarly bloated and inefficient suppliers of tanks, warships and missiles turned out to be.

The challenge facing Europe's crypto policymakers is a hard one. It is not just a matter of sacking a few thousand civil servants at agencies such as GCHQ, and letting a few CLEFS and equipment vendors go to the wall. It is the challenge of adapting to a major paradigm shift: from intelligence to evidence, from protecting lives to protecting money, from secrecy to authenticity, from classified to published designs, from tamper-proof hardware to freely distributed software, from closed to open systems, and from cosseted suppliers to the rough and tumble of the marketplace.

Every aspect of this change is likely to be alien and threatening to the signals security establishment. On past form, we expect that the securocrats will fail to adapt. They appear to be in a no-win situation; their attempts to retain control of the technology are probably doomed, and if they continue to fight market forces, then they risk humiliation, with further cuts in their organisations' budgets and influence.

Acknowledgement: The final version of this paper was written while the author was a guest of the Information Security Research Centre, Queensland University of Technology.

References

[And90]   RJ Anderson, "Solving a class of stream ciphers", in *Cryptologia* v **XIV** no 3 (July 1990) pp 285–288

[And92]   RJ Anderson, "UEPS — A Second Generation Electronic Wallet". in *Computer Security — ESORICS 92*, Springer LNCS **648**, pp 411–418

[And93]   RJ Anderson, "Why Cryptosystems Fail", in *ACM Conference on Computer and Communications Security* (Nov 1993) pp 215–227; journal version in *Communications of the ACM* v **37** no 11 (Nov 1994) pp 32–40

[And94]   RJ Anderson, "Liability and Computer Security: Nine Principles", in *Computer Security — ESORICS 94*, Springer LNCS v **875** pp 231–245

[And95]   RJ Anderson, "NHS-wide networking and patient confidentiality", in *British Medical Journal* v 311 (1 July 1995) pp 5–6

[ano94]   anonymous, "SESAME", posted to Internet newsgroup `sci.crypt` as message `<154315Z07111994@anon.penet.fi>`, 7th November 1994; and followup postings

[Art95]   C Arthur, news article in *New Scientist*, 11th March 1995; when accused by a labour spokesman of misquoting, he supplied the tape of his interview to the net. See 'Re: Britain to outlaw PGP - whats happened so far', posted as article `<D60F7L.KvH@exeter.ac.uk>` to `sci.crypt`, 25 Mar 1995.

[AB94]   RJ Anderson, SJ Bezuidenhout, "Cryptographic Credit Control in Prepayment Metering Systems", in *1995 IEEE Symposium on Security and Privacy, pp 15–23*

[AP94]   Associated Press, "BANKS-ATMS", wire item **1747**, 30 November 1994, New York

[Ban93]   KM Banks, *Kluwer Security Bulletin*, 4 October 93

[Bar93]   ED Bartlett, "RMS need to safeguard patient records to protect hospitals", in *Hospital Risk Management* v 15 (1993) pp 129–133

[Bau94]   MS Baum, *'Federal Certification Authority Liability and Policy — Law and Policy of Certificate-based Public Key and Digital Signatures'*, U.S. Department of Commerce Report Number NIST-GCR-94-654

[Ber94]   T Berson, *private communication*

[Boe95]   Bank of England, "Crest's security", in *Crest project newsletter, April 1995*

[Bor95]   S Bortzmeyer, "Data Encryption and the Law(s) — Results", *available from* `http://web.cnam.fr/Network/Crypto/survey.html` (15/12/94);

[Bro75]   FP Brooks, *'The mythical man-month: Essays on software engineering'* (Reading, Massachussetts, 1975)

[BT93]   *'RedCARE — The secure alarm networks'*, British Telecom, 1993

[Con94]   "Who's Reading Your Medical records?", in *Consumer Reports* (Oct 1994) pp 628–632

[CI94]   *Cards International* has country surveys about once a month; similar information can be found in *Banking Technology*

[CZ95]   Report of 4th Deutschen IT-Sicherheitskongreß, Bad Godesberg, 8–11 May 1995, in Computer Zeitung no 21 (25th May 1995) p 21

[Eis95]   S Eisvogel, posting about German 'Fernmeldeanlagen Ueberwachungs-Verordnung' of May 4th 1995 to tv-crypt mailing list

[ESO94]   Conference debate on security evaluation, ESORICS 94

[Gai92]   JL Gailly, "French law on encryption", *posted to Internet newsgroup* `sci.crypt` *as message* `<831@chorus.chorus.fr>`, *28 Oct 92 by* `jloup@chorus.fr` *(Jean-loup Gailly)*

[Gir93]   Y Girardot, "The Smart Option", in *International Security Review Access Control Special Issue* (Winter 1993/1994) pp 23–24

[Gor93]   J Gordon, "How to Steal a Car", talk given at 4th IMA Conference on Cryptography and Coding, December 1993

[Han94]   R Hanson, "Can wiretaps remain cost-effective?", in *Communications of the ACM v 37 no 12 (Dec 94) pp 13–15*

[Haw94]   N Hawkes, "How to find the money on lottery street", in *The Times* (8/10/94) weekend section pp 1 & 3

[Heu95]   A Heuser, writing on behalf of BSI to U Möller, copied at http://www.thur.de/ulf/krypto/bsi.html

[Inm95]   P Inman, "Bank of England share system 'open to fraud' ", in *Computer Weekly*, 23rd March 1995, pp 1 & 18

[ISO11166] 'Banking — Key management by means of asymmetric algorithms — Part 1: Principles, procedures and formats; Part 2: Approved algorithms using the RSA cryptosystem, *International Standards Organisation, 15th November 1994*

[Jac95]   L Jackson, "NHS Computer is 'Paparazzi's Dream' ", *Press Association* report 1520, 1st June 1995.

[Kri93]   HM Kriz, "Phreaking recognised by Directorate General of France Telecom", in *Chaos Digest 1.03 (Jan 93)*

[Lab95]   Labour party policy on the information superhighway, at URL http://www.poptel.org.uk/labour-party/content.html

[Llo94]   C Lloyd, "Place your bets while on the hoof", in *the Sunday Times* 2nd October 1994 section 2 p 11

[LB94]   N Luck, J Burns, "Your Secrets for Sale", in *Daily Express*, 16th February 1994 pp 32–33

[LKB+94] S Landau, S Kent, C Brooks, S Charney, D Denning, W Diffie, A Lauck, D Miller, P Neumann, D Sobel, "Codes, Keys and Conflicts: Issues in US Crypto Policy", *Report of the ACM US Public Policy Committee June 1994*

[Mads93] W Madsen, "NCIC criticised for open security and privacy doors", in *Computer Fraud and Security Bulletin (Oct 93) pp 6 - 8*

[Mad94]   W Madsen, "Norwegian encryption standard moves forward", in *Computer Fraud and Security Bulletin (Nov 94) pp 10–12*

[Moe95]   Ulf Moeller, "Kryptographie: Rechtliche Situation", at http://www.thur.de/ulf/krypto/verbot.html

[New94]   M Newman, "GSM moves past analog", in *Communications Week* issue **135** (28 November 1994) p 40

[Orl95]   S Orlowski, "Encryption and the Global Information Infrastructure, An Australian Perspective", *this volume*

[Pat95]   N Pattinson, Schlumberger, *personal communication*

[Rac88]   Racal Research Ltd., "GSM System Security Study", 10th June 1988

[Rai95]   "Counterfet Software Operations", press release no. OTC 06/27 1135 on CompuServe.

[Ran94]   J Randall, "BSkyB set for record £5bn stock market debut", in *The Sunday Times* (2nd October 1994) section 2 p 1

[Rem94]   MNR Remijn, "Tekst van de memorie van toelichting van de wet tegen crypto", posted to Internet newsgroup nlnet.cryptografie as message <1994Apr15.124341.20420@news.research.ptt.nl>

[Rob93]   D Robinson, "Cellular phones offer chip opportunity", in *Cards International* no **98** (24th November 1993) p 10

[Rya94]   I Ryan, "Market diversity points way forward", in *Cards International* no 111 (13th June 1994) p III

[Sin95]   Discussion at Singapore National Computer Board, 30th June 1995

[SCN94]   "German Motorway Toll Trial is GSM-Based", in *Smart Card News* v **3** no 3 (March 94) pp 41–44

[Tol93]   Discussions with staff of Tollpass Ltd., Edinburgh

[Tor94]   A Torres, "Commission wants black box, smart cards to enforce road safety", *Reuters RTec* 09/02 **0804**

[Yel95]   B Yeltsin, Decree no. 334, 3rd April 1995; English translation at http://www.eff.org/pub/Privacy/

Saving Dollars Makes Sense of Crypto Export Controls

by Mark Ames, Telstra Corporation
Melbourne, Australia
mames@vitg.exec.telecom.com.au

Life is an enigma. Dreams are the key.

Disclaimer:

This article does not represent the view, opinions, or intentions of the Telstra Corporation or any of its subsidiary or related companies. The views expressed are solely those of the author, who accepts full responsibility for content.

Abstract:

This analysis looks at the economic basis of US crypto export controls. The financial implications to policy makers and policy targets are considered. From an economic point of view, US crypto export controls are well founded, but risky for the US in the longer term. Prospects for change depend on the reactions of key players targeted by the policies and the perception of long term risk by the policy makers.

Introduction

US policies restricting export of information security products protect the economic intelligence gathering position of the United States. The principal driving this policy is cost control; the export of security products including standard US commercial encryption would greatly increase the cost of gathering economic and industrial intelligence. The US is not the only country imposing export controls on cryptographic hardware and software. It is, however, the most vigorous in enforcing them, and the availability of secure US hardware and software products is a significant factor in the operations of non-US organisations, both corporate and government. US export controls do not apply to Canada, which imposes parallel controls.

Most analysis of US crypto export controls has focused on emotive and ideological issues. Many commentators maintain this is a last ditch stand by cold warriors in the NSA who feel threatened by their loss of control over cryptography. This may be true, but it is not the driving force behind continued US Administration support for restrictions on the export of strong cryptographic systems.

Policy Targets, Real and Imaginary

One justification for export controls is to restrict the use of cryptography by criminals and foreign governments [ACM94] [NIST93]. Public policy instruments cannot do this. In the past, governments held a monopoly on cryptography, both in the US and elsewhere. Cryptographic tools were not readily available, and governments hoarded the expertise to develop them. It was a limited competition, as described in Sol-zhenitsyn's novel "The First Circle," to see which government could develop the best systems.

Now the situation is different. Cryptographic software, algorithms, and expertise are commodity items. Criminals and foreign governments have easy access to DES and a number of other tested algorithms free of charge on the Internet. Information on implementation, cryptanalysis, and software to cloak encrypted messages in binary media files is also in the public domain. There is a rapidly growing skill base of crypto-graphers in academic and corporate roles. Designing and implementing crypto systems for nefarious purposes was never easier than it is today. In this situation one of the primary justifications for cryptographic export controls – keeping the technology out of the hands of foreign governments, criminals, and terrorists – is simple propaganda playing on public fears.

US cryptographic export controls have the principle effect of limiting the use of strong cryptography by organisations targeted for strategic economic intelligence. The US policy is effective because those organisations buy industry standard US computer products with export grade security, and top management is not yet seriously concerned about the threat of economic intelligence gathering. While these attitudes persist, US crypto export policies will remain effective and are unlikely to change.

The Policy Makers

US export control policies maintain economic advantage and control the cost of main-taining that advantage. The role of the US in the world economy has slipped drama-tically over past decades. The US has lost its advantage in many areas of manufac-turing and technology. Its policies focus on supporting a few key leading industries, in-cluding computer software and advanced technology.

Strategic economic intelligence gathered by the communications intelligence (COMINT) program of the US National Security Agency (NSA) supports US eco-nomic advantage. The NSA is the single most significant contributor to US crypto-graphy policies. An intelligence requirements review in 1992 shifted the post-1947 emphasis on military and political targets to economic ones [Majo93]. In this New World Order, national security is about economic, not military advantage. The NSA is not the only significant COMINT agency in the world [Mads91], but it has the advantage in terms of technology and its global infrastructure.

US policy makers see the potential loss of cheap economic intelligence as the principal risk to manage. Unencrypted electronic information flows have grown exponentially over the past 20 years and with them an overflowing bounty of signals intelligence. The immediate availability of DES in exported US products would effectively close a window on cheap and plentiful communications intelligence that has widened to enormous proportions in that time. This is the current policy motivator. The US and their COMINT allies attempt to control research and software development, influence international standards, pressure foreign governments, and use other means at their disposal to restrict the spread of strong cryptography [Ande95]. The combined effects of international competitive pressures, shifting alliances based on economic rather than military interests, and growing demand for products with strong crypto must place the continued effectiveness of these efforts in doubt.

"Export Quality"

Outside the US and Canada, corporations and other organisations (including the Australian Department of Defence) cannot legally obtain standard international security features (56 bit DES and 155 digit RSA) in mass market software from US vendors. Until 1992, these products could not legally be sold or distributed outside the US unless vendors removed all cryptographic functionality. Otherwise, each sale required approval and licensing by the US State Department.

In response to pressure from software vendors, the US government instituted the Commodity Jurisdiction system, which allowed software meeting certain criteria to receive a general export licence [USDS92]. The government does not license DES or 155 digit RSA. Instead, proprietary (symmetric) algorithms with a maximum key length of 40 bits are licensed, as is RSA with a reduced modulus size, typically 64 digits [Lotus92]. Some products are still not available outside the US (ViaCrypt's commercial version of PGP). 'Export' standard security mechanisms are substituted in others (as in Lotus Notes and Microsoft Exchange). These products offer confidentiality options because customers want them and governments (especially the US government) encourage their use.

Businesses have to content themselves with crippled 'export' versions or do without. Alternatively, they can source non-US encryption products to add to commercial software to protect their information. The added costs are a huge disincentive. Export versions of US software are required to make 'add-on' security as difficult as possible, increasing integration and systems management costs. Top management resists these additional overheads, especially since the vendors of these export versions of US software tout their 'government approved', 'export quality' security features, even though the protection they offer is questionable.

The Bottom Line

What is the practical impact of limiting non-US organisations to export quality 'encryple-ware'? Using a one million dollar Wiener machine [Wien93], a 56 bit DES key can be recovered in a few hours. A 40 bit key can be recovered within a second. A 100 digit RSA modulus can be recovered in a similar time. This means that 40 bit traffic can be cryptanalysed and monitored on line. Using the same technology, 56 bit traffic must be recorded and broken off line. The added cost, over and above the required computer power, is significant. The imposition of 40 bit security in export software gives the NSA intelligence on the cheap. It simply could not afford to monitor the same amount of 56 bit traffic: the cost would be 10,000 times greater.

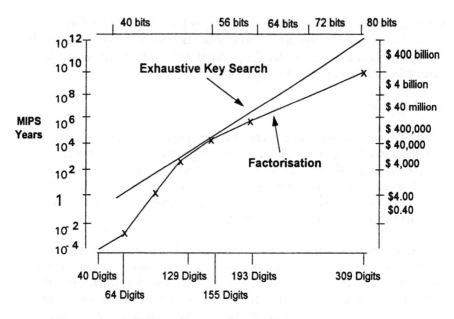

Fig 1. Relative Cryptanalysis Costs

Figure 1 draws on the work of Ron Rivest and his paper on the difficulty of factoring large numbers [Rive90]. Rivest introduces the concept of the *MIPS Year*, being 3.15×10^{13} operations, as a measure for the amount of processing power required to factor a large number. These are conservative figures assuming optimised design, configuration, programming, and operation of the processors used.

Here we have used his calculation for the quadratic sieve algorithm:

$$L(n) = \exp \sqrt{\ln(n) \bullet \ln\ln(n)}$$

for the factorisation of a modulus length of n digits. The processing power required to determine an asymmetric key is calculated as:

$$\frac{2^n}{2}(300)$$

where n is the number of bits in the key and 300 is taken as the number of operations required to test a key against a block of ciphertext. This is the average time to search the key space.

The relative costs are based on one MIPS Year in 1995 costing $4.00. A Wiener machine would appear to reduce the cost of key-search by two orders of magnitude. In any event, the process is not trivial. A DEC Alpha processor running at 1000 MIPS would still take the better part of a year to deliver a 40 bit key or factor a 100 digit RSA modulus. A million dollars worth of Alpha processors could find a 40 bit key in a minute; a Wiener machine could do the job in milliseconds.

Figure 1 shows clearly that factoring a 64 digit RSA modulus is a far easier and cheaper task. Factoring a public key modulus makes it possible to intercept session keys on the fly, eliminating the need for cryptanalysis by exhaustive key search. This task needs to be performed once for each user and server. (These keys are rarely changed.) This makes it possible then to impersonate users with forged digital signatures and forged 'secure' email. It also makes it possible to impersonate a server or high level user and freely browse information held in organisational databases. The competitive value of this information can be considerable in a product like Lotus Notes and its soon to be competitor, Microsoft Exchange.

US export controls drop entry level costs for government or private COMINT from billions or hundreds of millions to a shoe-string couple of million dollars. Measured against the value of a crucial patent or winning a major supply bid or contract tender, this is an attractive investment. Given that anyone with a million dollars could build a Wiener machine, 40 bit traffic can be recovered on line by any COMINT agency or private group targeting that type of intelligence.

In dropping the membership fee for entry into the COMINT club, the NSA must be confident that it can exploit 40 bit traffic more effectively than its competitors. Restricting export products to 40 bit crypto also focuses COMINT activities away from the US and Canada.

Policy Futures

If the perceptions of policy makers and the actions of policy targets remain unchanged, DES will be freely exportable in 2018. In the meantime, what factors are likely to influence the US to change the current policy? Making more and better algorithms freely available on the Internet will have little effect: this cryptographic technology does not usually find its way into business computer systems. Lawsuits, constitutional challenges, and congressional debates are unlikely to change the official view, considering the dollars and bureaucratic futures at stake.

US hardware and software vendors continue to put pressure on the US government to relax export controls, claiming that they are losing $9 billion a year in sales because of the restrictions [Clari94]. Nonetheless, the US has 70% of the global market share in software [Info95] (where the restrictions fall most heavily), and hardware exports continue to grow [EET95]. The lost sales could be seen as the price paid by the US computer industry for maintaining competitive advantage through economic intelligence.

What is really preserving the status quo and keeping the NSA's costs down is the persistent ignorance or lack of concern among international executives for the safety of their corporate information. Management guru Peter Drucker [Druc88] recognised years ago that modern organisations are critically dependent on their information. The Organisation for Economic Cooperation and Development confirms that dependency, extending it to social and economic infrastructures [OECD92]. The FBI [Majo93] and the NSA warn US business and research establishments of the threats posed by industrial and economic intelligence gathering by foreign countries. Similar messages are repeated in other countries.

Cryptographic export controls give the NSA a competitive edge. Non-US corporations also understand the need to have an edge against their competitors. If that edge can be gained with effective information security, ITAR or PKP licensing restrictions are unlikely to stand in the way. The US government will change its policies when they are no longer perceived to be effective. As non-US firms take steps to protect their information against communications intelligence gathering, COMINT costs increase, and crypto export restrictions cease to be effective. Given the increasing awareness of the value and the threats to corporate information, this is likely to happen well before 2018. However, this 'steady as she goes' course carries risks that US policy makers may be underrating.

Competition & The Risk of Computationally Secure Cryptography

Other players will enter the COMINT arena, including information 'entrepreneurs,' because of the lower entry cost and the high value of competitive information. The major COMINT agencies and their governments are very discreet in passing on intelligence 'product', taking care to conceal any indicator that the information was acquired through eavesdropping, systems penetration, or cryptanalysis. As more countries and private operations focus on commercial 40 bit traffic, information gleaned by these operators will be looking for a market. Sooner or later these smaller, inexperienced agencies (government or private) will slip up. Word will get out that someone using communications intelligence techniques obtained trade secrets from a competitor. This would seriously undermine confidence in the security and safety of electronic commerce, and in US export software in particular.

If management sees the profits or competitiveness of their organisation threatened by communications intelligence, we can expect them to respond vigorously. If a 56 or 64 bit key length will improve security, why not use 128 bit commercial crypto systems if more is better? Why stop at 155 digits for an RSA modulus when 300 digits can be implemented just as easily? DES and RSA 155 are computationally strong algorithms. Machinery to break them in reasonable time costs hundreds of millions of dollars. Only a select club of government COMINT agencies has these resources, which they must carefully allocate against high value targets. Stronger ciphers, such as the 128 bit International Data Encryption Algorithm (IDEA) or 300 digit RSA are computationally secure – it is impractical or impossible to build machines to break them now or in coming decades. Use of these secure algorithms can eliminate the possibility of effective COMINT or cryptanalytic attack. The short term benefits of cheap intelligence are clear. How closely have US policy makers looked at the longer term possibility of secure cryptographic algorithms frustrating key areas of their COMINT operations?

Conclusion and a Proposal

US crypto export policies will, over the next five years, cease to be cost effective as organisations based outside the US and Canada identify their sensitive information assets and take steps to protect them against communications intelligence gathering. The measures they implement are just as likely to include computationally secure algorithms in place of computationally resistant systems like DES and RSA 155. The US and other governments risk finding the situation uncontrollable unless they act together soon. If computationally secure algorithms are widely implemented, COMINT will lose its current position (and budget) to more traditional methods of clandestine information gathering.

Acting in their own interests, the US and other major COMINT nations could cooperate on a pragmatic basis and offer the international community strong crypto now. Countries accepting a strong international crypto standard would be protecting national information resources that contribute to the growth of their economies.

The major COMINT agencies would have to accept a short term loss in productivity in order to maintain their long term position. The COMINT playing field will be levelled and high entry costs maintained, limiting the number of players. The NSA would maintain a clear, if diminished advantage in the economic intelligence stakes.

Lifting crypto export controls would ease domestic pressure on the US government and its crypto policy position. Strong, computationally resistant crypto would find easy acceptance as an international standard, and the US government could focus its efforts (more successfully) against the spread of computationally secure cryptographic systems. US software vendors would have a wonderful opportunity to market international security upgrades, and new markets would open up for hardware vendors and specialist security companies.

We systems security professionals would be able to work more productively. We could use cryptography as a standard, effective tool for secure systems design without having to work around US export controls and a mixed bag of national restrictions.

References

[ACM94]   S. Kent, et al, "Codes, Keys and Conflicts: Issues in U.S. Crypto Policy," Report of a Special Panel of the ACM U.S. Public Policy Committee (USACM) June 1994

[Ande95]   Ross Anderson, "Crypto in Europe — Markets, Law, and Policy", Cryptography Policy and Algorithms Conference, Brisbane, Australia (Pre-Proceedings), July 1995.

[Clari94]   "Key Vote Nears On Encryption Exports", newsbytes@clarinet.com, 18 Jul 94.

[Druc88]   Peter Drucker, "The Coming of the New Organisation," Harvard Business Review, January-February 1988

[EET95]   "Exports propel the U.S. electronics industry to new heights," *Electronic Engineering Times,* May 29, 1995, v850 p46

[Info95]   "Vendors say U.S. encryption export restrictions threaten Internet Security," *InfoWorld,* Mar 06, 1995, v17, no10, p 10)

[Lotus92]   "Lotus Notes Security White Paper", Lotus development Corporation, 1992.

[Mads91]   W. Madsen, "Data Privacy: Legislation and Intelligence Agency Threats," *Computer Security and Information Integrity,* Elsevier, Amsterdam, 1991, (IFIP/Sec '90 Proceedings.)

[Majo93]   David G. Major, FBI Intelligence Division, "Economic Espionage and the Future of U.S. National Competitiveness," reprinted by Datapro Information Services Group, Delran, N.J.

[NIST93]   L. Hoffman, et al, "Cryptography: Policy and Technology Trends," National Institute of Standards, Gaitersberg, MD.

[OECD92]   Organisation for Economic Co-Operation and Development, Committee for Information, Computer, and Communications Policy, *Guidelines for the Security of Information Systems,* Paris, 1992.

[Rive90]   DR. RON RIVEST ON THE DIFFICULTY OF FACTORING, ftp://rsa.com/pub/ciphertext/vol1n1.txt.

[USDS92]   "GUIDELINES FOR PREPARING COMMODITY JURISDICTION REQUESTS", US Dept of State, Bureau of Politico-Military Affairs, Office of Defense Trade Controls, 17 Sep 1992

[Wien93]   Michael Wiener, "Efficient DES Key Search", presented at Crypto '93.

A Proposed Architecture for Trusted Third Party Services

Nigel Jefferies, Chris Mitchell, Michael Walker

Information Security Group
Royal Holloway, University of London
Egham Hill,
Egham, Surrey TW20 0EX
England

Abstract

In this paper we propose a novel solution to the problem of providing trusted third party services, such as the management of cryptographic keys for end-to-end encryption, in a way that meets legal requirements for warranted interception. Also included is a discussion of what might be a reasonable set of requirements for international provision of such services, as well as some analysis of the cryptographic properties of the architecture and how it might operate in practice.

1 Introduction

There has been much recent discussion on the question of how to meet users' requirements for security services, such as confidentiality and authentication. This has been largely prompted by the US government's Clipper proposals [FIPS185], as well as the increasing use of electronic means for transferring commercially sensitive data. On the one hand, users want the ability to communicate securely with other users, wherever they may be, and on the other hand, governments have requirements to intercept traffic in order to combat crime and protect national security. Clearly, for any scheme to be widely acceptable, it must provide the service users want, as well as meeting the legal requirements in the territories it serves.

To create a platform that can be used to provide user services, it is anticipated that solutions will be based on the use of trusted third parties (TTPs) from which users can obtain the necessary cryptographic keys with which to encrypt their data or make use of other security services. Law enforcement agencies' requirements will be focussed on the need to obtain the relevant keys from a TTP within their jurisdiction, so that they can decrypt precisely those communications that they are authorised to intercept.

In this paper we propose a novel mechanism that will enable TTPs to perform the dual rôle of providing users with key management services and providing law enforcement agencies with warranted access to a particular user's communications. Unlike other proposals, the mechanism allows users to update their keys according to their own internal security policies. We then list typical requirements for such a scheme, and consider how well the proposed mechanism meets these requirements. We conclude by considering possible variants of the basic method and how other proposed schemes for using TTPs in this way relate to the described method.

This paper was produced as part of the UK DTI/EPSRC-funded LINK PCP project 'Third-Generation Systems Security Studies'. Participants in this project are Vodafone Ltd, GPT Ltd and Royal Holloway, University of London.

2 The Mechanism

The proposed mechanism is based upon the Diffie-Hellman algorithm for key exchange [DH76]. To simplify our description, we consider the mechanism only in relation to one-way communication (such as e-mail). The adaptation of the scheme for two-way communication is very straightforward.

More specifically we present the mechanism in the context of a pair of users A and B, where A wishes to send B a confidential message and needs to be provided with a session key to protect it. We suppose that A and B have associated TTPs TA and TB respectively, where TA and TB may be different.

Prior to use of the mechanism, TA and TB need to agree on a number of parameters, and exchange certain information.

- Every pair of TTPs whose users wish to communicate securely must agree between them values g and p. These values may be different for each pair of communicating TTPs, and must have the usual properties required for operation of the Diffie-Hellman key exchange mechanism, namely that g must be a primitive element modulo p, where p is a large integer (satisfying certain properties). These values will need to be passed to any client users of TA and TB who wish to communicate securely with a client of the other TTP.
- Every pair of TTPs whose users wish to communicate securely must agree on the use of a digital signature algorithm. They must also each choose their own signature key/verification key pair, and exchange verification keys in a reliable way. Any user B wishing to receive a message from a user A by TTP TA must be equipped with a copy of TA's verification key (presumably by their own TTP TB) in a reliable way.
- Every pair of TTPs whose users wish to communicate securely must agree a secret key $K(TA,TB)$ and a Diffie-Hellman key generating function f. This function f shall take as input the shared secret key and the name of any user, and generate for that user a private integer b satisfying $1 < b < p-1$ (which will be a 'private receive key' assigned to that user—see immediately below). The secret key $K(TA,TB)$ might itself be generated by a higher-level Diffie-Hellman exchange between the TTPs.

Given that B is to be provided with the means to receive a secure message from A, prior to use of the mechanism A and B need to be provided with certain cryptographic parameters by their respective TTPs.

- Using the function f, the secret key $K(TA,TB)$ and the name of B, both TA and TB generate a private integer b satisfying $1 < b < p-1$. This key is known as B's *private receive key*. The corresponding *public receive key* is set equal to g^b mod p. The private receive key b for B needs to be securely transferred from TB to B (like the other transfers discussed here, this can be performed `off-line'). Note that B can derive its public receive key from b simply by computing g^b mod p. Note also that this key can be used by B to receive secure messages from any user associated with TA; however, a different key pair will need to be generated if secure messages need to be received from users associated with another TTP.
- A must also be equipped with a copy of B's public receive key. This key can be computed by TA using f, the name of B, and the key $K(TA,TB)$, and then is transferred in a reliable way from TA to A.
- Finally, by some means (perhaps randomly), TA generates another Diffie-Hellman key pair for use by A to send secure messages to any user associated with TB. Hence TA generates for A a *private send key*, denoted a (where $1 < a < p-1$). The corresponding public send key is equal to g^a mod p.

In addition TA signs a copy of A's public send key concatenated with the name of A using its private signature key. The public and private send keys, together with the signature, are then passed to A by some secure means. This process can be carried out as often as A wishes, according to its own internal security policy. So, for instance, A can have a new send key pair for each message sent to B.

Hence, prior to commencement of the mechanism, A possesses the following information:
- the private send key a for user A;
- a certificate for the public send key (g^a mod p) for user A, signed by TA;
- the public receive key (g^b mod p) for user B, and;
- the parameters g and p.

This information can be employed to generate a shared key g^{ab} mod p for the encryption between A and B. This key can be used as a session key, or, even better, as a key-encryption key (KEK). The KEK would then be used to encrypt a suitable session key. This would facilitate the sending of e-mail to multiple recipients, for instance, as well as allowing the use of a new key for each message. User A then sends the following information to user B:

- the message encrypted using the session key g^{ab} mod p;
- the public send key g^a mod p for user A (signed by TA);
- the public key g^b mod p for user B.

Once received, the public receive key g^b mod p allows user B to look up its corresponding private receive key b (there will be a different receive key for each TTP with whose users B communicates). User B can then generate the (secret) session key g^{ab} mod p by operating on the public send key g^a mod p for user A using its private key b, and thus can decrypt the received message.

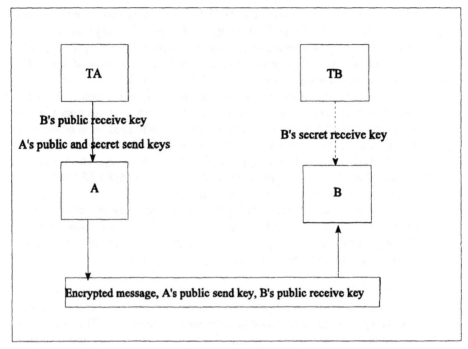

Fig. 1. Use of TTP Scheme for One-Way Communication

Should there be a warrant for legal interception of this communication, an intercepting authority can retrieve the private key of one of the users from the associated trusted third party within its jurisdiction and use this with the public key of the other user (which is transmitted along with the encrypted message) to find the session key for the encryption. There is no requirement for the intercepting authority to retrieve the private keys of both users.

3 A Typical Set of Requirements on a Trusted Third Party Scheme

Clearly, the definition and agreement of a set of requirements acceptable across a broad set of countries is largely a political process. However, we can give a set of typical or likely requirements on which to base an analysis of the suitability of the proposed mechanism.

1   *Use of the scheme should provide visible benefits for the user.* This means that the TTPs can offer their services to users on a commercial basis. The scheme described in this paper permits this to happen. By signing up to a licensed TTP, the user will be able to communicate securely with every user of every TTP with whom his TTP has an agreement. The user is able to choose from a number of TTPs in his home country, thus increasing his trust in the TTP.

2   *The scheme should allow national and international operation.* The proposed scheme achieves this by ensuring that the intercepting authority can obtain the required keys from a TTP within its jurisdiction.

3   *Details of the scheme should be public.* This is achieved for the proposed scheme by the publication of this paper!

4   *The scheme should be based on well known techniques,* and Diffie-Hellman certainly qualifies.

5   *All forms of electronic communication should be supported.* The proposed scheme can easily be adapted to include two-way communication such as voice telephony.

6   *The scheme should be compatible with laws and regulations on interception, and on the use, export and sale of cryptographic mechanisms.* This matter is the subject of further study, but no problems have yet been identified.

7   *Access must be provided to the subject's incoming and outgoing communication, where a warrant is held.* This is clearly achieved for the proposed scheme, as the subject's TTP can provide the appropriate send and receive private keys.

8   *The scheme should support a variety of encryption algorithms, in hardware and software.* As the proposed scheme deals solely with key management, any suitable encryption algorithm can be used, as long as it is available to the recipient and legitimate interceptors. The best way to achieve this may be to use a standard list of algorithms, such as the ISO register [ISO9979].

9   *An entity with a warrant should not be able to fabricate false evidence.* This is particularly applicable in countries where intercepted communications are admissible as evidence in court. The proposed scheme as it stands does not meet this requirement, but the provision of digital signatures as an additional service by the TTP will allow it to be met.

10   *Where possible, users should be able to update keys according to their own internal policies.* The proposed scheme allows a user to generate new send key pairs as often as wished (provided that he deposits them with his TTP) or have them generated by his TTP. The receive keys, which are generated deterministically based on the TTP shared key and the user's identity, are intended to be fairly permanent and change only if the TTPs' shared key or the user's identity changes.

11   *Abuse by either side should be detectable by the other.* We believe that this is the case for the proposed scheme, although abuse by collusion between the two sides may still be possible. The main disincentive to such abuse may be the `shrink-wrapped' provision of the software, which we would expect to be bundled in with, say, an email system or other telecommunications software.

12   *Users should not have to communicate with TTPs other than their own.* The only communication required in the proposed scheme is with the user's own TTP.

13   *Online communication between TTPs should not be required.* The independent generation of the receive keys in the proposed scheme means that no such communication is required for the proposed scheme.

4 Options and Other Issues

4.1 Trusting TTPs

The receiving party must trust the sending party's TTP, in order to verify the sending party's public key, and also because the sender's TTP can generate the receiver's private key. However, this trust only concerns communications between the receiver and senders belonging to that TTP. There will be a need for a certification hierarchy to identify a common point of trust for different TTPs.

4.2 The Choice of Values

There has been considerable discussion in the literature on the benefits of using a composite modulus for Diffie-Hellman, for instance. This, and other matters such as the length of the modulus p and the primitive element g, are beyond the scope of this paper.

4.3 Commercial Value

The proposed scheme relies entirely on its perceived value to users in order to be taken up. Service providers will want to recover the cost of setting up the service from their customers. Therefore the scheme must be able to provide value-added end-to-end services that users want. Further investigation is required to assess the level of demand for services such as:

- end-to-end encryption;
- end-to-end authentication;
- non-repudiation of sent message;
- message integrity.

Given that users will be paying for these services, they will expect a sufficient level of security. In the event of security failure with financial impact on the user, he will expect to be able to recover this, either via his insurers or from the organisation running the TTP. This makes running a TTP a potentially expensive business, unless the financial risks run by the TTP can be adequately protected against. If TTPs are not commercially viable, then the scheme is a non-starter.

4.4 Combined Two-Way Session Key

The two-way version of the proposed scheme provides two keys for communication: one for each direction. These could be combined to form a single session key, or just one of the keys could be used. The advantages and disadvantages of this are a matter for further study.

4.5 Sharing Keys Between Trusted Third Parties

In an environment where commercial TTPs will be looking to offer additional services to their users, it is possible that some users will want the extra reassurance offered by having their keys shared between a number of independent TTPs. The proposed protocol is easily adaptable to provide this feature. For instance, the ideas of Micali [Mica93] for adding secret sharing on top of existing schemes could be adopted.

4.6 Resistance to Burmester-type Attacks

The kind of attacks pointed out by Burmester [Burm94], whereby knowledge of a session key is used to obtain previously used session keys, are not effective against this scheme. This is because a user will have different keys for communication with users associated with different TTPs.

5 Other Published Schemes

5.1 The Goss Scheme

A scheme designed by Goss has been patented in the US [Goss90]. In this scheme, a shared secret key is established by combining two Diffie-Hellman exponentiations using fixed and varying (per session) parameters. At first sight, this appears to correspond to the receive and send keys in the proposed scheme. However, the Goss scheme uses a universal modulus and primitive element. If x, x' are A's fixed and variant keys, and y, y' are B's, then the shared key is calculated as

$$\alpha^{xy'} \oplus \alpha^{x'y}.$$

This could be viewed as a variant of the proposed two-way protocol whereby a universal modulus and primitive element are used and the two keys are combined by XOR-ing them.

5.2 Yacobi Scheme

This scheme [Yaco90] is almost identical to the Goss one, but uses a composite modulus and combines the session keys by modular multiplication rather than XOR-ing.

References

[Burm94]   Mike Burmester. `On the Risk of Opening Distributed Keys'. In *Advances in Cryptology—CRYPTO '94*, Springer-Verlag, Berlin (1994), pp.308–317.

[DH76]   Whitfield Diffie and Martin E. Hellman. `New Directions in Cryptography.' *IEEE Transactions in Information Theory* **IT-22** (1976) pages 644-655.

[FIPS185]   National Institute of Standards and Technology. FIPS Publication 185: *Escrowed Encryption Standard*, February 1994.

[Goss90]   US Patent 4956863. *Cryptographic Method and Apparatus for Public Key Exchange with Authentication*. Granted 11 September 1990.

[ISO9979]   ISO/IEC 9979:1991. *Data Cryptographic Techniques — Procedures for the Registration of Cryptographic Algorithms*. December 1991.

[Mica93]   Silvio Micali. *Fair Cryptosystems*. MIT Technical Report MIT/LCS/TR-579.b, November 1993.

[Yaco90]   Yacov Yacobi. `A key distribution paradox.' In *Advances in Cryptology — CRYPTO '90*, Springer-Verlag, Berlin (1991), pp.268–273.

A New Key Escrow Cryptosystem

Jingmin He and Ed Dawson
Information Security Research Center
Queensland University of Technology
Gardens Point, GPO Box 2434
Brisbane QLD 4001, Australia

Abstract

The Escrowed Encryption Standard (EES) proposed by U.S. government has gained much attention in the last two years. It was claimed that EES can provide cryptographic protection to unclassified, sensitive data, while at the same time, allow for the decryption of encrypted messages when lawfully authorized. Later, some criticism was proposed to reveal the weakness of the EES proposal.

In this paper we propose a new key escrow cryptosystem. The system eliminates many known weaknesses of the EES proposal. Particularly, the new system strengthens the weakest part of the EES proposal, i.e., the checksum generation, and provides other benefits that the original EES proposal did not have.

1   Introduction

In April 1993, the U.S. government announced a proposal for a new federal standard encryption system with key escrow capability [8]. The proposal, called the Escrowed Encryption Standard (EES), or commonly known as the "Clipper proposal", has two prominent features: a new symmetric encryption algorithm Skipjack and a key escrow system packed in a tamper-free hardware chip. It was claimed that the Skipjack algorithm is significantly stronger than those currently available to the public (e.g., DES). Furthermore, by escrowing the session key used by Skipjack, the government, once obtaining some kind of legal authorization (e.g., a court order), can gain access to the communications encrypted under the system. This will allow the government to enforce laws in special circumstances under appropriate legal safeguards.

Since the announcement of the Clipper proposal, there has been much controversy. One controversy is about the new block cipher Skipjack because it is a classified algorithm. A more controversial aspect of the Clipper proposal is its key escrow function.

Skipjack is a symmetric block cipher with a key length of 80 bits. Both the plaintext and ciphertext lengths are 64 bits. It is essentially a drop-in replacement for the well-known DES. Since Skipjack algorithm is classified, it has not been subjected to any serious assessment and criticism by civilian researchers and users. Instead, it is installed in and protected by a tamper-free hardware device (called the Clipper chip) and only the integrated hardware device as a whole is available to ordinary users.

The most important new feature of the EES proposal is its key escrow function. Whenever a Clipper chip is used, a "Law Enforcement Access Field (LEAF)" must be present for both the encryption and decryption operations. The LEAF contains an encrypted copy of the current session key. When lawfully authorized, a law enforcement agency can decrypt a LEAF and thus successfully wiretap a communication.

The general idea of key escrow cryptography is very attractive and promising. But the current EES proposal has some weaknesses, some of them have been pointed out by Blaze [4]. At the same time, many researchers have been trying to develop new key escrow systems, either in hardware or software [1][2][6].

In this paper, we propose another key escrow cryptosystem. The new system has many new features that the original EES proposal does not have. More specifically, the new system has the following new features:

1. As with the EES proposal, we assume there are two key escrow agents. Each of the two key escrow agents holds a unit subkey for each chip. But in our system, the two subkeys are never combined together or given out.

2. A longer checksum of 80 bits is used. The process for the generation of the checksum is well-defined and thus is not classified. This new generation will significantly enhance the integrity of the LEAF.

3. A logic timestamp is provided for each use of the chip. An investigator can only gain access to special designated communications.

4. The identity of a message receiver is embedded in the encryption process. This will allow a law enforcement agency to trace the identities of both the message sender and receiver. This feature is optional.

The paper is organized as follows. In Section 2 we describe the key escrow component of EES, including an overview of the security analysis of Blaze in [4]. In Section 3 we present our new key escrow cryptosystem, and give a detailed analysis on its security and efficiency. Section 4 contains some concluding remarks.

2 The Escrowed Encryption Standard

In this section we give a detailed review of the key escrow system in the Clipper proposal.

2.1 Hardware Chips

Skipjack is used as the basic encryption function. Some standard interface is provided for the use of Skipjack. Similar to DES, Skipjack is a 64 bit codebook block cipher, i.e., a 64-bit ciphertext is produced for a 64-bit plaintext. But Skipjack has a longer key of 80 bits, as opposed to DES's 56 bits. Hereafter, we use E to denote the Skipjack cipher and $E(k, M)$ to denote the ciphertext of the plaintext M when a 80-bit session key k is used.

The current proposed EES system is packed into a tamper-free (tamper-resistant) hardware and is popularly known as a "Clipper chip" (or EES processor, EES device). Each Clipper chip has a unique identification number and some other special parameters installed by the manufacturer. We will use $Chip_i$ to denote the Clipper chip of user i. Its unique identification number is denoted by ID_i.

Since the Clipper chip is tamper-free, no one, including the owner of a chip, can access its contents. Consequently, the Skipjack algorithm is well protected. A purchaser of a Clipper chip cannot learn anything about Skipjack resided inside the chip. Besides, a limited number of secrets can be safely stored in the Clipper chip.

But the most important reason to use a tamper-free hardware device is to ensure the proper function of the whole key escrow process. That is, whenever a user uses his/her Clipper chip, the pre-installed program will be executed in its integrity, and no steps can be bypassed. This is the key to the concept of key escrow, as we will see later.

At present, there are actually two EES devices available. One is the Clipper chip (MYK-78) which employs the Skipjack algorithm. The other is the Capstone chip (MYK-80) which adds built-in support for public key negotiation of session keys, digital signatures, random number generation, and other features. In this paper, we will mainly discuss the Clipper chip.

2.2 Chip Programming

When a user A wants to purchase a Clipper chip from an authorized product vendor, the vendor will supply a chip with a unique serial number (chip identification number) to a "key escrow" process for chip programming, conducted by a National Program Manager for Key Escrowing (hereafter called Key Escrow Programming Facility, or KEPF, or the manager). As a result of this process, some programs and secret parameters are installed inside the chip. Two escrow agents are involved in the escrowing process. One is NIST itself and the other is the U.S. Department of Treasure.

More specifically, the following secrets are installed inside the chip.

1. A unique device key (also called unit key) UK_A. It is generated and shared by the two independent escrow agents. For example, each of the two escrow agents generates a random seed and feeds it to a programming facility. As a result, two key components K_1 and K_2 are produced, where $UK_A = K_1 \oplus K_2$. K_1 is kept by one agent and K_2 is kept by another, together with the corresponding identification number ID_A.

2. A family key FK. This family key is common to all interoperable EES devices.

Besides, some programs (and of course the Skipjack algorithm) and other possible features are also installed. After all necessary installation, the completed chip is issued to user A.

2.3 LEAF and Encryption/Decryption

The Law Enforcement Access Field (LEAF) is the main feature of EES. Whenever a Clipper chip is used, a LEAF is produced to keep a track of that use.

For example, consider $Chip_A$ of user A. Suppose A wants to send a message M to some other people using his Clipper chip. To do this, A will first negotiate a session key k with the receiver (in whatever way). Then A simply feeds k and M to $Chip_A$. Two components will be produced: a ciphertext $E(k, M)$ and $LEAF(A, k)$, the value of the LEAF.

LEAF is a 128 bit structure. It contains a 32 bit unit identification number, the encrypted session key (80 bits), and a 16 bit checksum. For a Clipper chip $Chip_A$, when a session key k is used to do encryption/decryption, its LEAF is of the following form:

$$LEAF(A, k) = E(FK, D(A, k)),$$

where

$$D(A, k) = (ID_A, E(UK_A, k), f(A, k, IV)),$$

$f(A, k)$ is a 16-bit checksum generated by some classified function f, and IV is an initialization vector.

A will send both $E(k, M)$ and $LEAF(A, k)$ to the receiver. At the receiving side, B has a Clipper chip $Chip_B$. B knows the secret session key k because k was negotiated between A and B. Although B knows the session key k and the ciphertext $E(k, M)$, he does not know E since Skipjack is a classified algorithm. So he cannot decipher $E(k, M)$ without using his Clipper chip $Chip_B$.

On the other hand, $Chip_B$ has been pre-programmed in such a way that a decryption operation is not performed unless a valid LEAF is present. More specifically, the decryption process has been programmed in the following way:

1. The family key FK is used to decipher $LEAF(A, k) = E(FK, D(A, k))$. The result is $D(A, k) = (ID_A, E(UK_A, k), f(A, k, IV))$.

2. $f(A, k, IV)$ is computed and compared with the received checksum. If they do not match, stop.

3. The session key k is used to decrypt $E(k, M)$ to obtain the cleartext M.

As we can see, the checksum verification inside a Clipper chip is crucial to the application of EES. Since EES is packed inside a tamper-free hardware, no one can use a Clipper chip in such a way that some steps pre-programmed inside the chip are skipped. Particularly, the checksum verification cannot be bypassed. Only after this verification is passed successfully can the decryption operation be actually performed. In other words, if A wants his message to be received and decrypted by a receiver, he must provide a valid LEAF.

2.4 Wiretapping by Authorized Investigator

One of the two major goals of an escrowed key cryptosystem is to provide some facility for legal wiretapping. Suppose an investigator (e.g., FBI) wants to wiretap a message transmission from A to B. To do this, the investigator must first obtain some legal authorization, for example, a court order. He then presents this court order to the two key escrow agents. After verifying the court order, the two agents will give the investigator the necessary information (the two key components K_1 and K_2) to recover the unit key UK_A. The investigator is also given the family key FK. Using FK and UK_A, the investigator will be able to decrypt $LEAF(A, k)$ to obtain the session key k, and thus enable him to decrypt the ciphertext $E(k, M)$.

On the other hand, without a court order, no one (except a cooperation between the two escrow agents) can obtain the unit key of any Clipper chip, and therefore no one is able to find out the session key for any particular communication. As a result, the privacy of all users is protected.

2.5 Weakness of EES

Although there is a general controversy on the Clipper proposal, the main technical criticism comes from the work of Blaze [4]. In [4] Blaze analyzed the behaviour of EES devices and tried to figure out some of the structural properties of the internal EES implementation, particularly the generation of the checksum. He pointed out that it is possible for rogue applications to use the Skipjack algorithm integrated inside a Clipper chip to do encryption/decryption, without using the LEAF. In other words, the integration of the EES devices is not maintained at the highest level.

Recall that inside the Clipper chip, if a wrong IV or wrong session key is used, a different checksum will be generated and a mismatch is detected. Therefore, decryption will not be performed.

One of the main drawbacks of the EES proposal is the generation of the checksum. The checksum is a 16 bit string which is a function of the form $f(A, k, IV)$. Actually the identity of the message sender A is not used, so it should be written as $f(k, IV)$. Although the exact method of checksum generation is classified, for many known cryptographic modes (like cipher block chaining, cipher feedback, etc.), a checksum can be forged to "fool" the checksum verification. Furthermore, 16 bits are not long enough to be secure against exhaustive search.

Another serious drawback is that once an investigation is authorized by the court, the investigator will be able to decrypt everything transmitted by the suspect, without any time restriction.

We point out the following new observation not mentioned in [4].

In EES, only information about a message sender is included in the LEAF. One cannot trace the destination of a message. This will cause some danger in some applications. For example, in some criminal investigations, it is important to know who is going to take action after receiving a message from a particular message sender.

3 A New Key Escrow Cryptosystem

In this section we propose a new key escrow cryptosystem. The new system has the following added features:

1. Each of the two key escrow agents holds a unit subkey for each chip. The two subkeys are never combined together or given out.

2. A longer checksum of 80 bits is used. The process for the generation of the checksum is well-defined and thus is not classified. This new generation will significantly enhance the integrity of LEAF.

3. A logic timestamp is provided for each use of the chip. An investigator can only gain access to special designated communications.

4. The identity of a message receiver is embedded. This will allow a law enforcement agency to trace the identities of both the message sender and receiver.

We will need to use a one-way function F of length 80, i.e., the value of F is 80 bits long.

3.1 The Protocol

The overall mechanism of our protocol is the same as the current Clipper chip. We still assume two key escrow agents $Agent_1$ and $Agent_2$. For the chip issued to user A, the following are installed:

$$ID_A, K_1, K_2, FK, C,$$

where:

- ID_A is the unique 32 bit unit serial number for A's chip. This number should be made public so that any user who wants to send A secret messages will be able to get this identity number. For example, all serial numbers can be stored in a catalog like a phone book.

- K_1, an 80 bit long string, is part of the unit key held by $Agent_1$.

- K_2, also an 80 bit long string, is another part of the unit key held by $Agent_2$.

- FK is the common family key of 80 bits.

Besides, there are two registers of 32 bits, RID and C. Both are initially empty. RID will be used to store the identity of a message receiver, and C will be used to store a logic timestamp. Note that there is no single "unit key", like the one used in the Clipper chip (UK_A). Instead, two separate and independent keys are held by the two escrow agents respectively. They will never be simply joined together, say, by an exclusive or operation.

Encryption

Suppose A wants to send a message M to user B, using his chip. To do this, A will first negotiate a session key k with B. Then A feeds k, ID_B, M, and a random number (timestamp) C of 32 bits to $Chip_A$. ID_B is the current number stored in RID. Two components will be produced: a ciphertext $E(k, M)$ and a $LEAF(A, B, k, C)$, the value of the LEAF.

The Structure of LEAF

Our LEAF is a 256 bit structure. It contains the identification number of the sender (ID_A) and that of the receiver (ID_B), the encrypted session key (80 bits), the timestamp C, and a 80 bit checksum. That is,

$$LEAF(A, B, k, C) = E(FK, D(A, B, k, C)),$$

where

$$D(A, B, k, C) = <ID_A, ID_B, C, E(K', k), checksum>,$$

$$K' = F(K_1, C) \oplus F(K_2, C),$$

$$checksum = F(ID_A, ID_B, C, k, IV),$$

and IV is an initialization vector.

A will send both $E(k, M)$ and the $LEAF$ to the receiver.

Decryption

At the receiving side, the decryption process of $Chip_B$ has been programmed in the following way:

1. The family key FK is used to decipher $LEAF$. The result is

$$D(A, B, k, C) = <ID_A, ID_B, C, E(K', k), checksum>.$$

2. The second component is compared with the chip identification. If they do not match, stop.

3. $F(ID_A, ID_B, C, k, IV)$ is computed and compared with the received checksum. If they do not match, stop.

4. The session key k is used to decrypt $E(k, M)$ to obtain the cleartext M.

Legal Wiretapping

Suppose an investigator wants to wiretap a particular message transmission from A to B. To do this, the investigator first obtains a court order. He then uses the family key FK to decrypt $LEAF$ to obtain the timestamp C. He presents both C and the court order to the two key escrow agents. After verifying the court order, $Agent_1$ will compute $K_1' = F(K_1, C)$ and give K_1' to him. Similarly, $Agent_2$ will compute $K_2' = F(K_2, C)$ and give K_2' to him. The investigator computes $K' = K_1' \oplus K_2'$, decrypts $E(K', k)$ to obtain the session key k, and finally decrypts $E(k, M)$ to obtain the message M.

It may be argued that it is a severe limitation on the investigator to have to obtain a different court order for each time a different session key is selected. One method of overcoming this problem is to replace the logical timestamp C with a real timestamp which is extracted from $Chip_A$ using a time clock synchronised to Synchronous Universal Coordinated Time (UTC) stored on the chip. For example we could use the day of the year as the timestamp. In this fashion an investigator would obtain the capability of decrypting all intercepted messages for a single day from one court order.

3.2 Analysis

As with the original Clipper chip, the security of the encryption/decryption operations in using Skipjack is not compromised and is not our concern here. Instead, we mainly focus on the integrity of the whole system, i.e., we are mainly interested in the use of the LEAF.

Pre- or post-encryption is always a way to circumvent the LEAF in using any key escrow system. That is, before using the key escrow system, an attacker can always encrypt his message with another cipher. In [4], it was argued that

> "It is not clear, however, what the attacker gains from doing this, since if the second cipher is believed strong there is no need to use Skipjack in the first place, and if it is believed weak it does not protect the traffic from the government anyway."

So in this paper we will not consider the pre- or post-encryption as a threat to the use of the key escrow system.

LEAF Feedback

Since we have insisted on using a tamper-free hardware chip, we can assume that the pre-installed program will be executed in its integrity. That is, before decryption, a valid LEAF must be present. Note that in our system, we have explicitly defined the structure of the LEAF.

The main threat comes from using an incorrect LEAF for the current communication by the message sender. Note that in the Clipper proposal, the LEAF mainly depends on the session key k and the initialization vector IV. In a rogue application where both the sender and the receiver try to circumvent the LEAF, the receiver can generate an arbitrary $LEAF_B$ by himself. When he receives a transmission from the sender, he may ignore the $LEAF_A$ part and feed his own $LEAF_B$ and IV to his Clipper chip. If $LEAF_B$ matches part of $LEAF_A$, then for that part, an incorrect LEAF has been obtained and thus the LEAF is useless. This attack is called "LEAF Feedback" [4]. Simply speaking, LEAF Feedback relies on the partial synchronization of the initialization vector IV.

In our system, this kind of synchronization is almost impossible because we have used a one-way function F. Suppose user A (the message sender) intends to use the session key k to send a message M to user B. User A cannot simply change some bits of $LEAF(A, B, k, C)$ because this will cause unacceptable change in $D(A, B, k, C)$ and almost surely result in a rejection at the receiving side. So A can only try to produce a fake D' of the following form

$$D' = < u, v, w, x, y >$$

such that $E(FK, D') = E(FK, D(A, B, k, C))$, where all of u, v, w, x, y can be chosen arbitrarily by A and B. It is possible that A and B can produce incorrect values for u, v, w and x, due to the similar reason with the LEAF Feedback attack. But A or B cannot produce a correct y (the checksum) since y must be generated from the right session key. That is, by using a one-way function, it is impossible for any application to achieve any kind of "synchronization" needed in the LEAF Feedback attack. The more secure the one-way function is, the more unlikely the synchronization could be achieved. We believe that a one-way function F of 80 bits long should be secure enough.

Brute-Force LEAF Search

For the Clipper proposal, a Brute-Force LEAF Search attack is possible because the checksum is only 16 bit long. In our case, this attack is clearly infeasible.

Timestamp

In [2], several methods were mentioned to deal with the timestamp problem. That is, by providing a timestamp for each transaction, an investigator can only gain access to designated communications between users. Usually, a physical global clock is used.

In our scheme, a logic timestamp is provided for each use of the chip. The message sender should be responsible for using a different timestamp C for each different use of his chip. If he uses the same C in two different transactions, then the two transactions will be exposed even if the investigator has been given a court order to wiretap only one of the transactions. So the logic timestamp is totally controlled by the users themselves.

Tracing the Receiver

Tracing the identity of a message sender is important in many applications. For example, this helps identify traitors in a business [5]. In some applications, tracing the identity of a message receiver is also very important. In [6], a scheme based on discrete logarithms was proposed to trace the identity of a message receiver. In our system, this is easy because the identity of a message receiver is included in the LEAF.

4   Conclusion

We have proposed a replacement for the key escrow procedure in the Clipper chip. As was shown, this new method is secure from the attacks described by Blaze. In addition, this procedure allows law enforcement agencies in a legal wiretap to identify both the sender and receiver of a cryptogram. However, an investigator can only decrypt cryptograms for which he has been granted access to.

It should be noted that there should only be a small change in the complexity of the Clipper chip due to the addition of the one-way function F. This should not cause a significant degradation in the throughput of this device.

References

[1] D.M.Balenson, C.M.Ellison, S.B.Lipner, and S.T.Walker, A new approach to software key escrow encryption, manuscript, 1994.

[2] T.Beth, H.Knobloch,M.Otten,G.J.Simmons, and P.Wichmann, Towards acceptable key escrow systems, Proceedings of the 2nd ACM Conference on Computer and Communications Security, 1994, pp.51-58.

[3] T.Beth and M.Otten, eds, Proceedings of the E.I.S.S. Workshop on Escrowed Key Cryptography, Workshop held at the European Institute for System Security, University of Karlsruhe, June 22-24, 1994.

[4] M.Blaze, Protocol failure in the escrowed encryption standard, Proceedings of the 2nd ACM Conference on Computer and Communications Security, 1994, pp.59-67.

[5] B.Chor, A.Fiat and M.Naor, Tracing traitors, CRYPTO'94, pp.257-270.

[6] Y.Desmedt, How to secure the LEAF in Clipper Alternatives, in [3], pp.78-86.

[7] S.Micali, Fair public-key cryptosystems, CRYPTO'92, pp.

[8] National Institute for Standards and Technology, Escrowed Encryption Standard, Federal Information Processing Standards Publication 185, U.S. Dept. of Commerce, 1994.

[9] M.Smid, US government key escrow, talk outline in [3], pp. 45-55.

How to Fairly Reconstruct a Shared Secret

Jingmin He and Ed Dawson
Information Security Research Center
Queensland University of Technology
Gardens Point, GPO Box 2434
Brisbane QLD 4001, Australia

Abstract

In this paper we consider the secret reconstruction problem in a secret sharing scheme. We show how to use a slowly-information-revealing process to achieve a fair reconstruction of a shared secret. We give a detailed analysis on the advantage cheaters could gain in such a process.

1   Introduction

In a secret sharing scheme [12, 3], a secret is shared among many users in such a way that any qualified group of users can reconstruct the secret, but any unqualified group of users have absolutely no information about the secret.

Traditional secret sharing schemes cannot prevent cheating by themselves. For example, in a (t, n) secret sharing scheme, t users should be able to recover the secret. Suppose they pool their shares one by one, without using any *simultaneous pooling mechanism*. After being able to see all the shares of other $t - 1$ users, the last user might change his mind and either simply refuse to give his share or give out a fake share, without caring if his fake share could be detected. In the end, the cheater can recover the true secret, but the rest cannot. We call this an *unfair recovery of a shared secret*.

In this paper we consider the problem of fair reconstruction of a shared secret. This problem is a special case of the general problem of fair computation which has been investigated by quite a few people, all used the same idea, that is, the use of a process in which information is revealed slowly. But the implementations of this idea are not exactly the same in this case. Also, almost all discussions are very informal with regard to the characterization of the process itself. In this paper we give formal proofs on the behaviour of this process when adapted to our special case of secret sharing. Particularly, we give a concrete estimate on the rate with which information is revealed in this process.

The paper is organized as follows. Section 2 briefly reviews secret sharing schemes, and their two phases of secret sharing and secret reconstruction. Section 3 reviews previous related work. Section 4 contains some preliminary material. Our scheme of fairly reconstructing a shared secret is given in section 5. Its full analysis is given in section 6.

2 The Problem

2.1 Secret Sharing and Reconstruction

In a (t, n) secret sharing scheme (threshold scheme), a secret s is divided by a dealer into n shares $s_i, 1 \leq i \leq n$, such that knowledge of any t shares is sufficient to efficiently reconstruct s, but knowledge of any $t - 1$ or fewer shares provides absolutely no information about the value of s. The dealer will divide the secret and deliver all shares in a *secret sharing phase*. In another *secret reconstruction phase*, t (or more) users will pool their shares together and reconstruct the secret s. The parameter t is called the threshold value. Ito, Saito and Nishizeki [8], and Benaloh and Leichter [2] consider more general secret sharing schemes in which any monotone access structure is possible. In this paper we will concentrate on threshold schemes but all discussions are valid for generalized secret sharing schemes.

There are two phases in a secret sharing scheme. In the first phase, secret shares are delivered to all participants by secure channels. The use of these secure channels is necessary. In the secret reconstruction phase, secure channels are not necessarily needed among the involved participants. Note that there is always a concern about a possible wiretap in the reconstruction phase, i.e., an "outsider" might try to wiretap the conversations between "insiders" and obtain all secret shares of those participants whereas he himself has not actually participated in the reconstruction phase. To prevent such a wiretap, secure channels are always necessary. But we will only be concerned about the fairness of the reconstruction process and will assume that a wiretap is not a problem.

2.2 Cheaters

If all involved participants are honest in the reconstruction phase, then they just need to pool their shares together and reconstruct the secret according to the protocol. But serious problems arise when there might be some cheaters involved. What is a cheater in a secret sharing scheme? We might try to answer this question by first defining what is a honest user. A user is considered to be honest if he behaves according to the specified secret sharing scheme (protocol). If a user is not honest, then he is a cheater.

The above definition seems to be good, but it is incomplete. The main problem is that it does not characterize the behaviour of a user very properly. Since there are two separate phases in a secret sharing scheme, the behaviour of users must be specified very carefully. Let's explain why.

Suppose we have a (t, n) secret sharing scheme, where $t \geq 3$. The dealer has delivered a secret share privately to each of the n users. Suppose this first phase is finished at time T_1. At some later time $T_2 > T_1$, t users should be able to recover the shared secret. Suppose $(t - 1)$ users in a group A meet together at some time between T_1 and T_2 and each of them shows his share to other people. Suppose P_1 is one of them. Then at time T_2, when P_1 joins some other group B of $(t - 2)$ users, once P_1 gets one of their shares, he can recover the secret by himself and then he can reveal the secret to his old friends in group A, while no user in group B can recover the secret even if they have P_1's share.

One may argue that the above example is not good. This is true. But we must admit that it could happen in practical applications. In the above example, there are actually $(2t - 3)$ users involved, and we certainly should consider the $(t - 1)$ users in group A as cheaters.

Now we can define what is a cheater. In a secret sharing scheme, once a user obtains a share from the dealer, from that very time on, he must behave according to the specified protocol. Particularly, this requires him not to reveal his share to any other group of users before he formally joins them so that altogether they will form a qualified group of users to recover the shared secret. Otherwise he is considered to be a cheater. When we say h is an upper bound on the total number of cheaters, we mean there could be at most h cheaters among all users during the whole time.

In this paper we use the following computation model. There are a total of n users. Any pair of users are connected by a secure private channel. A secret is to be shared among them by a honest dealer using a (t, n) secret sharing scheme. When a group of users decide to cooperate to recover the secret, they send messages to each other via private channels. In other words, we use a secure asynchronous complete network as our computation model.

3   Previous Work

Tompa and Woll [15] first realize the problem of unfairness. They propose a probabilistic method to ensure a fair reconstruction of a shared secret, but they assume the use of a stronger computation model, that is, all involved users must pool their messages *simultaneously*. In other words, there is a central "message collector" in the network which will broadcast what it has collected in a specified period of time. Initiated by their work, many people have studied the problem of *cheating-detection/identification* extensively.

3.1   Cheating Detection/Identification and VSS

Chor, Goldwasser, Micali and Awerbuch [6] introduce the primitive of VSS (*verifiable secret sharing*). Rabin and Ben-Or [10] give an excellent implementation of VSS schemes which can be used to identify cheaters and their schemes are unconditional secure. Brickell and Stinson's method [4] can identify cheaters with very high probability. Their method has similar properties as those of Rabin and Ben-Or's method. For other results, see [4] and its references.

It might be argued that a VSS scheme can be used directly to prevent unfair recovery of a shared secret. Unfortunately this is not true. Let us elaborate on this further.

We use the VSS scheme of Rabin and Ben-Or as an example. Their scheme is a $(t + 1, n)$ scheme, where there are n users and $n \geq 2t + 1$. It is assumed that there are at most t cheaters. Each user has a set of messages, among them some are used for the reconstruction of the shared secret, and some are used to check the validity of information provided by other users. Suppose all the n users submit their shares (and some associated information). Since the scheme has a threshold value $(t + 1)$, t users cannot obtain any information about the secret. Since there are at least $(t + 1)$ honest users among all the n users, the true shared secret can be reconstructed correctly.

Now we can see that the VSS scheme is actually an error-tolerant scheme in which all n users are required to submit their shares. If only $(t+1)$ users are involved, and one of them is a cheater, then only he can reconstruct the true secret if he does not submit his true share, although his cheating can surely be detected by all others. This is exactly the problem we want to deal with in this paper.

We are considering a (t, n) scheme where t users, if all are honest, should be sufficient to recover the secret. Among the t involved users, some might be cheaters. Our goal is to achieve a fairness among these t users and still maintain the threshold value t. Since the threshold value must remain to be t, it is possible that the t involved users cannot reconstruct the secret due to the existence of cheaters. However, if the secret can be obtained by anyone, then (at least) the honest users must be able to obtain it as well. Of course a VSS is a very powerful tool and as we will show later, a multi-stage use of a VSS will be sufficient to solve our problem. But as we have shown, a direct use of a VSS does not solve our problem.

3.2 Fair Multiparty Computation

Chaum, Damgard, and van de Graaf [5], Beaver and Goldwasser [1], and Goldwasser and Levin [7] have considered the general problem of fair multiparty computation. Earlier, the two-party case has been investigated extensively by Tedrick [13, 14], Luby, Micali and Rackoff [9], Vazirani and Vazirani [16], and Yao [17]. All of these are based on a computational model, i.e., some computational assumptions are used, like the existence of an oblivious transfer protocol, or even stronger, the Quadratic Residuosity Assumption.

The idea used in all the above work is the same, i.e., a process of slowly revealing information is used. But the revealing rate of the process is discussed in a very informal way. In this paper, we use the same idea to show how to fairly recover the shared secret in a secret sharing scheme. We give rigorous proofs for all results. Particularly, we give an explicit upper bound on the revealing rate of the information-revealing process which is also an upper bound on the advantage cheaters could gain in our scheme. In our case, we do not need to use any assumption because we have assumed a honest dealer.

4 Some Preliminaries

We use $P[A]$ to denote the probability of event A.

Definition 1 *Let* $0 \le \epsilon \le 1/2$. *A coin with bias* ϵ *towards bit* b *is a random variable* x *such that* $P[x = b] = 1/2 + \epsilon, P[x = 1 - b] = 1/2 - \epsilon$. *We also say that* x *is* ϵ-biased *towards* b. $\square$

Let ϵ be given. Then a coin with bias ϵ towards a bit b can be constructed easily. For example, suppose G is a random number generator with uniform distribution in the interval $[0, 1]$. Run G to obtain a random point a. If a lies in $[0, 1/2 + \epsilon)$, the outcome is set to $x = b$; if a lies in $[1/2 + \epsilon, 1]$, the outcome is set to $x = 1 - b$. Then x is the required biased coin.

The law of large numbers is very useful in proving the convergence of a probabilistic process. We will use the following stronger version of the law of large numbers which is derived from *Bernstein*'s improvement on the famous *Chebyshev* inequality.

Theorem 1 ([11],page 388) *Let* $x_1, x_2, \ldots, x_n$ *be random variables which are independent and identically distributed,*

$$P[x_i = 1] = p, P[x_i = 0] = q, i = 1, 2, \ldots, n,$$

where $p > 0, p + q = 1$. *Define the* sum *variable* $S_n = x_1 + x_2 + \cdots + x_n$. *Let* ϵ, δ *be any positive numbers such that* $0 < \epsilon \leq pq$. *Let* $n(\epsilon, \delta) = 9\ln(2/\delta)/(8\epsilon^2)$. *Then for any* $n \geq n(\epsilon, \delta)$, *we have*

$$P[|\ S_n/n - p\ | \leq \epsilon] \geq 1 - \delta. \square$$

In this paper, we will use the VSS of Rabin and Ben-Or as a basic building block, although we can use any other unconditional secure VSS as well. We call it scheme $\mathcal{A}$. Scheme $\mathcal{A}$ has the property that whenever a user sends a fake share to some other user, his cheating can be detected almost surely (i.e., with a probability of at least $1 - 2^{-k}$ where k is a security parameter). For details of the Rabin–Ben-Or scheme, see [10].

5 The Scheme

The idea is simple. We share a secret bit by bit. For each bit b, we expand it into a long binary sequence $L(b)$ by repeatedly flipping a coin biased towards b. We share each bit of the sequence independently, using the VSS scheme $\mathcal{A}$.

Let k be a security parameter which will characterize all involved quantities. Let P, Q be any fixed polynomials such that

$$1/P(k) + 1/Q^2(k) \leq 1/4. \tag{1}$$

Let T be a fixed polynomial and

$$N = (9\ln 2/8)(T(k) + 1)P^2(k), \epsilon = 1/Q(k), \alpha = 1/P(k), p = 1/2 + \epsilon, \delta = 2^{-T(k)}. \tag{2}$$

Let $s = s_1 s_2 \cdots s_m$ be the secret to be shared among n users $P_i, 1 \leq i \leq n$. The dealer will share s in m stages, one stage for each bit. Since all stages are exactly the same, we only describe one stage for one secret bit b.

The dealer does the following.

1. Constructs a coin with bias ϵ towards b, and produces a list of N independent outcomes by flipping this coin N times. Let the list be $L(b) = c_1, c_2, \ldots, c_N$.

2. Shares each outcome c_i independently by using the VSS scheme A.

This completes the dealer's part.

Suppose t users plan to work together and reconstruct the secret bit b. They will reconstruct the N outcomes hidden in L, one by one. Once the whole list has been reconstructed without any cheating being detected, they will output bit b' as their guess about the true b as follows.

1. Let n_1 and n_0 be the number of reconstructed 1s and 0s respectively.

2. If $| n_1/N - p | \leq \alpha$, then $b' = 1$.

3. If $| n_0/N - p | \leq \alpha$, then $b' = 0$.

4. Otherwise output a random bit as b'.

Remark: Note that in most previous work, the output is determined by a simple majority voting. In our case, that will not work at all. As we can see, a margin of α is necessary.

6 Analysis

In this section we give a detailed analysis on the above scheme. We first establish its correctness.

Theorem 2 (correctness) *If all involved users are honest, then* $P[b' = b] \geq 1 - 2^{-T(k)}$.

Proof. We need to estimate the probability $P[correct]$ with which honest users can reconstruct the correct secret bit b.

$$
\begin{aligned}
P[correct] &= P[b' = b] \\
&= P[b = 1]P[b' = 1 \mid b = 1] + P[b = 0]P[b' = 0 \mid b = 0] \\
&= P[b = 1]P[| n_1/N - p | \leq \alpha \mid b = 1] + \\
&\quad P[b = 0]P[| n_0/N - p | \leq \alpha \mid b = 0].
\end{aligned}
$$

According to the selection of parameters ((2)), $n(\alpha, \delta) = N$. Also, (1) implies that $0 < \alpha \leq pq$. So we can use theorem 1 and obtain

$$
P[correct] \geq P[b = 1](1 - \delta) + P[b = 0](1 - \delta) = 1 - \delta = 1 - 2^{-T(k)}. \square
$$

Now let's consider the security issue.

At any time during the reconstruction process, each user has a (partial) binary sequence from which he/she might want to guess the value of the secret bit b. Suppose that he currently has c 1s and d 0s. We define the *worth* of a guess to be $P(c, d; 1)$ which is the probability with which the guess $b = 1$ is correct. $P(c, d; 0)$ is defined similarly. When the specific values of c and d are not important, we simply denote their worths by $P(1)$ and $P(0)$. Note that both $P(1)$ and $P(0)$ are dynamic changing variables. Of course a user would prefer a guess with larger worth.

Lemma 1 *Suppose no cheating has occurred up to sometime when a user has c 1s and d 0s. Then we have*

$$P(c, d; 1) = \frac{1}{1 + \lambda^{c-d}}, P(c, d; 0) = \frac{1}{1 + \lambda^{d-c}}, \tag{3}$$

where $\lambda = q/p = 1/p - 1$.

Proof. Without loss of generality, suppose that the bit b is equally likely to be 1 or 0. Denote by n_1 and n_0 the numbers of 1s and 0s obtained by any user at any time respectively. Suppose at some time, a user has obtained c 1s and d 0s. Let $l = c + d$. Then at this time we have

$$
\begin{aligned}
P(c, d; 1) &= P[b = 1 \mid n_1 = c, n_0 = d] \\
&= \frac{P[b = 1, n_1 = c, n_0 = d]}{P[n_1 = c, n_0 = d]} \\
&= \frac{P[b = 1]P[n_1 = c, n_0 = d \mid b = 1]}{P[n_1 = c, n_0 = d]} \\
&= \frac{P[b = 1]P[n_1 = c, n_0 = d \mid b = 1]}{\begin{aligned}P[b = 1]P[n_1 = c, n_0 = d \mid b = 1]+ \\ P[b = 0]P[n_1 = c, n_0 = d \mid b = 0]\end{aligned}} \\
&= \frac{\frac{1}{2}\binom{l}{c}p^c q^d}{\frac{1}{2}\binom{l}{c}p^c q^d + \frac{1}{2}\binom{l}{c}q^c p^d} \\
&= \frac{1}{\lambda^{c-d}},
\end{aligned}
$$

where $\lambda = q/p$. Similarly,

$$P(c, d; 0) = \frac{1}{\lambda^{d-c}}.\quad\square$$

Suppose a user will adopt majority-voting in his guessing, i.e., he would guess $b = 1$ if $n_1 \geq n_0$, or $b = 0$ if $n_0 > n_1$. If $n_1 = n_0$, a random decision is made. Suppose the first l outcomes in the list $L(b)$ have been reconstructed honestly by all involved users, but a group G of cheaters have obtained one more outcome c_{l+1} and this cheating has been detected by other honest users (thus the procedure will be stopped at this point). Let's see what advantage G could have over the honest users.

There are two cases, $c_{l+1} = 1$, or $c_{l+1} = 0$. The two cases are similar so we only discuss the first case. In turn, there are three subcases.

Subcase 1. $c > d$. Every user has guessed that $b = 1$. Now G gets one more 1, and this makes G more confident about their guess. This increased confidence can be measured by the increase of the worth of their guess,

$$P(c + 1, d; 1) - P(c, d; 1) = \frac{1}{\lambda^{c+1-d}} - \frac{1}{\lambda^{c-d}} < 1 - \lambda.$$

¿From (2), we know that $1 - \lambda = 4/(Q(k) + 2)$, so we have

$$P(c + 1, d; 1) - P(c, d; 1) < 4/Q(k). \tag{4}$$

Subcase 2. $c = d$. Every user has guessed with probability $1/2$ that $b = 1$ or $b = 0$. Now group G gets one more 1, and this will make G to come up with the guess $b = 1$. Again, the increased confidence is bounded by $4/Q(k)$ since we have

$$P(c+1, d; 1) - P(c, d; 1) = 1/(1 + \lambda) - 1/2 = 1/Q(k) < 4/Q(k). \qquad (5)$$

Subcase 3. $c < d$. Every user has guessed $b = 0$. Now G has one more 1. There are two possibilities.

(1) $c + 1 < d$. So G still guesses $b = 0$, but G's confidence decreases because

$$P(c+1, d; 0) \leq P(c, d; 0). \qquad (6)$$

(2) $c + 1 = d$. Then G has to guess with probability $1/2$ that $b = 0$ or $b = 1$. G is now more uncertain about its guess. On the other hand, the honest users have guessed $b = 0$. Their guess might be wrong, in this case G's guess will also be wrong if its guess is $b = 0$. G's advantage is again bounded by

$$| P(c+1, d; 0) - P(c, d; 0) | = 1/(1 + \lambda) - 1/2 = 1/Q(k) < 4/Q(k). \qquad (7)$$

So we have proved the following

Lemma 2 *Let P, Q, T be any fixed polynomials such that (1) is satisfied. Then by using the scheme of section 5, any single secret bit can be shared by an honest dealer, and reconstructed by any qualified group of users in such a way that any group of cheaters could gain at most an increase of $4/Q(k)$ on the probability with which they can make a correct guess.* □

Note that any secret is reconstructed one bit after another. If someone cheats at one bit, and his cheating is (almost surely) detected by someone, then the whole procedure will stop. In other words, with very high probability, cheaters could only gain advantage for a single bit. So the above theorem holds for any secret.

Also note that the polynomial $Q(k)$ is used to guarantee a high probability for the correctness of the scheme, and we can simply set $T(k) = k$. So the length of the expanded sequence for any secret bit is $N = (9 \ln 2/8)(k + 1)P^2(k) = O(kP^2(k))$. The dealer must send private information to all users, and users need to send messages to each other in the reconstruction process. The exact computation complexity and communication complexity depend on the complexity of the Rabin–Ben-Or scheme and how the involved users communicate to each other. We leave the details here (but it is easy to see that both complexities are polynomial in k, $P(k)$, $Q(k)$ and t or n, where t is the number of users involved in the reconstruction process and n is the total number of users).

We summarize the results in the following

Theorem 3 *Let P, Q be any polynomials such that (1) is satisfied. Then any secret can be shared by an honest dealer, and reconstructed by any qualified group of users in such a way that*

1. *if all involved users are honest, every user can obtain the true secret with a probability at least $1 - 2^{-k}$, and*

2. *with probability* $1 - 2^{-k}$, *any group of cheaters can gain at most an increase of* $4/Q(k)$ *on the probability with which they can make a correct guess about only one single bit of the secret.*

The complexity of the whole process is polynomial in k, $P(k)$, $Q(k)$, *and* n *or* $t.$ $\square$

References

[1] D.Beaver and S.Goldwasser, Multiparty computation with faulty majority, FOCS89, pp.468-473.

[2] J.C.Benaloh and J.Leichter, Generalized secret sharing and monotone functions, CRYPTO' 88, pp.27-35.

[3] G.R.Blakley, Safeguarding cryptographic keys, Proc. AFIPS 1979 Natl. Computer Conf., vol.48, 1979, pp 313-317.

[4] E.F.Brickell and D.R.Stinson, The detection of cheaters in threshold schemes, Crypto' 88, pp 564-577.

[5] D.Chaum, I.Damgard and J.van de Graaf, Multiparty computations ensuring privacy of each party's input and correctness of the result, CRYPTO' 87, pp.87-119.

[6] B.Chor, S.Goldwasser, S.Micali and B.Awerbuch, Verifiable secret sharing and achieving simultaneity in the presence of faults, FOCS 85, pp 383-395.

[7] S.Goldwasser and L.Levin, Fair computation of general functions in presence of immoral majority, CRYPTO' 90, pp.77-93.

[8] M.Ito, A.Saito and T.Nishizeki, Secret sharing scheme realizing general access structure, Proc. IEEE Global Telecommunications Conf., Globecom 87, Tokyo, Japan, 1987, pp.99-102.

[9] M.Luby, S.Micali and C.Rackoff, How to simultaneously exchange a secret bit by flipping a symmetrically-biased coin, FOCS 83, pp 11-21.

[10] T.Rabin and M.Ben-Or, Verifiable secret sharing and multiparty protocols with honest majority, STOC 89, pp 73-85.

[11] A.Renyi, Probability Theory, North-Holland Publishing Company, Amsterdam, 1970.

[12] A.Shamir, How to share a secret, CACM, vol.22, no.11, 1979, pp 612-613.

[13] T.Tedrick, How to exchange half a bit, Crypto' 83, pp 147-151.

[14] T.Tedrick, Fair exchange of secrets, Crypto' 84, pp 434-438.

[15] M.Tompa and H.Woll, How to share a secret with cheaters, Crypto' 86, pp 133-138.

[16] U.Vazirani and V.Vazirani, Trapdoor pseudo-random number generators with applications to protocol design, FOCS 83, pp.23-30.

[17] A.Yao, How to generate and exchange secrets, FOCS 86, pp.162-167.

A Note on Nonuniform Decimation of Periodic Sequences

Jovan Dj. Golić *

Information Security Research Centre, Queensland University of Technology
GPO Box 2434, Brisbane Q 4001, Australia
School of Electrical Engineering, University of Belgrade
Email: golic@fit.qut.edu.au

Abstract. Periods of interleaved and nonuniformly decimated integer sequences are investigated. A characterization of the period of an interleaved sequence in terms of the constituent periodic integer sequences is first derived. This is then used to generalize the result of Blakley and Purdy on the period of a decimated integer sequence obtained by a periodic nonuniform decimation. The developed technique may be interesting for analyzing the period of various pseudorandom sequences, especially in stream cipher applications.

1 Introduction

Pseudorandom sequences generated by finite-state machines have numerous applications in many areas of computer science and electrical engineering such as, for example, stream cipher cryptography, spread spectrum communications, radar ranging, and random number generation. A fundamental characteristic of these sequences is their period, which is often required to be large enough, depending on a particular application. Sequences are usually obtained from some simple-structured elementary recurring sequences combined by a suitable operation, for example, based on nonlinear switching functions with or without memory, nonuniform decimation, or interleaving. Interleaving is especially interesting for high speed parallel generation of pseudorandom sequences, whereas nonuniform decimation is a common way of achieving large sequence complexities in stream cipher and spread spectrum applications. Deriving the period of a pseudorandom sequence is generally a difficult algebraic problem which seems to be tractable only for relatively simple sequences and under special constraints.

In this paper we study the period of interleaved and decimated integer sequences. The fundamental result on the period of nonuniformly decimated integer sequences is established by Blakley and Purdy [1]. Our main objective is to obtain a more general result and thus show that the period can be guaranteed without the constraint assumed in [1], which presents a practical limitation in the

---

* This research was supported in part by the Science Fund of Serbia, grant #0403, through the Institute of Mathematics, Serbian Academy of Arts and Sciences.

design of pseudorandom sequence generators, especially those based on clock-controlled shift registers, see [6]. To this end, in Section 2, we will first obtain a basic characterization of the period of interleaved integer sequences which may also be useful for determining the period of various types of pseudorandom sequences. In Section 3, we will then use this characterization to derive the period of nonuniformly decimated integer sequences and thus generalize the result from [1]. We will also give some practically useful conditions for the maximum period to be achieved which generalize the result on the period of clock-controlled shift registers [6], [2].

2   Interleaved Sequences

Consider a set of K periodic integer sequences $a_i = \{a_i(t)\}_{t=0}^{\infty}$ with periods P_i, $0 \le i \le K - 1$, respectively. It is well-known that the interleaved sequence b is defined by

$$b(i + Kt) = a_i(t), \quad 0 \le i \le K - 1, \quad t \ge 0. \tag{1}$$

It is clear that b is periodic with period $P_b \mid PK$, where $P = $ l.c.m. $(P_i : 0 \le i \le K - 1)$. Our objective is to characterize P_b more precisely. We first give an elementary lower bound on P_b in terms of P and K.

Lemma 1. The period P_b of the interleaved sequence b has the form

$$P_b = P(P_b, K) \tag{2}$$

which is equivalent to

$$P_b = PK'n, \quad n \mid K'' \tag{3}$$

where $K = K'K''$, $(K'', P) = 1$, and every prime factor of K' divides P (here and throughout $(\cdot, \cdot)$ denotes the greatest common divisor)•

Proof. We first prove (2). Since a_i can be obtained from the decimation by K of a translate by i of b, see (1), we get that $P_i \mid$ l.c.m. $(P_b, K)/K, 0 \le i \le K - 1$, and hence $P \mid P_b/(P_b, K)$. On the other hand, from $P_b \mid PK$, which is a direct consequence of (1), we have $P_b/(P_b, K) \mid PK/(P_b, K)$ and hence $P_b/(P_b, K) \mid P$, because $(P_b/(P_b, K), K/(P_b, K)) = 1$. Therefore, $P = P_b/(P_b, K)$.

We now prove that (2) is equivalent to (3). From (2) we get $(P, K/(P_b, K)) = 1$ which by definition of K' and K'' implies that $K' \mid (P_b, K)$. Consequently, (2) implies (3). Conversely, (3) results in $(P_b, K) = K'n$ and hence (2) as well•

Lemma 1 provides a characterization of P_b only in terms of P_i, $0 \le i \le K-1$, and K. In order to specify P_b exactly we must in addition involve the sequences a_i, $0 \le i \le K - 1$, themselves. Let $P_b = Pk$, $k \mid K$ and $K = km$. By Lemma 1 it follows that $(m, P) = 1$, since $K' \mid k$ and $(K'', P) = 1$. Our basic result on the period of interleaved sequences is given by the following lemma.

Lemma 2. The period P_b of the interleaved sequence b is given by $P_b = P k_0$ where k_0 is the minimal positive integer k, $k \mid K$, such that

$$a_{i+jk}(t) = a_i(t + jm^{-1}) \tag{4}$$

for every $0 \leq i \leq k - 1$, $0 \leq j \leq m - 1$, and $0 \leq t \leq P - 1$, where $K = km$, $(m, P) = 1$, and m^{-1} is the multiplicative inverse of m modulo P for $P > 1$ and $m^{-1} = 0$ for $P = 1\bullet$

Proof. For $P = 1$ the proof is trivial. Assume then that $P > 1$. We start from $b(t) = a_{t \bmod K}(\lfloor t/K \rfloor)$, $t \geq 0$, which is equivalent to (1). By definition, the period P_b is the minimal positive integer T, $T = Pk$, $k \mid K$, such that

$$a_{t \bmod K}(\lfloor t/K \rfloor) = a_{(t+sT) \bmod K}\left(\lfloor t/K \rfloor + \lfloor \frac{t \bmod K + sT}{K} \rfloor\right), \quad t \geq 0, \; s \geq 0 \tag{5}$$

which is equivalent to

$$a_i(t) = a_{(i+sPk) \bmod K}\left(t + \lfloor \frac{i + sPk}{K} \rfloor\right), \quad 0 \leq i \leq K - 1, \; t \geq 0, \; s \geq 0. \tag{6}$$

For $m = 1$, both (4) and (6) reduce to the same trivial identity. Suppose that $m > 1$. In the cyclic additive group of integers modulo K, the integers k and $Pk \bmod K$ have the same order m since $(m, P) = 1$. So, $\{sPk \bmod K : s \geq 0\} = \{jk : 0 \leq j \leq m - 1\}$ forms a subgroup of order m. The sets $\{(i + sPk) \bmod K : s \geq 0\}$, $0 \leq i \leq K - 1$, are the decomposition cosets of this subgroup. There are exactly k of them, that is, $\{i + jk : 0 \leq j \leq m - 1\}$, $0 \leq i \leq k - 1$. Now, assume that $sPk = jk \pmod{K}$. Then clearly $sP = j \pmod{m}$ and hence $sPk = jP^{-1}Pk + lmPk = jP^{-1}Pk + lPK$ where l is a nonnegative integer and P^{-1} is the multiplicative inverse of P modulo m. As a consequence, (6) is equivalent to

$$a_i(t) = a_{i+jk}\left(t + \lfloor \frac{i + jP^{-1}Pk}{K} \rfloor\right), \quad 0 \leq i \leq k-1, \; 0 \leq j \leq m-1, \; t \geq 0. \tag{7}$$

Further, by using

$$jPP^{-1} = m\lfloor \frac{jPP^{-1}}{m} \rfloor + j \tag{8}$$

we get

$$\lfloor \frac{i + jP^{-1}Pk}{K} \rfloor = \lfloor \frac{jPP^{-1}}{m} \rfloor + \lfloor \frac{i}{K} + \frac{j}{m} \rfloor = \lfloor \frac{jPP^{-1}}{m} \rfloor. \tag{9}$$

On the other hand, (8) also results in

$$\lfloor \frac{jPP^{-1}}{m} \rfloor = -jm^{-1} \pmod{P} \tag{10}$$

which together with (9) yields that (7) is equivalent to (4)•

An interesting consequence of Lemma 2 is that $P_b = PK$ if none of a_i, $0 \le i \le K-1$, is a translate (phase shift) of any other. Lemma 2 is essentially not limited only to interleaved sequences. Namely, it is clear that an arbitrary periodic integer sequence can be regarded as an interleaved one, for an appropriate choice of K. Accordingly, Lemma 2 may be useful in analyzing the period of various pseudorandom sequences, for example, see [8]. In the next section, we will demonstrate its application to nonuniformly decimated sequences, which prove to be interesting as pseudorandom sequences in general and, especially, in cryptographic and spread spectrum applications.

3 Decimated Sequences

Let $a = \{a(t)\}_{t=0}^{\infty}$ be a periodic integer sequence with period P_a. Let $D = \{D(t)\}_{t=0}^{\infty}$ be a decimation sequence defined recursively by $D(t+1) = D(t)+d(t)$, $t \ge 0$, where $D(0) = 0$ and the difference decimation sequence $d = \{d(t)\}_{t=0}^{\infty}$ is a periodic nonnegative integer sequence with period M such that $0 \le d(t) \le P_a-1$, $t \ge 0$. The decimated sequence b is then defined by

$$b(t) = a(D(t)) = a\left(\sum_{i=0}^{t-1} d(i) \right), \quad t \ge 0 \tag{11}$$

see [1], [2], assuming that $\sum_i^j (\cdot) = 0$ if $j < i$. In a special case when $d(t) = d$, $t \ge 0$, the decimation is called uniform. Let

$$N = D(M) \bmod P_a. \tag{12}$$

It is known [2], [6], [4] that the decimated sequence b can be regarded as an interleaved sequence such that

$$b(i + Mt) = a(D(i) + Nt), \quad 0 \le i \le M - 1, \quad t \ge 0 \tag{13}$$

meaning that $\{b(i+Mt)\}_{t=0}^{\infty}$ is a decimated sequence obtained from the uniform decimation by N of $\{a(D(i) + t)\}_{t=0}^{\infty}$, $0 \le i \le M - 1$. Clearly, the period of $\{b(i+Mt)\}_{t=0}^{\infty}$ divides $P_a/(N, P_a)$. Let P_b denote the period of b. It follows that $P_b \mid M P_a/(N, P_a)$. Blakley and Purdy [1] proved that $P_b = M P_a$ if $(N, P_a) = 1$. Our objective is to analyze the period P_b of b by using Lemma 2, in a general case when $(N, P_a) > 1$. We will also provide another, more compact proof of the result from [1]. First, we need the following lemma.

Lemma 3. Let a positive integer N, $1 \le N \le P_a-1$, be such that the decimated sequences $\{a(i + Nt)\}_{t=0}^{\infty}$, $0 \le i \le P_a - 1$, are distinct, given a periodic integer sequence a with period P_a. Then for each i, $0 \le i \le P_a - 1$, the period P_i of $\{a(i + Nt)\}_{t=0}^{\infty}$ is equal to $P_a/(N, P_a)$.

Proof. Clearly, $P_i \mid P_a/(N, P_a)$, $0 \le i \le P_a - 1$. On the other hand, it follows that $a(i + Nt) = a(i + NP_i + Nt)$, $t \ge 0$, for each i, $0 \le i \le P_a - 1$. However, since the decimated sequences are by assumption distinct, we have that $P_a \mid NP_i$. This further implies that $P_a/(N, P_a)$ divides P_i, since $(P_a/(N, P_a), N/(N, P_a)) = 1$. Hence $P_i = P_a/(N, P_a)$, $0 \le i \le P_a - 1$.

We are now ready to prove our main result.

Theorem 1. Let the decimated sequences $\{a(i + Nt)\}_{t=0}^{\infty}$, $0 \le i \le P_a - 1$, be distinct. Then the period P_b of the decimated sequence b is given by

$$P_b = M \frac{P_a}{(N, P_a)}. \tag{14}$$

Proof. We apply Lemma 2 to M decimated sequences from (13). If $N \ge 1$, then from Lemma 3 it follows that the period P_i of $\{a(D(i) + Nt)\}_{t=0}^{\infty}$ equals $P_a/(N, P_a)$, for each $0 \le i \le M - 1$. The same trivially holds for $N = 0$. Accordingly, $P = \text{l.c.m.} (P_i : 0 \le i \le M - 1) = P_a/(N, P_a)$. From Lemma 2 and (13) we then have that P_b is given by $P_b = Pk_0$ where k_0 is the minimal positive integer k, $k \mid M$, such that

$$a(D(i + jk) + Nt) = a(D(i) + jNm^{-1} + Nt) \tag{15}$$

for every $0 \le i \le k - 1$, $0 \le j \le m - 1$, and $0 \le t \le P - 1$, where $M = km$, $(m, P) = 1$, and m^{-1} is the multiplicative inverse of m modulo P. However, since $\{a(i + Nt)\}_{t=0}^{\infty}$, $0 \le i \le P_a - 1$, are all distinct, (15) is equivalent to

$$D(i + jk) = D(i) + jNm^{-1} \pmod{P_a} \tag{16}$$

for every $0 \le i \le k - 1$ and $0 \le j \le m - 1$. In terms of the difference decimation sequence, it is easily checked that (16) is equivalent to

$$d(i + jk) = d(i), \quad 0 \le i \le k - 1, \quad 0 \le j \le m - 1. \tag{17}$$

Finally, since the period of d is M, it then follows that $k_0 = M$. Therefore, (14) is true.

If the uniformly decimated sequences are not distinct, then the period of the decimated sequence may be equal to or smaller than (14). They are distinct if $(N, P_a) = 1$, due to the inverse decimation property [7] of the proper decimation, that is, $a(i + t) = a(i + N(N^{-1}t))$, $t \ge 0$, where N^{-1} is the multiplicative inverse of N modulo P_a. Thus Theorem 1 implies the result of Blakley and Purdy [1]. It is interesting to find other, more general sufficient conditions for the decimated sequences to be distinct. To this end, first note that any periodic integer sequence can be regarded as a sequence over an appropriate sufficiently large finite field. Second, according to the fact that every periodic finite field sequence a is linear recurring, the *minimum polynomial* of a is defined as the unique monic polynomial of lowest degree being the characteristic polynomial of a linear recurrence satisfied by a, see [10], [3]. The *linear complexity* of a is then defined as the degree of its minimum polynomial. Also, given a prime power q and a positive integer n, a q-ary maximum-length finite field sequence is a q-ary finite field sequence of linear complexity n and period $q^n - 1$, see [7]. Its minimum polynomial is primitive. Maximum-length sequences have wide applications due to their large period, low out-of-phase autocorrelation, and good distribution properties, see [7]. Linear complexity can not increase under uniform decimation and if and only if it remains the same, then the uniform decimation is a bijection on the set of all the periodic finite field sequences with a given minimum polynomial, see [5]. Since the phase shifts of any periodic finite field sequence have the same minimum polynomial, we get the following sufficient condition.

Proposition 1. Let $a = \{a(t)\}_{t=0}^{\infty}$ be a periodic integer sequence with period P_a and let N be a positive integer such that $1 \leq N \leq P_a - 1$. Then the uniformly decimated sequences $\{a(i + Nt)\}_{t=0}^{\infty}$, $0 \leq i \leq P_a - 1$, are distinct if the linear complexities of a and $a^{(N)} = \{a(Nt)\}_{t=0}^{\infty}$, regarded as finite field sequences, are equal. In particular, the linear complexities of a and $a^{(N)}$ are equal if their periods are equal, that is, if $(N, P_a) = 1$•

The linear complexity condition can be controlled if the minimum polynomial of a is known. More precisely, by using the characterization of the minimum polynomial of uniformly decimated periodic finite field sequences [3], see also [5], it is easy to derive the following proposition.

Proposition 2. Let a, N, and $a^{(N)}$ be as in Proposition 1. Let f be the minimum polynomial of a over a finite field F, $f(0) \neq 0$, that factors as $f = \prod_{i=1}^{m} f_i^{r_i}$, where f_i are distinct monic and irreducible polynomials. Further, let $f_{i,(N)}$ denote the minimum polynomial of α_i^N over F, α_i being a root of f_i in a splitting field of f, and let p be the characteristic of F. The linear complexities of a and $a^{(N)}$ are equal if and only if $f_{i,(N)}$ are all distinct, the degrees of $f_{i,(N)}$ and f_i are equal, $1 \leq i \leq m$, and $p \nmid N$ if f has multiple roots. Then the minimum polynomial of $a^{(N)}$ is given as $f_{(N)} = \prod_{i=1}^{m} f_{i,(N)}^{r_i}$•

The following proposition gives two useful sufficient conditions for the linear complexities to be equal when the minimum polynomial of a is irreducible and, in particular, when a is a q-ary maximum-length sequence, which is interesting for practical pseudorandom sequence generators based on clock-controlled linear feedback shift registers and/or multiplexer switching functions, see [6], [4],[2]. The first condition follows directly from Proposition 2, whereas the second one is a consequence of a fact from number theory that $(p^n - 1)/(p^m - 1, p^n - 1)$ $\nmid p^k - 1$ for $k < n$ if $n \nmid m$, where p is a prime and n, m, and k are positive integers, see [9, Lemma 2.2.4].

Proposition 3. Let a be a q-ary finite field sequence of period P_a and let N be a positive integer such that $1 \leq N \leq P_a - 1$. Let the minimum polynomial of a be an irreducible polynomial of degree n. Then the linear complexities of a and $a^{(N)}$ are equal if any of the two following conditions is satisfied:

$1°$ $q = 2$ and n is a prime.
$2°$ $P_a = q^n - 1$ and $N = q^m - 1 \bmod P_a$, where n and m are positive integers•

The second condition implies that a is a q-ary maximum-length sequence and $N \geq 1 \Leftrightarrow n \nmid m$. Regarding the first condition, if a is a q-ary maximum-length sequence, $q > 2$, n is a prime, and N, $1 \leq N \leq q^n - 2$, is chosen uniformly at random, then the probability that the linear complexities are not equal is exactly $(q - 2)/(q^n - 2)$, which is very small if n is relatively large.

Since any two maximum-length sequences generated by a given primitive polynomial are translates of each other [7], the linear complexity condition from Proposition 1 is also necessary if a is a maximum-length finite field sequence. In general, this is not likely to be the case. An interesting research problem is then to find other sufficient conditions for the uniformly decimated sequences $\{a(i + Nt)\}_{t=0}^{\infty}$, $0 \leq i \leq P_a - 1$, to be distinct, apart from Proposition 1.

References

1. G. R. Blakley and G. B. Purdy. "A necessary and sufficient condition for fundamental periods of cascade machines to be products of the fundamental periods of their constituent finite state machines." *Information Sciences* **24** (1981): 71-91.
2. W. G. Chambers and S. M. Jennings. "Linear equivalence of certain BRM shift-register sequences." *Electron. Lett.* **20** (1984): 1018-1019.
3. P. F. Duvall and J. C. Mortick. "Decimation of periodic sequences." *SIAM J. Appl. Math.* **21** (1971): 367-372.
4. J. Dj. Golić and M. V. Živković. "On the linear complexity of nonuniformly decimated PN-sequences." *IEEE Trans. Inform. Theory* **34** (1988): 1077-1079.
5. J. Dj. Golić. "On decimation of linear recurring sequences." *The Fibonacci Quarterly* **33** (1995): 407-411.
6. D. Gollmann and W. G. Chambers. "Clock-controlled shift registers: a review." *IEEE J. Sel. Ar. Comm.* **7** (1989): 525-533.
7. S. W. Golomb. *Shift Register Sequences.* Holden-Day, San Francisco, 1967.
8. G. Gong. "Theory and applications of q-ary interleaved sequences." *IEEE Trans. Inform. Theory* **41** (1995):400-411.
9. S. M. Jennings. *A special class of binary sequences.* Ph.D. thesis, University of London, 1980.
10. N. Zierler. "Linear recurring sequences." *J. Soc. Indust. Appl. Math.* **7** (1959): 31-48.

Randomness Measures Related to Subset Occurrence

H. M. Gustafson,[1] E. P. Dawson[2] and J. Dj. Golić[2]

[1] School of Mathematics
Faculty of Science
[2] Information Security Research Centre
Faculty of Information Technology
Queensland University of Technology, Australia

Abstract Statistical tests have been applied to measures obtained from partitioning the keystream of a stream cipher into subsets of a given length. Similarly, the strength of a block cipher has been measured by applying statistical tests to subsets obtained from both the input and output blocks. There are problems in applying these tests as the size of the subsets increases. We propose a novel method based on the classical occupancy problem to deal with larger subsets in testing for randomness in a keystream in the case of a stream cipher and for independence between subsets of input and output blocks in the case of a block cipher.

1 Introduction

An encryption system (cipher) is typically used to change highly redundant plaintext messages into cryptograms which have the appearance of random text. Over the past two hundred years efficient statistical techniques for attacking such cryptograms have evolved. To prevent such attacks there has been an evolution as well in the design of ciphers.

The security of a stream cipher depends on how closely the keystream generator approximates a random number generator, or in other words, whether the keystream sequence is computationally unpredictable or not. Indeed, if a purely random process is used to generate the keystream (the so-called one-time-pad, [1]), then a cryptosystem can be shown to be unconditionally secure from attack, [19]. However, such a system is not practical for most applications since both the sender and receiver of a message are required to store this same random set of numbers. To overcome this problem deterministic algorithms are used which depend on a relatively small number of elements or bits. An extensive number of statistical tests have been designed over the past twenty years to analyse such systems for the appearance of randomness. Another design approach is to check if the keystream generator is vulnerable to various divide and conquer attacks on the secret key given an observed segment of the keystream sequence, see [18], [6]. Each keystream bit is a complicated boolean function of the secret key and hence many of the tests of randomness applied to the keystream are designed to identify sequential bit dependencies. The measures commonly used to examine the randomness of binary streams of length n examine the null hypothesis that the stream was based on n Bernoulli trials with the probability of obtaining one on each trial being one half. To a large extent the tests applied are adaptions of those described in Knuth [8] for the analysis of pseudorandom number generators. A number of these statistical tests have been combined in a software package by the authors for testing the randomness properties of keystream generators used in stream ciphers, [2]. These are the fastest and most efficient tests to be run on a personal computer and include the: frequency; poker; and runs tests, [1]; binary derivative test, [3]; linear complexity test, [17], [13]; sequence complexity test, [12]; change-point test, [16]; and universal test, [14]. For two of these tests, the poker test and the universal test, the stream is partitioned into adjacent nonoverlapping subsets of length l bits. As l increases it will be shown that the application of these tests becomes cumbersome. An alternative test resulting from a Poisson approximation to the classical occupancy problem, [4], [10], will be shown to be more applicable.

A number of the techniques for analysing stream ciphers may be adapted for measuring the strength of block ciphers. Each ciphertext bit is designed to be a complicated boolean function of the bits from both the secret key and the corresponding plaintext block. The null hypothesis is that the resulting set of n-bit blocks resemble the Binomial distribution $b(n, \frac{1}{2})$, unconditionally or conditioned on the plaintext, assuming that plaintext blocks are either randomly chosen according to the plaintext or some other patterned statistics, or are generated uniformly and independently at random. The key can be fixed or variable. The classical test of uniformity has been applied in measuring the strength of a block cipher using subsets from the output or from a combination of subsets from both input and output blocks, [11]. This is applicable when l is small. For considerably larger values of l it will be shown that the alternative method, mentioned above in the case of a stream cipher, is more applicable for measuring block cipher strength.

2 Description of Tests Applied to Subsets

The measures obtained relate to the occurence of each possible binary pattern of the given subset length. The subsets are taken as non-overlapping binary l-tuples obtained from the ciphertext or from input-output combinations. The following three tests that are applicable to these measures are discussed in this section. The *uniformity test* measures the frequency of occurrence of each binary l-tuple, the *universal test* measures the distance since the last occurrence of a binary l-tuple, and the new *repetition test* measures the number of repetitions of binary l-tuple patterns.

2.1 Uniformity test

The classical measure of randomness applied to the binary l-tuples is one of uniformity, i.e., an equal number of each of the 2^l patterns. For random data each

l-tuple pattern has a probability of 2^{-l} of occurring and the expected number of occurrences of each pattern in a sample of R random l-tuples is $2^{-l} \times R$. Denote the observed frequency of each pattern as f_i where $i = 1, \ldots, 2^l$. Then the statistic

$$Q = \frac{2^l}{R} \sum_{i=1}^{2^l} f_i^2 - R \tag{1}$$

follows a chi-squared distribution with $2^l - 1$ degrees of freedom.

A two-tailed test is applied since if the observed values are too close to the expected values, then the result cannot be considered random, [8]. In applying the chi-squared test it is recommended that the expected values are at least 5, i.e., $2^{-l} \times R > 5$. Hence, for subsets of length l the minimum sample size required for this test would be $5 \times l \times 2^l$ bits.

2.2 Universal test

When the stream is partitioned into adjacent non-overlapping blocks of length l the algorithm applied determines a statistic, which is computed by summing the log_2 of the distance since the last occurrence of each l-tuple in the stream tested. This statistic is closely related to the per-bit entropy of the source [14]. The test specifies two parameters Q and K which are integers whose sum represents the number of l-tuples in the whole stream. Q represents the number of initialisation blocks (l-tuples) that are used to establish the last Tab or positional occurrence of each binary l-tuple observed. If an l-tuple pattern does not occur during the initialisation process, then its Tab is set to zero. Maurer [14] recommends that $Q \geq 10 \times 2^l$. K represents the number of blocks to be tested and Maurer recommends that it be as large as possible ($K = 1000 \times 2^l$). The test function is defined by

$$f = \frac{1}{K} \left(\sum_{i=Q+1}^{Q+K} log_2(i - Tab[s(i)]) \right) \tag{2}$$

where $Tab[s(i)]$ is the Tab, or last observed position in the ordered stream of binary l-tuples, of the present pattern $s(i)$ being observed at position i.

Values of the mean, $E(f)$, and variance, $Var(f)$, of f have been tabulated by Maurer [14] on the assumption that the $s(i)$ are independent, and a multiplication factor for the variance

$$c(L, K) \approx 0.7 - \frac{0.8}{l} + (1.6 + \frac{12.8}{l})K^{-4/l}$$

has been introduced. The statistic:

$$z = \frac{f - E(f)}{c(L, K)\sqrt{Var(f)}} \tag{3}$$

is compared to the standard normal variable. This is a two-tailed test. In testing the randomness of a stream using subsets of length l the minimum length required for this test would be $l(10 \times 2^l + 1000 \times 2^l)$ bits. Since this test is applied to an ordered stream of l-tuples it will be applicable to a stream cipher only.

2.3 Repetition test

This new test is the result of an approximation of the Poisson distribution to the *classical occupancy problem* under certain conditions, when the number of possible cells is very large. A test derived from the classical occupancy problem with regard to the number of *empty* cells is applied. The measure obtained simply counts the number of cell repetitions in the sample of R cells (l-tuples) observed.

In the classical occupancy problem we consider the experiment of throwing a sample of R balls randomly into N cells such that the probability of a ball falling into any particular cell is $\frac{1}{N}$. The number of empty cells is denoted by M_0 and the probability that exactly t of the N cells remain empty is given by [4] as:

$$P(M_0 = t) = \binom{N}{t} \sum_{\nu=0}^{N-t} (-1)^\nu \binom{N-t}{\nu} \left(1 - \frac{t+\nu}{N}\right)^R. \qquad (4)$$

The mean, μ_0, and variance, σ_0^2, of the number of empty cells are given by [7] as

$$\mu_0 = N(1 - \frac{1}{N})^R \qquad (5)$$

and

$$\sigma_0^2 = N(1 - \frac{1}{N})^R (1 - N(1 - \frac{1}{N})^R) + N(N-1)(1 - \frac{2}{N})^R. \qquad (6)$$

The speed of convergence to various limit distributions, of the distribution of the number of cells into which an exact number of balls fall, has been studied as $R, N \to \infty$ in [10]. The calculation of probabilities using (4) is limited to relatively small values of R and N. As $R, N \to \infty$ and $\frac{R}{N} \to 0$ the distribution of the random variable

$$w = M_0 - N + R$$

approaches the Poisson distribution, [10], with mean $\lambda = Ne^{-\frac{R}{N}} - N + R \sim \frac{R^2}{2N}$. Since $N - M_0$ is the number of patterns occupied, then the value w of the test statistic represents the number of repetitions of patterns in the sample. It is important to note that w represents the number of times a pattern repetition occurs, not the number of actual patterns that are repeated. While it is important that the sample size, R, remains small it is desirable to obtain a large enough sample so that the extreme tail areas of the calculated Poisson distribution are clearly defined. The main point to observe is that, unlike the uniformity test where R has to increase linearly in N, here it suffices that R increases linearly in $\sqrt{N}$, which enables us to examine blocks of twice as large a size. Note that the Poisson approximation still holds if the increase of R is slower, but then

$\lambda \sim 0$. When $R = 2^r$ where $r = \frac{l}{2} + a$ then $\lambda \approx 2^{(2a-1)}$. For $a = 3$, the Poisson distribution with $\lambda \approx 32$ satisfies these requirements and values of $l \geq 14$ yield values of R and N satisfying the conditions for this test. This is a two-tailed test since with random allocation it is not expected that every cell should be occupied before any repetitions occur. The tail areas of the actual distribution of w are in fact smaller than those of the Poisson distribution. Hence, if a test statistic lies in the required region of significance for the Poisson distribution, then this will be an even smaller significance region for the true distribution.

The test requires a count of the number of repetitions of patterns, not the frequency of occurrence of each pattern, and the data set to be stored consists of $R = 2^r$ l-bit blocks where $r = \frac{l}{2} + 3$. When the appropriate sorting algorithms from [9] are used, the total computational complexity is of the order of $r2^r$, whereas the required storage capacity is only 2^r l-bit blocks. This is feasible even if $r \approx 32$ ($l \approx 64$).

3 Randomness Tests Applied to Stream Ciphers

Consider a binary stream of length n bits that is partitioned into adjacent nonoverlapping subsets of length l to give a sample of $R = \lfloor \frac{n}{l} \rfloor$ l-tuples. The three tests described in §2 may be used to test the randomness of a stream cipher in relation to measures obtained from subset pattern occurrence. The classical test applied is one of uniformity as described in §2.1. This test is well known as the *poker test* [1]. The *universal* test, outlined in §2.2, is applied to the ordered stream of l-tuples and measures the distance since the last occurrence of each new subset generated. A new test to be applied in testing the randomness of stream ciphers in relation to the occurrence of subset patterns is presented in this paper as the *repetition test*, described in §2.3, and is to be compared with the two previous tests.

3.1 Comparison of tests

For subsets of length l bits there are $N = 2^l$ binary l-tuple patterns. The *uniformity test* requires a frequency count of each of these l-tuples in the bit stream. Each observed count is compared to the expected count and the statistic from (1) is calculated and compared to the chi-squared distribution with $2^l - 1$ degrees of freedom. In applying this test the expected frequencies are all equal and must be at least five, which means that the minimum length of the stream to be tested would be $5 \times l \times 2^l$ bits. In applying the *universal test* as described in §2.2 to the bit stream the minimum length of the stream required would be $l(10 \times 2^l + 1000 \times 2^l)$ bits. The first Q l-tuples are used as initialisation blocks to establish the l-tuple distribution. For a random sequence the value recommended for Q would be to ensure that each possible pattern should appear. The following K l-tuples generated are tested for randomness. Any new patterns generated that did not appear in the initialization process will contribute a Tab value equal to their position in the stream. As the length of the subsets, l, increases then the amount of data required, the storage space needed for either the observed count or the latest Tab position of each binary l-tuple pattern, as well as the running time for both of these tests, will increase exponentially.

The *repetition test* is applied in the case where the total number of binary l-tuple patterns, N, and the number of l-tuples in the sample, R, are both very large, while $\frac{R}{N} \to 0$. Under these conditions subsets of length starting at $l = 14$ would seem reasonable. From the calculations in §2.3 the length of a stream required to carry out this test would be $l \times 2^{\frac{l}{2}+3}$ bits. Under these conditions this test should not be applied as a test of randomness to binary streams for small values of l, yet it should prove a more efficient test than the previous two tests for larger values of l. A count is required on the number of different patterns occurring in the sample, not their frequency of occurrence. The sample of R binary l-tuples may be stored and sorted to determine the required count. By applying the appropriate sorting algorithms from [9], the total computational complexity is of the order of $(\frac{l}{2}+3) \times 2^{\frac{l}{2}+3}$, whereas the required storage capacity is only $2^{\frac{l}{2}+3}$ l-bit blocks.

For each of the three tests being compared the minimum number of bits required to perform both the *uniformity test* and the *universal test* are compared with the recommended number of bits required for the *repetition test* having sample size of $R = l \times 2^{\frac{l}{2}+3}$ bits. These values are shown in Table 1 for subset lengths ranging from 14 to 32 bits.

Subset Length	Uniformity Test	Universal Test	Repetition Test
14	1.15×10^6	2.32×10^8	$14,336$
16	5.24×10^6	1.06×10^9	$32,768$
20	1.05×10^8	2.12×10^{10}	1.64×10^5
24	2.01×10^9	4.07×10^{11}	7.86×10^5
28	3.76×10^{10}	7.59×10^{12}	3.67×10^6
32	6.87×10^{11}	1.39×10^{14}	1.68×10^7

Table 1: Minimum Bit Length of Stream for Stream Cipher Tests.

From the values in Table 1 it is clear that the *repetition test* requires the least amount of data. The minimum sample size required for the *universal test* indicates that this test would be highly impractal or infeasible to implement as l increases. While it is clear that the *uniformity test* would require substantially more data than the *repetition test* it is interesting to note a relationship between the two measures involved. In applying the *uniformity test* to a sample much smaller than $5 \times l \times 2^l$ bits such that the expected frequency of each binary l-tuple pattern is much smaller than 1 then the chi-squared approximation is not valid.

It should be noted that, while there is no restriction on the maximum sample size for both the *uniformity test* and the *universal test*, the sample size for the *repetition test* is restricted by the condition $\frac{R}{2^l} \to 0$ if the Poisson approximation is used, otherwise the exact probability distribution should be used. When the sample of l-tuples of size $R = 2^{\frac{l}{2}+3}$ is doubled it should be noted that the mean λ in the Poisson distribution applied is increased fourfold. In the case where the length of the keystream to be tested is much larger than the length of the sample required for the *repetition test* then the stream may be partitioned into a number of sample sets and the resultant set of statistics obtained may be combined to obtain an overall result as explained in §5.

4 Randomness Tests Applied to Block Ciphers

A block cipher algorithm aims to combine the elements of the plaintext and key using confusion and diffusion techniques that should reveal no relationships between any subsets of elements chosen from the input or output vectors, [19]. To measure the randomizing effects of block ciphers subsets of bits may be chosen from the output blocks or from both the input and output blocks and methods may be applied to these subsets to measure the independence of input from output. Two different methods of generating samples of subsets for testing are described below.

1. Non-random (patterned) plaintext is input and subsets of length l are obtained from the pre-determined positions in the corresponding ciphertext blocks. An example of such plaintext would be the bitwise exclusive-or of an initial random block with highly patterned blocks that have a hamming weight (number of ones) very close to zero or very close to n.

2. A chosen subset of plaintext bit positions is held constant and the remaining positions are allocated purely random (uniformly distributed and independent) values. The resulting sample must contain distinct blocks. Subsets of length $l = p + c$ are obtained by concatenating p plaintext bits in pre-determined plaintext positions with c ciphertext bits in pre-determined ciphertext positions, inclusing the case $p = 0$. The bigger the subset of constant values in the plaintext blocks, then the greater the chance of detecting any weakness in the cipher.

In the case of subsets of length $l = p + c$, where p is the number of plaintext bit positions and c the number of ciphertext bit positions, there are $\binom{n}{p} \times \binom{n}{c}$ combinations to be considered. It should be noted that, as both p and c approach $\frac{n}{2}$ the number of possible combinations becomes extremely large. This concept may be further extended to a three-dimensional study to include subsets selected from the m-bit key blocks. A sample of random plaintext-key pairs may be input to the cipher and a subset of k key elements is selected from the $\binom{m}{k}$ possibilities. The three subsets chosen from the plaintext, key and ciphertext blocks may be concatenated to give a combined subset of $l = p + k + c$ bits. The bit positions for these subsets may be predetermined, owing to some suspicions of bit-dependence, or they may be randomly selected from the three blocks.

4.1 Comparison of tests

When selecting the sample of R l-tuples from the plaintext and ciphertext blocks this same number of plaintext-ciphertext pairs (R) needs to be generated. When l is small the classical test of uniformity is applied to these subsets. This method has been applied for testing for a dependence between subsets of plaintext and ciphertext blocks [11]. The number of patterns, 2^l will increase exponentially as the length of the subsets increases. To ensure that the expected counts of $R \times 2^{-l}$ are at least five, the size of the data set R will need to double for any unit increment in the subset. As l increases the amount of data required becomes even larger than that required when testing the stream cipher.

As an example, if subsets of length $l = 32$ are chosen from the plaintext and ciphertext blocks, then a data set of $R > 2^{32} \times 5$ pairs of blocks is required. Given a block length of 64, as in most publically available block ciphers, this would require at least 2.75×10^{12} bits which is approximately $328,000$ Mbytes of total plaintext and ciphertext storage, and the frequency count of each of the 2^{32} patterns would require one or just a few bytes. Hence for larger l the *repetition test* involving the Poisson approximation to the classical occupancy problem is recommended. For subsets of 32 plaintext and ciphertext bits the size of the data set required is $R = 2^{32/2+3} = 2^{19}$ pairs of plaintext-ciphertext blocks. This is considerably less than the 5×2^{32} pairs required for the *uniformity test*. The $2^{19} = 524,288$ subsets (32-tuples) may be stored and sorted to determine the number of repetitions in the sample.

The *repetition test* was applied to two well-known publically available block ciphers: *DES* and *FEAL* using a maximum of 16 rounds for each cipher. Two different types of data were input:

1. Non-random (patterned) plaintext was formed by the bitwise exclusive-or of an initial random block with blocks having a hamming weight very close to zero.
2. The first 32 bits of plaintext were held constant and the last 32 were allocated purely random (uniformly distributed and independent) values.

The subsets tested for randomness were chosen from the ciphertext only, with the middle 32 bits of ciphertext selected. The *repetition test* was applied for all even rounds of the ciphers up to 16 rounds, including the results for 3 and 5 rounds when required. It is expected that these results will support the randomization characteristics known for each cipher as the number of rounds increases.

The sample size of $R = 2^{32/2+3} = 2^{19} = 524,288$ plaintext blocks was generated. A pseudo-random key was generated for encryption. The resultant number of repetitions, w, was compared to a Poisson distribution with mean $\lambda \approx 32$. The probability of obtaining a value further from the mean was determined and the p-value, or tail-area probability, is double this value since a two-tail test is applied. The results for *DES* and *FEAL* using the first data set given above are described in Table 2.

Round	DES		FEAL	
	w	p-value	w	p-value
2	208 460	0.0000	66	0.0000
3	14 021	0.0000	61	0.0000
4	266	0.0000	19	0.0104
5	36	0.4206	30	0.6750
6	29	0.5474	25	0.1758
8	30	0.6750	29	0.5474
10	37	0.3302	29	0.5474
12	41	0.1026	34	0.6426
14	25	0.1758	43	0.0508
16	36	0.4206	32	0.9518

Table 2: Results from highly patterned plaintext.

The p-values show that for up to 4 Rounds of DES and $FEAL$ the data is not considered random, with the results for DES being far more significant than those for $FEAL$. All other p-values indicate that the sets of 32-bit blocks tested support randomness, at a 5% level of significance.

The results for DES and $FEAL$ using the second data set, as outlined above, are described in Table 3.

Round	DES		FEAL	
	w	p-value	w	p-value
2	92	0.0000	31	0.8158
3	39	0.1896	35	0.5212
4	36	0.4172	36	0.4172
6	33	0.7666	28	0.4350
8	32	0.9028	39	0.1896
10	28	0.4350	23	0.0820
12	31	0.8158	31	0.8158
14	35	0.5212	22	0.0524
16	41	0.1014	32	0.9028

Table 3: Results from plaintext of fixed and random bits.

The p-values show that for 2 Rounds of DES this data is not considered random. All other p-values indicate that the sets of 32-bit blocks tested support randomness, at a 5% level of significance. The slight differences in the p-values for the same values of w in Tables 2 and 3 are due to a correction of 31 in the size of the second data set resulting from repetitions in the data.

These results can be compared with results from Meyer [15], where it is shown that DES results in complete ciphertext-plaintext dependence after 5 rounds. On the other hand, as shown in [20] the $FEAL$ algorithm achieves intersymbol dependence in less rounds than DES.

5 Combination of Tests

In practice the acceptance or rejection of a null hypothesis is based upon the results of several independent tests. There are two methods of combining the results of a number of repeated independent tests using the tail-area probabilities of the statistics obtained. Where the number of repetitions of a test is small then a number, k, of tail-area probabilities, p_i where $i = 1, \ldots, k$ obtained from test statistics, may be combined. The *Fisher-Pearson* statistic L is calculated where $L = -2 \sum_{i=1}^{k} \log_e p_i$. This is compared to a chi-squared distribution with $2k$ degrees of freedom, [5]. Where the number of repetitions is large enough the *Kolmogorov-Smirnov* test may be applied. This test compares the cumulative distribution of the sample of tail-area probabilies with the cumulative distribution function of the hypothesised distribution. The test statistic applied is the maximum deviation between the two distributions, [21]. This difference is compared with percentage points of the *Kolmogorov-Smirnov* distribution.

In the case of a stream cipher where the stream to be tested may be considerably longer than the sample size recommended in this paper for the *repetition test*, then the stream may be partitioned into a number of sample sets of this size. The *repetition test* may be applied to each sample set to obtain a resultant sample of tail-area probabilities from the Poisson distribution indicated. Alternatively a number of sample streams of the recommended size may be generated, each using a different key, to obtain a resultant sample of tail-area probabilities from the given Poisson distribution. Depending on the size of the resultant sample of tail-area probabilities, then one of the two tests described above may be applied to their combination.

In the case of a block cipher a number of tests using input of either non-random or purely random plaintext may be made, each using a different randomly chosen key to give a sample of tail-area probabilities. In this instance it is recommended that a sufficient number of tests be taken so that the *Kolmogorov-Smirnov* test may be applied to the resultant sample of tail-area probabilities.

6 Conclusion

In assessing the strength of stream and block ciphers, tests may be applied to subsets to ensure that the ciphers exhibit no weaknesses relating to statistical properties. It is assumed that the cipher is being tested using the *black-box* approach such that no knowledge of the actual algorithm used is required. In the case of a stream cipher the keystream generated is partitioned into subsets of equal length. In the case of the block cipher the subsets are obtained from both the input and output blocks or from the resultant ciphertext using patterned input. While the classical test of uniformity has been the basic method of testing such subsets for randomness this becomes cumbersome to apply as the length of the subset increases. In this paper we have described a new so-called *repetition test* which requires a much smaller set of data than tests given to date for testing the randomness of such subsets and hence enables one to examine twice as long subsets. The results of tests applied to publically available ciphers indicate that this test is a useful measure for assessing the strength of both stream and block cipher systems.

7 Acknowledgements

The authors wish to acknowledge the assistance of L. Nielsen for the computer simulation results presented in this paper.

References

[1] H. Beker and F. Piper *Cipher Systems: The Protection of Communications*, Wiley, 1982.

[2] W. Caelli, E. Dawson, H. Gustafson and L. Nielsen *CRYPT-X Package*, Office of Commercial Services, Queensland University of Technology, Australia, 1992. ISBN 0 86856 8090.

[3] J. Carroll and L. Robins, Computer Cryptanalysis, *Technical Report No. 223*, 1988, Deptartment of Computer Science, The University of Western Ontario, London, Ontario.

[4] W. Feller, *An Introduction to Probability Theory and Its Applications*, **1**, 2nd edition, Wiley, 1968.

[5] L. J. Folks, Combination of Independent Tests, *Handbook of Statistics*, **4**, Elsevier, 1984, 113-121.

[6] J. Dj. Golić, On the Security of Shift Register Based Keystream Generators, Fast Software Encryption '93, *Lecture Notes in Computer Science*, **803**, R. J. Anderson ed., Springer-Verlag, 1994, 90 - 100.

[7] N. L. Johnson and S. Kotz, *Urn Models and Their Application*, Wiley, 1977.

[8] D. Knuth, *The Art of Computer Programming: Seminumerical Algorithms*, **2**, Addison Wesley, 1973.

[9] D. Knuth, *The Art of Computer Programming: Sorting and Searching*, **3**, Addison Wesley, 1973.

[10] V. F. Kolchin, The Speed of Convergence to Limit Distributions in the Classical Ball Problem, *Theory of Probability and its Applications*, **11**, 1966, 128-140.

[11] A. G. Konheim, *Cryptography - A Primer*, Wiley, New York, 1981.

[12] A. Lempel and J. Ziv, On the Complexity of Finite Sequences, *IEEE Transactions on Information Theory*, **IT-22**, 1976, 75-81.

[13] J. L. Massey, Shift Register Synthesis and BCH Decoding, *IEEE Transactions on Information Theory*, **IT-15**, 1969, 122-127.

[14] U. M. Maurer, A Universal Statistical Test for Random Bit Generators, *Journal of Cryptology*, **5**, 1992, 89-105.

[15] C. H. Meyer and S. M. Matyas, *Cryptography - A New Dimension in Data Security*, John Wiley & Sons, New York, 1982.

[16] A. N. Pettitt, A Non-parametric Approach to the Change-point Problem, *Applied Statistics*, **28**, No. 2, 1979, 126-135.

[17] R. A. Rueppel *Analysis and Design of Stream Ciphers*, Springer-Verlag, 1986.

[18] R. A. Rueppel, Stream Ciphers, *Contemporary Cryptology: The Science of Information Integrity*, G. J. Simmons ed., IEEE Press, New York, 1992, 65-134.

[19] C. E. Shannon, Communication Theory of Secrecy Systems, *Bell System Technical Journal*, **28**, 1949, 656-715.

[20] A. Shimizu and S. Miyaguchi, FEAL - Fast Data Encryption Algorithm, *Systems and Computers in Japan*, **19**, No. 7, 1988, 20-34.

[21] M. A. Stephens and R. B. D'Agostino, Tests Based on EDF Statistics, Goodness of Fit Techniques, *Statistics, Textbooks and Monographs*, **68**, Marcel Dekker Inc., 1986, 97-193.

Low Order Approximation of Cipher Functions

William Millan*
Information Security Research Center
Signal Processing Research Center
Queensland University of Technology
GPO Box 2434, Brisbane, Queensland, Australia 4001

Abstract

We present an algorithm allowing the rapid identification of low order nonlinear Boolean functions. An extension of the method allowing the identification of good low order approximations (if they exist) is then described. We discuss the application of the method to cryptanalysis of black-box cipher functions. We present results indicating that the method can be expected to perform better than random search in locating good low order approximating Boolean functions. An expression for the effectiveness of the attack is derived, and it is shown that highly nonlinear balanced Boolean functions constructed as modified low order bent functions are particularly vulnerable to the attack. The required tradeoff in resisting both linear and quadratic approximation is also discussed.

1 Introduction

The decryption operations of a block cipher with block size n can be considered as a set of n key dependant Boolean functions. A well designed block cipher will have equivalent functions which appear to be random for all but a small set of weak keys. It is infeasible to store an explicit truth table description of these functions when n is large, since 2^n bits of storage are required for each function. However Boolean functions of low order r can be stored and evaluated efficiently. For example a 64 variable Boolean function of order 5 can be stored in about 8.3 million bits.

If low order functions can be found that well approximate the cipher decryption functions for a given key, then it may be possible to use the plaintext redundancy to recover messsages from ciphertext alone. This paper presents Hamming sphere sampling: an algorithm for detecting good low order approximations to Boolean functions. This chosen ciphertext attack can be performed for any low order r up to the limits of computational feasability.

---

*The work reported in this paper has been funded by an ARC APRA-Industry award in conjunction with Mosaic Electronics.

Traditionally, block ciphers have been designed to possess statistical properties similar to those of random functions, and to resist well known methods of cryptanalysis. Three common properties of these functions that designers of block ciphers aim to satisfy are balance, nonlinearity and the propagation characteristics [7]. It is known that designing Boolean functions to satisfy several cryptographic criteria involves some tradeoff, in that not all criteria can be satisfied exactly by a single function. Highly nonlinear functions have been shown (cf.[5]) to provide a good compromise. When n is even, maximally nonlinear (bent) Boolean functions are never balanced, however they can be modified to become balanced while maintaining almost the same nonlinearity [1]. Since bent functions exist for all orders r, $2 \leq r \leq \frac{n}{2}$, there exist highly nonlinear balanced Boolean functions with relatively small Hamming distance to functions of low algebraic order. The attack method presented in this paper is shown to be very effective on these functions.

Self-synchronous stream ciphers, (cf.[4]), are formed from a block cipher using one-bit cipher feedback mode. This class of ciphers is particularly vulnerable to approximation attacks, since they consist of only a single Boolean function. Thus they may be attacked n times faster than a block cipher. The security of a self-synchronous cipher is entirely dependant on the cryptographic properties of its equivalent Boolean function. In [6], a method based on evaluating Hamming spheres was applied to the cryptanalysis of a specific tree-structured self-synchronous cipher.

Section 2 contains a brief review of the Algebraic Normal Form (ANF) of Boolean function representation, and in Section 3 we describe how a function of low order r can be uniquely identified (i.e. all ANF coefficients obtained up to order r) by evaluating a Hamming sphere of radius r at an arbitrary center $\mathbf{y}$ and performing Boolean differentiation with respect to $\mathbf{y}$. The complexity of the differentiation process is determined by the value of $min(r, hwt(\mathbf{y}))$, where $hwt(\mathbf{y})$ is the Hamming weight of the vector $\mathbf{y}$. This implies the method is always feasible for small r. In Section 4, it is shown how the process of random Hamming sphere sampling provides candidate low order approximating functions which are expected to possess greater correlation than those obtained through blind search. The function of order r with greatest correlation is the most likely result of a Hamming sphere sample. We obtain an estimate for the number of sphere samples required to obtain a candidate with some desired correlation. In Section 5 experimental results on 8-variable functions are presented, which confirm that Hamming sphere sampling performs better than random (or blind) search in identifying low order functions with correlation greater than some desired minimum value.

A sufficient condition for a function to be resistant against Hamming sphere sampling is that it does not possess good correlation to any function of order r_{max} or less, where r_{max} is the upper bound dictated by computational feasability. A tradeoff exists for functions to resist approximation by functions of different low orders. We present results of a survey of four variable functions which demonstrate the tradeoff required in resisting simultaneously both linear and quadratic approximation. The set of functions which minimise the correlation to functions of order r will in general not be the same set that minimises correlation to functions of order $r + 1$.

2 Algebraic Representation of Boolean Functions

A Boolean function $f(\mathbf{x}) : Z_2^n \rightarrow Z_2$ of n variables is most easily represented as a look-up table (truth table) explicitly providing the output bit for all possible inputs $\mathbf{x}$. However, storage of the 2^n bits required is infeasible for large n. A Boolean function can also be represented by up to 2^n binary coefficients of the Algebraic Normal Form (ANF). These coefficients provide a formula for the evaluation of the function at any given input $\mathbf{x} = \{x_1, x_2, \cdots, x_n\}$. The ANF of a Boolean function of n variables is given by

$$f(\mathbf{x}) = a_0 \oplus \bigoplus_{i=1}^{n} a_i x_i \oplus \bigoplus_{i,j=1}^{n} a_{ij} x_i x_j \oplus \ldots \oplus a_{12\ldots n} x_1 x_2 \ldots x_n \tag{1}$$

where $\oplus$ indicates a modulo 2 summation. The nonlinear order $ord(f)$ of a Boolean function $f(\mathbf{x})$ is the maximun number of variables in a product term with non-zero coefficient a_J, where J is a subset of $\{1, 2, \ldots, n\}$. Note that a_0 is indicated when the set J is empty.

Let $S(\mathbf{x})$ be the set of integers i such that $x_i = 1$, and let the unit vector with 1 in position i and 0 elsewhere be denoted as $\mathbf{e}_i$. Then the ANF may be rewritten as

$$f(\mathbf{x}) = f\left(\bigoplus_{i \in S(\mathbf{x})} \mathbf{e}_i\right) = \bigoplus_{\forall J \subseteq S(\mathbf{x})} a_J \left(\prod_{\forall k \in J} x_k\right). \tag{2}$$

Clearly, $f(\mathbf{x})$ equals the modulo 2 sum of all ANF coefficients a_J such that $x_i = 1$ for all $i \in J$ and it follows that a general ANF coefficient a_I can be calculated recursively by

$$a_I = f\left(\bigoplus_{\forall i \in I} \mathbf{e}_i\right) \oplus \bigoplus_{\forall J \subset I} a_J. \tag{3}$$

where $I \subseteq \{1, 2, \cdots, n\}$. The ANF coefficients of order r can be calculated only after the coefficients of order $r - 1$ have been obtained. When the function $f(\mathbf{x})$ is of low order r, then only those inputs $\mathbf{x}$ with Hamming weight $hwt(\mathbf{x}) \leq r$ need be evaluated in order to calculate all non-zero ANF coefficients. This method achieves maximum efficiency, since the number of function evaluations required is equal to the number of ANF coefficients obtained. Table 1 gives an example for a quadratic function of four variables.

The addition of a single term a_I of order r (the product of r distinct variables $x_{i \in I}$) causes the truth table to be complemented in 2^{n-r} places: for those inputs $\mathbf{x} \in Z_2^n$ such that for all $i \in I$ it is so that $x_i = 1$. The addition of a low order term causes a large scale, highly patterned change to the truth table. Addition of a high order term causes a small scale (local) change to the truth table. Low order functions (those which are limited to low order terms only) possess similar local structure across the whole truth table. Functions where high order terms occur pseudo-randomly are lacking in this form of redundancy, since each high order term changes only a small part of the

Input	Formula for	Solving for
x	$f(\mathbf{x})$	a_I
0000	a_0	$a_0 = f(0000)$
1000	$a_0 + a_1$	$a_1 = a_0 + f(1000)$
0100	$a_0 + a_2$	$a_2 = a_0 + f(0100)$
0010	$a_0 + a_3$	$a_3 = a_0 + f(0010)$
0001	$a_0 + a_4$	$a_4 = a_0 + f(0001)$
1100	$a_0 + a_1 + a_2 + a_{12}$	$a_{12} = a_0 + a_1 + a_2 + f(1100)$
1010	$a_0 + a_1 + a_3 + a_{13}$	$a_{13} = a_0 + a_1 + a_3 + f(1010)$
1001	$a_0 + a_1 + a_4 + a_{14}$	$a_{14} = a_0 + a_1 + a_4 + f(1001)$
0110	$a_0 + a_2 + a_3 + a_{23}$	$a_{23} = a_0 + a_2 + a_3 + f(0110)$
0101	$a_0 + a_2 + a_4 + a_{24}$	$a_{24} = a_0 + a_2 + a_4 + f(0101)$
0011	$a_0 + a_3 + a_4 + a_{34}$	$a_{34} = a_0 + a_3 + a_4 + f(0011)$

Table 1: Finding low order ANF coefficients.

truth table. Those functions with only a few high order terms (and those which are in the same equivalence classes under affine transformation) will have relatively small Hamming distance to some low order function, since each high order term changes only relatively few truth table positions.

The Hamming distance $d(f, g)$ between two functions f and g is the number of function values in which they differ:

$$d(f, g) = hwt(f \oplus g) = \sum_{\mathbf{x}} f(\mathbf{x}) \oplus g(\mathbf{x}).$$

where $\sum$ indicates a sum of real numbers. The correlation between two functions f and g is related to their Hamming distance by (eg. [5])

$$c(f, g) = 1 - \frac{d(f, g)}{2^{n-1}}$$

so that $-1 \leq c(f, g) \leq 1$.

2.1 Hamming Sphere Evaluation

Definition 1 *Given a Boolean function $f(\mathbf{x})$ of n variables, the Hamming sphere of radius r at center $\mathbf{y}$ is defined as the set of all input vectors with Hamming distance r or less from $\mathbf{y}$.*

In particular, the Hamming sphere of radius r centered at $\mathbf{y} = 0$ is the set of all inputs $\mathbf{x}$ with Hamming weight r or less. We define evaluation of a Hamming sphere to be the evaluation of the function at all inputs $\mathbf{x}$ in the sphere. When $ord(f) = r$, these function values yield ANF coefficients of $f(\mathbf{x})$ using equation (3), proving the following proposition:

Proposition 1 *Evaluating the Hamming sphere of radius r, at center $\mathbf{y} = 0$ is sufficient to identify a Boolean function of order r.*

Evaluating a Hamming sphere of radius r provides one equation for each of the message symbols in the Reed-Muller code $\mathcal{R}(r, n)$ [3]. These symbols are the ANF coefficients. Thus Hamming sphere sampling is an algorithm for probabilistic decoding of Reed-Muller codes, using only a small fraction of the received codeword.

Evaluating a Hamming sphere centered at some non-zero vector $\mathbf{y}$ will yield the ANF coefficients of the rth order function $f(\mathbf{x} \oplus \mathbf{y})$, which is a transformed version of the actual function $f(\mathbf{x})$. The transformation $T_{\mathbf{y}}(f(\mathbf{x})) = f(\mathbf{x} \oplus \mathbf{y})$ is the same as input masking with vector $\mathbf{y}$. It is (conceptually) easy to find the explicit truth table representation of $f(\mathbf{x})$ from that of $f(\mathbf{x} \oplus \mathbf{y})$, by swapping 2^{n-1} pairs of truth table values. However, such an operation is infeasible for large n due to storage problems. These problems are avoided using the ANF representation of low order functions since relatively few coefficients need be stored. For example, a fifth order function of 64 variables may be stored in about 8.3 million bits. Lower order functions require much less space.

In the next section we describe a way of obtaining the correct ANF coefficients of $f(\mathbf{x})$ from the calculated coefficients of $f(\mathbf{x} \oplus \mathbf{y})$. This allows an unknown r^{th} order function to be identified by the evaluation of any Hamming sphere of radius r.

3   Boolean Differentiation

In this section we present some results on Boolean differentiation which are essential for the efficient implementation of the sphere sampling method. We concentrate on differentiation in the ANF domain, since this is of practical interest. For a more generalised framework and some additional results, see [2]. It should be noted that the results in this section were obtained independantly of [2], which did not become known to us until after the conference presentation. The following definition appears in [3].

Definition 2 *The derivative of a Boolean function $f(\mathbf{x})$ with respect to a vector $\mathbf{y}$ is defined as*

$$d_{\mathbf{y}} f(\mathbf{x}) = f(\mathbf{x}) \oplus f(\mathbf{x} \oplus \mathbf{y}).$$

From this definition and the identity $T_{\mathbf{y}}(f) = f(\mathbf{x} \oplus \mathbf{y})$, we obtain

$$d_{\mathbf{y}} f(\mathbf{x}) = d_{\mathbf{y}} T_{\mathbf{y}}(f(\mathbf{x})).$$

It follows that

$$f(\mathbf{x}) = d_{\mathbf{y}} T_{\mathbf{y}}(f(\mathbf{x})) \oplus T_{\mathbf{y}}(f(\mathbf{x})). \tag{4}$$

The Hamming sphere sample $f(\mathbf{x})$ (a candidate approximating function) is the sum of an evaluated Hamming sphere $T_{\mathbf{y}}(f(\mathbf{x}))$ and its derivative with respect to the center of the sphere, $\mathbf{y}$. From the definition it is clear that Boolean differentiation is easy in the truth table domain, however in the ANF domain it is less clear how to proceed. In the next section we show how Boolean differentiation can be efficiently performed in the ANF domain, provided the function $f(\mathbf{x})$ has low order.

3.1 Boolean Differentiation of the ANF

Firstly, consider the case where the vector $\mathbf{y}$ has Hamming weight one ($\mathbf{y} = \mathbf{e}_i$). For any variable x_i, a function $f(\mathbf{x})$ may be expressed as

$$f(\mathbf{x}) = x_i f_i(\mathbf{x}) \oplus g_i(\mathbf{x}),$$

where both f_i and g_i are degenerate in x_i so that $f_i(\mathbf{x}) = f_i(\mathbf{x} \oplus \mathbf{e}_i)$ and $g_i(\mathbf{x}) = g_i(\mathbf{x} \oplus \mathbf{e}_i)$ are true for all $\mathbf{x}$. This can be done by separating those terms of the ANF of f which contain x_i from those which do not.

Proposition 2 $d_{\mathbf{e}_i} f(\mathbf{x}) = f_i(\mathbf{x})$.

Proof: From Definition 2 and the degeneracy of f_i and g_i, we have

$$
\begin{aligned}
d_{\mathbf{e}_i} f(\mathbf{x}) &= f(\mathbf{x}) \oplus f(\mathbf{x} \oplus \mathbf{e}_i) \\
&= x_i f_i(\mathbf{x}) \oplus g_i(\mathbf{x}) \oplus (x_i \oplus 1) f_i(\mathbf{x} \oplus \mathbf{e}_i) \oplus g_i(\mathbf{x} \oplus \mathbf{e}_i) \\
&= x_i f_i(\mathbf{x}) \oplus g_i(\mathbf{x}) \oplus x_i f_i(\mathbf{x}) \oplus f_i(\mathbf{x}) \oplus g_i(\mathbf{x})
\end{aligned}
$$

so that

$$d_{\mathbf{e}_i} f(\mathbf{x}) = f_i(\mathbf{x})$$

□

Since every term in $d_{\mathbf{e}_i} f(\mathbf{x})$ has order less than $ord(f)$ we have

Corollary 1 Let $ord(f(\mathbf{x})) = r$ then $ord(d_{\mathbf{e}_i} f(\mathbf{x})) \le r - 1$.

Proposition 2 allows differentiation with respect to a single variable to be performed in a single pass. This is feasible for low order functions of many variables. Next we show how this process may be extended to the general case of differentiation with respect to $\mathbf{y}$ of any Hamming weight, $1 \le hwt(\mathbf{y}) \le n$. For this we need some results on multiple derivatives.

Definition 3 Let $\mathbf{y} = \bigoplus_{i \in S} \mathbf{e}_i$ for some set $S \subseteq \{1, 2, \cdots, n\}$. Let $h = hwt(\mathbf{y}) = |S|$ so that we identify $\mathbf{y}$ with $S = \{i_1, i_2, \cdots, i_h\}$. Let the multiple derivative of $f(\mathbf{x})$ with respect to the set of variables $\{x_i | i \in S\}$ be denoted by

$$d_{\{\mathbf{y}\}} f(\mathbf{x}) = d_{\mathbf{e}_{i_1}} d_{\mathbf{e}_{i_2}} \cdots d_{\mathbf{e}_{i_h}} f(\mathbf{x}).$$

Proposition 3 Let $ord(f) = r$, then $ord\left(d_{\{\mathbf{y}\}} f(\mathbf{x})\right) \le max(0, r - h)$.

Proof: Follows from recursive application of corollary 1.

□

Proposition 4 For any two vectors $a, b \in Z_2^n$, we have

$$d_{a \oplus b} f(\mathbf{x}) = d_a f(\mathbf{x}) \oplus d_b f(\mathbf{x}) \oplus d_a d_b f(\mathbf{x}).$$

Proof: Firstly consider that

$$d_a f(\mathbf{x}) \oplus d_b f(\mathbf{x}) \quad = \quad f(\mathbf{x}) \oplus f(\mathbf{x} \oplus a) \oplus f(\mathbf{x}) \oplus f(\mathbf{x} \oplus b)$$
$$= \quad f(\mathbf{x} \oplus a) \oplus f(\mathbf{x} \oplus b).$$

Now,

$$d_a d_b f(\mathbf{x}) \quad = \quad d_a \left(f(\mathbf{x}) \oplus f(\mathbf{x} \oplus b) \right)$$
$$= \quad f(\mathbf{x}) \oplus f(\mathbf{x} \oplus a) \oplus f(\mathbf{x} \oplus b) \oplus f(\mathbf{x} \oplus a \oplus b)$$
$$= \quad d_{a \oplus b} f(\mathbf{x}) \oplus d_a f(\mathbf{x}) \oplus d_b f(\mathbf{x})$$

and the proposition follows. □

Theorem 1 *Let* $\mathbf{y} \in Z_2^n$ *have* $hwt(\mathbf{y}) = h$. *Let* G *denote the* h-*dimensional linear subspace of* Z_2^n *with basis the set of unit vectors* $\mathbf{e}_i$ *which sum to give* $\mathbf{y}$. *Let* G^* *contain all non-zero elements of* G *and let the multiple derivative notation defined above be used. Then*

$$d_\mathbf{y} f(\mathbf{x}) = \bigoplus_{z \in G^*} d_{\{z\}} f(\mathbf{x}). \tag{5}$$

Proof: By recursion on Proposition 4. Let $a = \mathbf{e}_i$ and $b = \mathbf{e}_2 \oplus \mathbf{e}_3 \oplus \cdots \oplus \mathbf{e}_h$ where $h = hwt(\mathbf{y})$. From Proposition 4, Theorem 1 is true when $h = 2$. In each level of recursion (adding one to h), the number of distinct non-zero terms being summed in equation (5) is doubled and another one added. Observing that there exist $2^h - 1$ distinct elements in G^* and $2(2^h - 1) + 1 = 2^{h+1} - 1$ we conclude by induction that $d_\mathbf{y} f(\mathbf{x})$ equals the sum of all multiple derivatives corresponding to elements of G^*. □

Note that, from Proposition 3, when $hwt(z) = ord(f) = r$ the multiple derivatives $d_{\{z\}} f(\mathbf{x})$ are all constant functions equal to the value of the ANF coefficient a_z. These terms affect only the constant ANF coefficient a_0. When $hwt(z) > r$, $d_{\{z\}} f(\mathbf{x}) = 0$, so that these terms in equation (5) need not be calculated. To obtain the ANF coefficients other than a_0, only those derivatives $d_{\{z\}} f(\mathbf{x})$ for which $hwt(z) < r$ need be calculated. From this argument, we see that the complexity of performing (5) is dictated by the value of $ord(f) = r$, rather than $hwt(\mathbf{y}) = h$. Thus the process of performing Boolean differentiation is feasible in the ANF domain for functions $f(\mathbf{x})$ of low order.

For example, when $f(\mathbf{x})$ is quadratic only single variable derivatives need be determined. In this case (5) becomes:

$$d_\mathbf{y} f(\mathbf{x}) = \left[\bigoplus_{j=1}^{h} d_{i_j} f(\mathbf{x}) \right] \oplus \left[\bigoplus_{j=1}^{h-1} \bigoplus_{k=j+1}^{h} a_{i_j, i_k} \right].$$

where $h = hwt(\mathbf{y})$ and a_{i_j, i_k} is a quadratic ANF coefficient of $f(\mathbf{x})$. This allows the ANF of the derivative (with respect to a vector $\mathbf{y}$) of a quadratic function $f(\mathbf{x})$ to be obtained from the ANF of f and the ANFs of a number (equal to $hwt(\mathbf{y})$) of simple (single variable) derivatives of f. The final complementation is determined as a sum of some quadratic coefficients of the function $f(\mathbf{x})$. When storage of the quadratic function f is feasible, then the storage of all simple derivatives is also feasible.

4 Analysis

The method of obtaining good low order approximation functions by Hamming sphere sampling may be described by the following algorithm.

(i) Given a fixed function $f(\mathbf{x})$, evaluate a Hamming sphere of radius r and center $\mathbf{y}$ using equation (3). This gives the ANF of a function $g(\mathbf{x} \oplus \mathbf{y})$ of order r.

(ii) Obtain the candidate approximation

$$\tilde{g}(\mathbf{x}) = g(\mathbf{x} \oplus \mathbf{y}) \oplus d_\mathbf{y} g(\mathbf{x} \oplus \mathbf{y})$$

by obtaining the derivative of the evaluated Hamming sphere. Now we have the ANF of a low order function $\tilde{g}$ which is indistinguishable from f on the sphere.

(iii) Estimate the correlation between f and $\tilde{g}$ by comparing their values on a set of random inputs $\mathbf{x}$. If the value of correlation is clearly less than required: go to (i), else if unsure: make more comparisons, else end.

The process of Hamming sphere sampling may be repeated with different centers $\mathbf{y}$ and radii r, until a function $\tilde{g}(\mathbf{x})$ with a suitably high correlation is obtained, or until a sufficient number of sphere samples have been taken to provide reasonable assurance that no good low order approximation exists.

Consider a fixed cipher function f and suppose there exists a low order approximating function $g(\mathbf{x})$ such that $c_{f,g} = c > 0$. We assume a probabilistic model in which g is randomly chosen so that

$$Pr(g(\mathbf{x}) = f(\mathbf{x})) = \frac{1-c}{2}, \qquad \mathbf{x} \in Z_2^n.$$

Those $\mathbf{x}$ for which $f(\mathbf{x}) \neq g(\mathbf{x})$ are the positions of the errors in the approximation, and may be considered to be scattered randomly in the space Z_2^n. Let

$$B(r,n) = \sum_{i=0}^{r} \binom{n}{i}$$

denote the volume of an n-dimensional Hamming sphere of radius r. For a Hammimg sphere chosen at random the probability that the sphere is error-free is then

$$p = \left(\frac{1+c}{2}\right)^{B(r,n)}$$

and the probability that at least one from N independantly chosen spheres is error-free is

$$P = 1 - (1-p)^N.$$

This means that the number, $N(c)$, of sphere samples required to achieve an error-free sphere with probability P is given as

$$N(c) = \frac{\ln(1-P)}{\ln\left(1 - \left(\frac{1+c}{2}\right)^{B(r,n)}\right)}.$$

c_{min}	$n = 128$	64	32	16
$r = 1$	0.68	0.39	0	0
2	0.995	0.981	0.921	0.694
3	0.9999	0.9992	0.993	0.94

Table 2: Minimum correlation required for feasible sphere sampling attack.

We may consider $P = 1 - e^{-1} \approx 0.63$ to be an acceptably good probability of obtaining $g(\mathbf{x})$ as a sphere sample in N trials. This allows

$$N(c) = \frac{1}{p} = \left(\frac{2}{1+c}\right)^{B(r,n)} .$$

This should be compared with $2^{B(r,n)-1}$, the total number of possible g required by the blind search method of obtaining candidate approximating functions. Hence we may define the gain factor of sphere sampling over blind search as

$$G(c) = \frac{1}{2}(1+c)^{B(r,n)} .$$

We observe that the gain factor exceeds 1.4 when $c > \frac{1}{B(r,n)}$. This implies that the method of Hamming sphere sampling can be expected to be superior to blind search for a wide range of correlations. However, the blind search method is infeasible for large n, so we have not yet shown the condition for feasible success of Hamming sphere sampling. For feasibility we require that the total number of chosen inputs is no more than about $2^{40} \approx 10^{12}$. The number of chosen inputs is the number of spheres times the volume of a sphere, so we expect the method to be successfull with probability at least 63% when there exists a function $g(\mathbf{x})$ of order r and correlation c which satisfies the inequality

$$N(c)B(r,n) < 10^{12}.$$

From this we obtain a lower bound on c:

$$c > \left(\frac{2 - \exp\left(\frac{\ln\left(\frac{10^{12}}{B(r,n)}\right)}{B(r,n)}\right)}{\exp\left(\frac{\ln\left(\frac{10^{12}}{B(r,n)}\right)}{B(r,n)}\right)}\right) .$$

Table 2 shows the value of c_{min} for some values of r and n.

Highly nonlinear balanced Boolean functions can be obtained by modification of low order bent functions [1]. Such balanced functions have correlation $1 - 2^{-\frac{n}{2}}$ with a low order function. These functions are particularly vulnerable to the sphere sampling attack, since this value of c exceeds the lower bound required for feasibility.

5 Test Results

Table 3 presents results of a survey of 8-variable balanced Boolean functions. The test compared the effectiveness of first order (linear) sphere sampling with that of random

Minimum c	% Random	% Same	% SS	Average Ratio
0.2344	2.5	94.7	2.8	1.04
0.2188	6.2	87.3	6.5	1.07
0.2031	12.2	73.7	14.1	1.13
0.1875	23.8	47.2	29.0	1.23
0.1719	33.2	25.3	41.5	1.26
0.1563	31.2	14.4	54.4	1.27
0.1406	32.1	11.0	56.9	1.20
0.1250	29.6	7.3	63.1	1.16
0.1094	28.8	5.1	66.1	1.13
0.0938	28.6	4.5	66.9	1.10
0.0781	28.8	3.9	67.3	1.07
0.0625	29.8	3.6	66.6	1.05
0.0469	32.7	3.7	63.6	1.03
0.0313	36.0	3.9	60.1	1.02
0.0156	38.2	6.5	55.3	1.01

Table 3: Comparison of Linear Sphere Sampling Vs Blind Search for n=8.

(or blind) search for selection of candidate approximation functions. For each function examined, the Walsh-Hadamard transform (for example, see [7]) was performed, allowing the exact number, N_R, of affine functions possessing correlation at least equal to a certain minimum c to be counted. In addition, a linear sphere sample was obtained at all 256 centers. The number of sphere samples, N_{SS}, that yeilded functions with correlation at least equal to c was determined. The table shows the percentage of functions for which $N_R > N_{SS}$, $N_R = N_{SS}$, and $N_R < N_{SS}$. The column headed Average Ratio shows the mean value of $\frac{N_R}{N_{SS}}$. For each cutoff c, a total of 3200 functions were tested. These results confirm the previous analysis that sphere sampling is more likely than blind search to result in a function with correlation exceeding some minimum value.

We now present results of a survey of 4 variable functions which clearly show the trade off between resisting linear approximation and quadratic approximation simultaneously. Table 4 shows the number of 4-variable functions with the indicated distances to the set of affine functions and the set of quadratic functions. For $n = 4$ the 896 quadratic bent functions attain the maximum nonlinearity of 6 (62.5% correlation). The maximum distance to quadratic functions is 4 (75% correlation), attained by the 32 affine functions. The maximum distance to the union of affine and quadratic functions is only 2 (87.5% correlation), attained by cubic functions.

6 Conclusions

It is well known that a cipher function with small distance to an affine function must be considered cryptographically weak. We propose that this notion be extended to include functions with small Hamming distance to any low order function, since the method

	Distance to Quadratic functions				
Nonlinearity	0	1	2	3	4
0	0	0	0	0	32
1	0	0	0	512	0
2	0	0	3840	0	0
3	0	17920	0	0	0
4	1120	0	26880	0	0
5	0	14336	0	0	0
6	896	0	0	0	0

Table 4: Number of functions with various distances to 1st and 2nd order functions, n=4.

of Hamming sphere sampling will indicate the best approximate function quickly, with high probability. To remain secure from these attacks, cipher functions must resist approximation by functions of all low orders. It has been shown that a tradeoff occurs in resisting both affine and quadratic approximation.

Acknowledgement

We would like to thank Jovan Golić for several helpful suggestions, especially regarding the analysis of the sphere sampling method.

References

[1] H. Dobbertin, "Construction of Bent Functions and Balanced Boolean Functions with High Nonlinearity", *presented at K.U.Leuven Workshop on Cryptographic Algorithms*, 1994.

[2] X. Lai, "Higher Order Derivatives and Differential Cryptanalysis", in *Communications And Cryptography, Two Sides of One Tapestry*, pp.227-233, Kluwer Academic Publishers, 1994.

[3] F.J. MacWilliams, N.J.A. Sloane "The Theory of Error Correcting Codes", North Holland Publishing Company, 1977.

[4] U.M. Maurer, "New Approaches to the Design of Self-Synchronizing Stream Ciphers", *EUROCRYPT '91*, Lecture Notes in Computer Science, vol.547, pp.458-471, Springer-Verlag, Berlin, New York, Tokyo, 1991.

[5] W. Meier, O. Staffelbach "Nonlinearity Criteria for Cryptographic Functions", *EUROCRYPT '89*, Lecture Notes in Computer Science, vol.434, pp. 549-562, Springer-Verlag, Berlin, Heidelberg, New York, 1990.

[6] W. Millan, E.P. Dawson, L.J. O'Connor, "Fast Attacks on Tree-Structured Ciphers", *Proceedings of Workshop on Selected Areas in Cryptography (SAC '94)*, pp.148-158, Queens University, Kingston, Canada, May 1994.

[7] B. Preneel, W. Van Leekwijck, L. Van Linden, R. Govaerts, J. Vanderwalle, "Propagation Characteristics of Boolean Functions", *EUROCRYPT '90*, Lecture Notes in Computer Science, vol.473, pp.161-173, Springer-Verlag, Berlin, New York, Tokyo, 1991.

Multiple Encryption with Minimum Key

Ivan B. Damgård and Lars Ramkilde Knudsen*

Aarhus University, Denmark

Abstract. In this paper we consider multiple encryption schemes built from conventional cryptosystems such as DES. The existing schemes are either vulnerable to variants of meet in the middle attacks, i.e. they do not provide security of the full key or there is no proof that the schemes are as secure as the underlying cipher. We propose a new variant of two-key triple encryption which is not vulnerable to the meet in the middle attack and which uses a minimum amount of key. We can prove a connection between the security of our system and the security of the underlying block cipher.

1 Introduction

Since its introduction in the late seventies, the American Data Encryption Standard (DES) has been the subject of intense debate and cryptanalysis. Like any other practical cryptosystem, DES can be broken by searching exhaustively for the key.

One natural direction of research is therefore to find attacks that will be faster than exhaustive search, measured in the number of necessary encryption operations. The most successful attack on DES known of this kind is the linear attack by Matsui [3, 4]. This attack requires about 2^{43} known plaintext blocks. Although this is less than the expected 2^{55} encryptions required for exhaustive key search, the attack is by no means more practical than exhaustive search. There are two reasons for this: first, one cannot in practice neglect the time needed to obtain the information about the plaintext; secondly, when doing exhaustive key search the enemy is free to invest as much in technology as he is capable of to make the search more efficient, in a known plaintext attack he is basically restricted to the technology of the legitimate owner of the key, and to the frequency with which the key is used. In virtually any practical application, a single DES key will be applied to much less than 2^{43} blocks, even in its entire life time. The difference between the two kinds of attacks is illustrated in a dramatic way by the results of Wiener [10] who shows by concrete design of a key search machine that if the enemy is willing to make a one million dollar investment, exhaustive key search for DES is certainly not infeasible.

As a result, we have a situation where DES has proved very resistant over a long period to cryptanalysis and therefore seems to be as secure as it can be in the sense that by far the most practical attack is a simple brute force search for

---

* Postdoctoral researcher sponsored by the Danish Technical Research Council.

the key. The only problem is that the key is too short given today's technology, and that therefore, depending on the value of the data you are protecting, plain DES may not be considered secure enough anymore.

What can be done about this problem? One obvious solution is to try to design a completely new algorithm. This can only be a very long term solution: a new algorithm has to be analysed over a long period before it can be considered secure; also the vast number of people who have invested in DES technology will not like the idea of their investments becoming worthless overnight. An alternative is to devise a new system with a longer key using DES as a building block. This way existing DES implementations can still be used.

We are in the situation, where we have a block cipher, that has proved to be very strong, the only problem being that the keys are too small and a simple brute-force attack has become possible. Thus, this section is motivated by the following general question: Given cryptosystem $\mathcal{X}$, which cannot in practice be broken faster than exhaustive key search, how can we build a new system $\mathcal{Y}$, such that

1. Keys in $\mathcal{Y}$ are significantly longer than keys in $\mathcal{X}$ (e.g. twice as long)
2. Given an appropriate assumption about the security of $\mathcal{X}$, $\mathcal{Y}$ is provably about as hard to break as $\mathcal{X}$ under any natural attack (e.g. ciphertext only, known plaintext, etc.).
3. It can be convincingly argued that $\mathcal{Y}$ can in fact not be broken faster than exhaustive key search, and is therefore in fact much stronger than $\mathcal{X}$.

Possible answers to this question have already appeared in the literature. The most well known example is known as two-key triple encryption, where we encipher under one key, decipher under a second key, and finally encipher under the first key. Van Oorschot and Wiener [9] have shown, refining an attack of Merkle and Hellman [6], that this construction is not optimal: under a known plaintext attack, it can be broken significantly faster than exhaustive key search. We propose a new variant of two-key triple encryption, which has all the properties we require above.

The security of a block cipher (in bits) is defined to be the logarithm base 2 of the number of encryptions an attack needs in order to be successful.

2 Multiple Encryption

In this section, we look at methods for enhancing cryptosystems based on the idea of encrypting plaintext blocks more than once. Following the notation of the introduction, we let $\mathcal{X}$ be the original system, and we let E_K and D_K denote encryption respectively decryption in $\mathcal{X}$ under key K. We assume that the key space of $\mathcal{X}$ consists of all k-bit strings, and that the block length of $\mathcal{X}$ is m. In a *cascade of ciphers* it is assumed that the keys of the component ciphers are independent. The following result was proved by Even and Goldreich.

Theorem 1 [2]. *A cascade of ciphers is at least as hard to break under any attack as any of the component ciphers in the cascade under a chosen plaintext attack.*

As seen, the result establishes a connection between the cascade and the component ciphers under a chosen plaintext attack on the latter. The following result covering all attacks was proved by Maurer and Massey.

Theorem 2 [5]. *A cascade of ciphers is at least as hard to break as the first cipher.*

A special case of a cascade of ciphers is when the component ciphers are equal, also called multiple encryption. In the following we consider different forms of multiple encryption.

2.1 Double Encryption

The simplest idea one could think of would be to encrypt twice using two keys K_1, K_2, i.e. let the ciphertext corresponding to P be $C = E_{K_2}(E_{K_1}(P))$. It is clear (and well-known), however, that no matter how K_1, K_2 are generated, there is a simple meet-in-the middle attack that breaks this system under a known plaintext attack using 2^k encryptions and 2^k blocks of memory, i.e. the same time complexity as key search in the original system. The memory requirements can be reduced heavily using the techniques of Quisquater and Delescaille, see [7], and it is clear that this system is not a satisfactory improvement over $\mathcal{X}$.

2.2 Triple Encryption

Triple encryption with three independent keys K_1, K_2, and K_3, where the ciphertext corresponding to P is $C = E_{K_3}(E_{K_2}(E_{K_1}(P)))$, is also not a satisfactory solution for a similar reason as for double encryption. A simple meet-in-the-middle attack will break this in time about 2^{2k} encryptions and space 2^k blocks of memory. Thus we do not get full return for our effort in tripling the key length - as stated in demand 3 in the introduction, we would like attacks to take time close to 2^{3k}, if the key length is $3k$. In addition to this, if $\mathcal{X} = \text{DES}$, then a simple triple encryption would preserve the complementation property, and preserve the existence of weak keys.

It is clear, however, that no matter how the three keys in triple encryption are generated, the meet-in-the-middle attack mentioned is still possible, and so the time complexity of the best attack against *any* triple encryption variant is no larger than 2^{2k}. It therefore seems reasonable to try to generate the three keys from two independent $\mathcal{X}$-keys K_1, K_2, since triple encryption will not provide security equivalent to more than 2 keys anyway.

2.3 Two-key Triple Encryption

One variant of this idea is well-known as two-key triple encryption, proposed by W. Tuchmann [8] The ciphertext C corresponding to the plaintext P is $C = E_{K_1}(D_{K_2}(E_{K_1}(P)))$. Compatibility with a single encryption can be obtained by setting $K_1 = K_2$. As one can see, this uses a particular, very simple way of generating the three keys from K_1, K_2. For two-key triple encryption there is a result similar to Th. 1.

Theorem 3. *Under a chosen plaintext/ciphertext attack two-key triple encryption is at least as hard to break as the underlying cipher.*

Proof: Assume that we have an algorithm A, which on input n chosen plaintexts, breaks a two-key triple encryption scheme, $\mathcal{Y}$, where $\mathcal{X}$ is the underlying cipher. Choose one key $K_{1,3}$ at random. Whenever A asks for the encryption of plaintext P, we encrypt P using the key $K_{1,3}$, yielding PP. Then we get the decryption CC of PP in the chosen ciphertext setting from $\mathcal{X}$. Now encrypt CC using again the key $K_{1,3}$ yielding C, which is input to A. Since by assumption A breaks the two-key triple scheme, it will output a candidate for the key in the second round, i.e. for the secret key of $\mathcal{X}$. □

Even though this result establishes some connection between the security of two-key triple encryption with the underlying cipher, it holds only for a chosen plaintext/ciphertext attack and still does not meet our second demand.

For the two-key triple encryption scheme, each of K_1 and K_2 only influences particular parts of the encryption process. Because of this, variants of the meet-in-the-middle attack are possible that are even faster than exhaustive search for K_1, K_2. In [6] Merkle and Hellman describes an attack on two-key triple DES encryption requiring 2^{56} chosen plaintext-ciphertext pairs and a running time of 2^{56} encryptions using 2^{56} words of memory. This attack was refined in [9] into a known plaintext attack on the DES, which on input n plaintext-ciphertext pairs finds the secret key in time $2^{120}/n$ using n words of memory. The attacks can be applied to any block cipher. Therefore two-key triple encryption does not meet our third demand.

We therefore propose what we believe to be stronger methods for generating the keys. Our main idea is to generate them *pseudorandomly* from 2 $\mathcal{X}$ keys using a generator based on the security of $\mathcal{X}$. In this way, an enemy trying to break $\mathcal{Y}$ either has to treat the 3 keys as if they were really random which means he has to break $\mathcal{X}$, according to Th. 2; or he has to use the dependency between the keys - this means breaking the generator which was also based on $\mathcal{X}$! Thus, even though we have thwarted attacks like Merkle-Hellman and van Oorschot-Wiener by having a strong interdependency between the keys, we can still, if $\mathcal{X}$ is secure enough, get a connection between security of $\mathcal{X}$ and $\mathcal{Y}$. In the following we concentrate on triple encryption schemes and in Sect. 3.4 generalize our results to any n-fold schemes.

3 Multiple Encryption with Minimum Key

3.1 General Description of $\mathcal{Y}$

Let a block cipher $\mathcal{X}$ be given, as described above. The key length of $\mathcal{X}$ is denoted by k. By $E_K(P)$, we denote $\mathcal{X}$-encryption under K of block P, while $D_K(C)$ denotes decryption of C. We then define a new block cipher $\mathcal{Y}$ using a function G:

$$G(K_1, K_2) = (X_1, X_2, X_3)$$

which maps 2 $\mathcal{X}$-keys to 3 $\mathcal{X}$-keys. We display later a concrete example of a possible G-function. This is constructed from a few $\mathcal{X}$-encryptions. Keys in $\mathcal{Y}$ will consist of pairs (K_1, K_2) of $\mathcal{X}$-keys. Encryption in $\mathcal{Y}$ is defined by

$$E_{K_1, K_2}(P) = E_{X_3}(E_{X_2}(E_{X_1}(P))),$$

where $(X_1, X_2, X_3) = G(K_1, K_2)$. Decryption is clearly possible by decrypting using the X_i's in reverse order.

3.2 Relation to the Security of $\mathcal{X}$

We would like to be reasonably sure that we have taken real advantage of the strength of $\mathcal{X}$ when designing $\mathcal{Y}$. One way of stating this is to say that $\mathcal{Y}$ is at least as hard to break as $\mathcal{X}$. By Th. 2, this would be trivially true if the three keys used in $\mathcal{Y}$ were statistically independent. This is of course not the case, since the X_i's are generated from only 2 keys. But if the generating function G has a pseudorandom property as stated below, then the X_i's are "as good as random" and we can still prove a strong enough result.

Definition 4. Consider the following experiment: an enemy B is presented with three k-bit blocks X_1, X_2, X_3. He then tries to guess which of two cases has occurred:

1. The X_i's are chosen independently at random.
2. The X_i's are equal to $G(K_1, K_2)$, for randomly chosen K_1, K_2.

Let p_1 be the probability that B guesses 1 given that case 1 occurs, and p_2 the probability that B guesses 1 given that case 2 occurs. The generator function G is said to be *pseudorandom*, if for any B spending time equal to T encryption operations,

$$|p_1 - p_2| \leq \frac{T}{V},$$

where V is the total number of keys in $\mathcal{X}$.

The intuition we want to express with this definition is that the generator function G should be at least as hard to break as it is to do exhaustive search for a key in system $\mathcal{X}$. Clearly, if the total number of keys is V, and you have resources for testing T randomly chosen keys, then the probability of finding the correct one is T/V. We therefore say that this also should be the maximum amount of success you can achieve against G using time T. Def. 4 is inspired by the complexity theoretic definition of a strong pseudorandom generator introduced by Blum and Micali [1].

In the rest of this subsection we consider attacks against $\mathcal{X}$ and $\mathcal{Y}$ in a fixed scenario with a given plaintext distribution and a given form of attack, such as known plaintext, chosen plaintext, etc. We do not specify these things further, because the reasoning below will work for any such scenario. The time unit will be encryption operations in system $\mathcal{X}$.

The next theorem shows the promised connection between security of $\mathcal{X}$ and $\mathcal{Y}$, i.e. in a given amount of time, an attack cannot do much better against $\mathcal{Y}$ than what is possible against $\mathcal{X}$.

Theorem 5. *Let p be the success probability of the best attack against $\mathcal{X}$ running in time T. Assume now that an attacker A against our new system $\mathcal{Y}$ runs in time T and has success probability $p + \epsilon$. If the function G used to construct Y is pseudorandom, then*

$$\epsilon \leq \frac{T}{V}$$

where V is the total number of keys in $\mathcal{X}$.

Proof: Let $\mathcal{Y}_0$ be the same system as $\mathcal{Y}$, but with independent keys X_i. By Th. 2, using A against $\mathcal{Y}_0$ leads to an attack against $\mathcal{X}$ with the same success probability. Hence by assumption, A's success probability against $\mathcal{Y}_0$ will be at most p. But then we can use A to make an algorithm B that fits Def. 4: Given X_1, X_2, X_3, B uses these as keys in the triple encryption system and simulate A's attack. If A is successful, B will guess that the X_i's are generated from K_1, K_2, if not, B will guess that they are independent. Since in one case A will be attacking $\mathcal{Y}$, and in the other case $\mathcal{Y}_0$, it is clear that for this B, we have by Def. 4

$$\epsilon \leq |p_1 - p_2| \leq \frac{T}{V}$$

$\square$

As an example of what the statement of the theorem means, consider an ideal case, where the best an attack against $\mathcal{X}$ can do, is to spend its time choosing random keys and test whether they fit with the information available. The success probability for time T would then be $T/2^k$ assuming a key can be tested in 1 encryption. Then the above theorem says that if G is pseudorandom, the success probability of any attack against $\mathcal{Y}$ running in time T can be at most $T/2^k + T/2^k$. This is larger than the original success probability against $\mathcal{X}$ by a factor of only 2.

3.3 A Concrete Two-key Triple Encryption Construction

We propose here a new construction for triple encryption, called **3-MAK** for triple encryption with **M**inimum **A**mount of **K**ey. In this construction X_1, X_2, X_3

are all used as keys for encryption. We define this construction of $G(K_1, K_2) = (X_1, X_2, X_3)$ by:

$$X_1 = E_{K1}(E_{K2}(IV_1))$$
$$X_2 = E_{K1}(E_{K2}(IV_2))$$
$$X_3 = E_{K1}(E_{K2}(IV_3))$$

where IV_i are three different initial values, e.g. $IV_i = C+i$, where C is a constant. In the following we will show that our construction meets our three demands from the introduction. So let us assume that the underlying block cipher $\mathcal{X}$ cannot in practice be broken faster than an exhaustive search. Our first demand is met since the keys in our scheme are twice as long as in the underlying block cipher. It is seen that double encryption is used to generate the three keys. Therefore, based on Th. 2, we conjecture that the generator G just defined is pseudorandom according to Def. 4.

Theorem 6. *If the security of $\mathcal{X}$ is k (bits), and the generator G is pseudorandom, then the security of 3-MAK with $\mathcal{X}$ is at least $k - 1$.*

Note that although we have proved a lower bound on the complexities of attacks (of any kind) against our new scheme in terms of the complexities of attacks against $\mathcal{X}$, it does not mean at all that there exists an attack with that complexity. We conjecture that the fastest attack against 3-MAK encryption is a brute force attack of complexity 2^{2k}. Attacks like the ones from [6, 9] are applicable to ciphers, for which the first and the third keys are equal. Since in our case every one of the three keys defined above are dependent of both master keys K_1 and K_2 in a complicated way we conclude that these attacks are not possible.

In Table 1 a schematic overview of our result is given. The first line in the table (i.e. for "Block cipher") is by assumption. For two-key triple encryption there exists no lower bound on the complexities of all attacks to our knowledge and the upper bound results from the meet-in-the-middle attacks described earlier in this paper. For three-key triple encryption the lower bound is by the result of Maurer and Massey and the upper bound is a simple meet in the middle attack again, believed as being the best attack. For 3-MAK encryption the lower bound is proved in this paper and the upper bound is our conjecture of the best attack. We invite the readers to come up with attacks violating this upper bound.

The key scheduling in our construction is slower than for the two-key triple encryption. In most software applications of the DES the key scheduling takes about twice the time of a single encryption. The key scheduling in the 3-MAK DES encryption scheme takes about the time of 5 key schedules and 6 encryptions of the DES, i.e., using the above estimate, a total time of about 16 DES-encryptions. For comparison the key schedules for two-key triple DES and triple DES with three independent keys take 4 and 6 encryptions, respectively. In encryption with our new construction the key schedule should be performed once and the three round keys stored. In that way encryption with 3-MAK DES is as

Scheme	Key size	Lower bound (all attacks)	Upper bound (best known attack)
Block cipher $\mathcal{X}$	k	k	k
Two-key triple $\mathcal{X}$	$2k$	?	k
Three-key triple $\mathcal{X}$	$3k$	k	$2k$
3-MAK $\mathcal{X}$	$2k$	$k-1$	$2k$

Table 1. Bounds on the complexities of attacks on the proposed scheme and the existing ones.

Scheme	Key size	Key-sched.	Weak keys	Complementation property
DES	56	2	Yes	Yes
Two-key triple-DES	112	4	Yes	Yes
Three-key triple-DES	168	6	Yes	Yes
3-MAK DES	112	16	No	No

Table 2. Comparison of the proposed scheme and the existing ones, all used with DES.

fast as for other triple encryption schemes with fixed keys. Finally we note that the absence of weak keys are guaranteed, since the three round keys are never equal and the complementation property does not hold. In Table 2 we give a schematic overview of the differences between our proposed scheme and the existing ones when used with DES. 'Keysched.' is the total number of encryptions in the key schedules using the above estimate in the case of DES. Finally we state if weak keys exist and if the complementation property holds.

3.4 Extensions

In the preceding sections we focused on triple encryption schemes. It is clear that our ideas can be extended to quadruple, quintuple, ..., n-fold schemes. Let $\mathcal{X}$ be a component cipher with key size k. In general a $2i$-fold encryption scheme based on $\mathcal{X}$ is vulnerable to a meet in a middle attack using 2^{ik} words of memory taking time about 2^{ik}. Similarly, a $(2i+1)$-fold encryption scheme based on $\mathcal{X}$ is vulnerable to a meet in a middle attack using 2^{ik} words of memory taking time about $2^{(i+1)k}$. Therefore, one does not get the security of the full key length. It is obvious that by generating the $2i$ respectively $2i+1$ keys pseudorandomly, defined in a similar manner as Def. 4, from i respectively $i+1$ keys, one can prove results similar to that of Th. 5.

4 Conclusion

We considered multiple encryption schemes built from conventional cryptosystems. We showed how to build triple encryption schemes with a security strongly connected to the security of the underlying block cipher and not vulnerable to the special meet in the middle attacks on existing schemes. Furthermore, a brute force meet in the middle attack has the same complexity as a brute force exhaustive key search, thus our scheme can be seen as an example of an optimal triple encryption scheme. Also, a generalisation to n-fold schemes was given.

References

1. M. Blum and S. Micali. How to generate cryptographically strong sequences of pseudorandom bits. *SIAM Journal on Computing*, pages 856–864, 1984.
2. S. Even and O. Goldreich. On the power of cascade ciphers. *ACM Trans. on Computer Systems*, 3:108–116, 1985.
3. M. Matsui. Linear cryptanalysis method for DES cipher. In T. Helleseth, editor, *Advances in Cryptology - Proc. Eurocrypt'93, LNCS 765*, pages 386–397. Springer Verlag, 1993.
4. M. Matsui. The first experimental cryptanalysis of the Data Encryption Standard. In Y. G. Desmedt, editor, *Advances in Cryptology - Proc. Crypto'94, LNCS 839*, pages 1–11. Springer Verlag, 1994.
5. U. Maurer and J.L. Massey. Cascade ciphers: The importance of being first. *Journal of Cryptology*, 6(1):55–61, 1993.
6. R. Merkle and M. Hellman. On the security of multiple encryption. *Communications of the ACM*, 24(7):465–467, 1981.
7. J.-J. Quisquater and J.-P. Delescaille. How easy is collision search. New results and applications to DES. In G. Brassard, editor, *Advances in Cryptology - Proc. Crypto'89, LNCS 435*, pages 408–413. Springer Verlag, 1990.
8. W. Tuchman. Hellman presents no shortcut solutions to DES. *IEEE Spectrum*, 16(7):40–41, July 1979.
9. P.C. van Oorschot and M.J. Wiener. A known-plaintext attack on two-key triple encryption. In I.B. Damgård, editor, *Advances in Cryptology - Proc. Eurocrypt'90, LNCS 473*, pages 318–325. Springer Verlag, 1990.
10. M.J. Wiener. Efficient DES key search. Technical Report TR-244, School of Computer Science, Carleton University, Ottawa, Canada, May 1994. Presented at the Rump Session of Crypto'93.

A One–Key Cryptosystem Based on a Finite Nonlinear Automaton

Marc Gysin

Department of Computer Science, The University of Wollongong, Wollongong NSW
2522, Australia, e-mail: marc@cs.uow.edu.au

Abstract. A finite nonlinear automaton for a one–key blockcipher cryptosystem is presented. The key defines the automaton. The mapping from the "keyspace" to the "automaton–space" is one to one. Similar to other cryptosystems, encryption and decryption is done by repeating a number of simple steps called "rounds". The number of rounds and the blocksize can be variable and does not depend on the key. The statistical properties measured on the ciphertext are satisfactory and in the same range as the properties of DES (Data Encryption Standard).

1   Introduction

Finite automata have always been of the interest in cryptography. Linear and nonlinear shift registers are a special case of finite so called "autonomous automata" and have been widely studied [1]. One–key and public key cryptosystems based on finite non–autonomous automata where the next state function depends on some input have been studied and presented in [2] and [3]. Some cryptosystems proposed are based on cellular automata [6]. In many of these cryptosystems, the key defines the initial state of the automaton. In [7], cryptosystems are outlined which are based on finite automata and with the key defining the automaton in some manner.

A lot of work has been done in grammatical syntax recognizers and finite acceptors [4], [5]. An acceptor is a machine that reads a certain input and then goes into an "accept" or "reject" state. Other kinds of automata are called Mealy machines (named after G.Mealy, 1955) which read an input character, move onto a next state (depending on the current state and the character read) and output a character. Hence, Mealy machines can be seen as converters from an input alphabet to an output alphabet. The way the input text is translated to the output text depends on the automaton and the next state function. In this paper, a cryptosystem is presented which is an extension of a Mealy machine and in which the key defines the automaton.

2   Mealy Machines and Extended Mealy Machines

Definition 1 Mealy Machine.
Let X, Y and S be finite sets. A *Mealy machine* is a quintuple $M = < X, Y, S, \delta, \lambda >$

where $\delta : S \times X \to S$ and $\lambda : S \times X \to Y$ are functions. At moment i, the automaton is in an internal state $s_i \in S$, accepts an input $x_i \in X$, goes to the next state s_{i+1} and puts out a character $y_i \in Y$ according to the next–state function δ and output function λ, respectively.

Thus, on an initial state $s_0 \in S$ of M, an input sequence $x_0, x_1, \ldots$ of M produces an output sequence $y_o, y_1, \ldots$ of M according to

$$s_{i+1} = \delta(s_i, x_i), y_i = \lambda(s_i, x_i), i = 0, 1, \ldots.$$

A simplified version of the cryptosystem presented in this paper can be seen as follows. Let $X = Y = \{0,1\}$ be a set of two elements. During encryption the plaintext $P = P_0 = x_{0,0}, x_{0,1}, \ldots, x_{0,8 \times blksize-1}$ is read and an output sequence $C_1 = x_{1,0}, x_{1,1}, \ldots, x_{1,8 \times blksize-1}$ is generated. C_1 is then reversed and $P_1 = C_1^{*} = x_{1,8 \times blksize-1}, \ldots, x_{1,1}, x_{1,0}$ is taken as the input sequence. The next output sequence is then reversed and so on. The encryption process stops after a specified number, r, of rounds and the ciphertext $C = C_r$ is the output sequence after r rounds and $r - 1$ times reversing the output sequence before feeding again it as an input sequence into the automaton. At the beginning and after each round the automaton (re)starts in its initial state s_0. The next–state function δ and the output function λ are defined by the key and by the inherent conditions imposed on the cryptosystem. One of these conditions is that for any key k, the λ function must be reversible in order to perform decryption properly. More precisely, for each k and s_i and each $x_i \neq x_k$, $\lambda(s_i, x_i) \neq \lambda(s_i, x_k)$ must be guaranteed. Deciphering can now be done in the following manner. We start in s_0 and take the cryptogram $C = x_{r,0}, x_{r,1}, \ldots, x_{r,8 \times blksize-1}$, reverse it if necessary and regain the previous sequence that generated C. This is possible due to the requirement that λ is reversible and that after each round we restart at the initial state s_0. The decryption process stops after r rounds and $r - 1$ times reversing the sequence obtained by operating the automaton in the decryption mode. The reversing of intermediate results is done in order to ensure that a single bit change in the plaintext (or ciphertext) affects all the bits *before* and after the position where the alteration was made.

Therefore, in this simplified cryptosystem, one round of encryption corresponds exactly to operating a Mealy machine M on a two elements (0 and 1) input and output alphabet starting at an initial state s_0. In each round, deciphering starts with the same corresponding bit and in the same state of the automaton as enciphering. Therefore, the requirement of λ being reversible is sufficient and the quite sophisticated theory about inversion of automata can be omitted.

We now slightly extend our definition of the encryption and decryption automaton.

Definition 2 Extended Mealy Machine.
An *extended Mealy machine* is an octuple $M_{ext} = < X, Y, S, T, \Theta, \beta, \delta, \lambda >$ where

X, Y, S are as above and $T = \{0, 1\}^n$ can be looked as additional "internal" state variables and $\beta : \{0, 1\}^n \to \{0, 1\}$ is a function. $\Theta : S \times \{0, 1\}^n \to \{0, 1\}^n$ is a function that "updates" the internal state variables and is independent on the input $x \in X$. $\delta : S \times X \times \{0, 1\} \to S$ and $\lambda : S \times X \times \{0, 1\} \to Y$ are functions similar to above but they also take the output of β into consideration.

At moment i, the automaton is in an internal state $s_i \in S$ with the t_i's $\in T$ being 0 or 1. The automaton accepts an input $x_i \in X$, goes to the next state s_{i+1} and puts out a character $y_i \in Y$ according to the next–state function δ and output function λ, respectively. The additional internal variables t_i are involved in both of these functions. β firstly returns a 0 or 1 value according to the values of the t_i's, and then δ and λ take the value of β as an additional argument together with the internal state s_i and the x_i read in order to go to the next state s_{i+1} and output a y_i, respectively.

After some basic considerations about extended Mealy machines with k states and n additional internal variables, it becomes quite obvious that an extended Mealy machine can be seen as a "normal" Mealy machine with $2^n \times k$ states. Nevertheless, for the remainder of this paper we will follow Definition 2 whenever possible because the "real–time" implementation can be easily explained when looking at it as an extended Mealy machine.

3   An Example

We have given the formal definition of a Mealy machine and an extended Mealy machine and we have outlined how a Mealy machine can define a cryptosystem. It remains to show how the key defines the automaton and how encryption and decryption is performed with extended Mealy machines.

Suppose the key is n bytes long. Then the number of internal states s_i is typically n or $n/2$ (in the latter case n must be even). That is, each byte (or each two bytes) of the key describe entirely one state s_i of the automaton. In the real–time application, the functions δ_{real} and λ_{real} go from $\{0, 1\}$ to S and Y, respectively. This works as follows. β firstly takes all the internal variables t_i and returns a 0 or 1 value, then $x \in X$ (recall $x \in \{0, 1\}$) is added to β modulo 2. This last value modulo 2 forms a $\{0, 1\}$–input condition which is taken by δ_{real} and λ_{real} in order to go to the next state and output a 0 or 1.

The algorithm defines the function β. For any state $s_i \in S$, one of the two successor states is always equal to $s_i + 1$ (reduced modulo the total number of states if necessary). This state is called the "plus–one" state. The other state will be referred to as the "alternative" state. The initial state is $s_0 = 0$ and all the t_i's are set to 0.

It is left for the key to describe (for each $s_i \in S$):

- an "update–mechanism" for the internal variables t_i;
- the next, so called "alternative" state;
- under which input condition 0 or 1 the automaton moves to the "alternative" state; and
- which value, 0 or 1, is output when moving to the "alternative" state.

Since λ is reversible, the input condition and the output bit for the "plus–one" state are now clearly defined. If, for example, the system moves from s_i to the "alternative" state with input condition 1 and puts out 0 then the input condition for the "plus–one" state is clearly 0 and the output will be 1.

Let us now give a simple example in detail. Suppose the key is 8 bytes long. That is, we have 8 states s_i. Each byte describes exactly one of these 8 states. For each state s_i we have the first 3 bits describing the "alternative" state, the next 2 bits describing the input condition and the output bit for the "alternative" state and 3 bits left over for bitwise addition modulo 2 to the internal state variables t_i.

For the key

$$k = A0B3FF253C49B061_{hex},$$

we get the automaton in Table 1. (I/O refers to the input condition and output bit for the corresponding state transition)

Table 1. The automaton for the key $k = A0B3FF253C49B061_{hex}$

state	key–byte	"plus–one" state	I/O	"alt" state	I/O	$\oplus t_i$'s
0	101 00 000	1	1/1	101 = 5	0/0	000
1	101 10 011	2	0/1	101 = 5	1/0	011
2	111 11 111	3	0/0	111 = 7	1/1	111
3	001 01 001	4	1/0	001 = 1	0/1	001
4	001 11 100	5	0/0	001 = 1	1/1	100
5	010 01 001	6	1/0	010 = 2	0/1	001
6	101 10 000	7	0/1	101 = 5	1/0	000
7	011 00 001	0	1/1	011 = 3	0/0	001

One round of encryption now works as follows. For $\beta = t_0 t_1 \oplus t_0 \oplus t_2$, the plaintext $P = 10110111$, and starting in $s_0 = 0$ with $t_0 = t_1 = t_2 = 0$, the automaton goes through the states 0, 1, 2, 3, 1, 2, 7, 3 and puts out 11011100. The details are given in Table 2.

Clearly, this process can be reversed for decryption. We start in $s_0 = 0$ again and $t_0 = t_1 = t_2 = 0$. From the corresponding output bit we immediately know the input condition and the next state of the automaton. We also know the

Table 2. One round of encryption

p_i	state	$t_0t_1t_2$	β	input cond.	next state	c_i
1	0	000	0	1	1	1
0	1	000	0	0	2	1
1	2	011	1	0	3	0
1	3	100	1	0	1	1
0	1	101	0	0	2	1
1	2	110	0	1	7	1
1	7	001	1	0	3	0
1	3	000	0	1	4	0

return value of β. By a simple addition modulo 2 of β and the input condition, we recover the corresponding plaintext bit. We move to the next state of the automaton and repeat this process until the plaintext is recovered completely. Observe that the states s_i which the automaton goes through and the internal values of the variables t_i are exactly the same for encryption and decryption.

4 The Real Implementation

In the real implementation, the number of rounds and the blocksize is variable and can be entered by the user of the system. As mentioned above, the output text is reversed after each round and the automaton starts again in its initial state and the t_i's are reset to zero.

The experiments carried out have been made with an automaton where two bytes of the key define one state. Out of these two bytes, the bits have been allocated in the following way:

- 2 bits for the input condition and the output bit of the "alternative" state;
- 10 bits for the internal variables $t_0, \ldots, t_9$; and
- 4 bits as an *offset* for the "alternative" state. That is, the "alternative" state is calculated as follows: $altstate = currstate - offset + 8$, where $offset$ is the corresponding decimal value of the 4 keybits and $currstate$ is the current state.

The total number of the states $s \in S$ does not have to be a power of 2, it just has to be greater or equal to 16. If it is greater than 16 we know in advance that for any state $s_i \in S$, some other states $s_j \in S$ cannot be successor states of s_i. However, the key length must be even.

β is a bent function on the 10 variables t_i:

$$\beta = t_0t_1 \oplus t_2t_3 \oplus t_4t_5 \oplus \bar{t_6}t_7 \oplus \bar{t_8}t_9 \oplus t_0 \oplus \bar{t_2} \oplus t_4 \oplus \bar{t_5} \oplus t_6 \oplus \bar{t_9}.$$

A bent function is taken because we want the output of β to appear like a random function of the input variables t_i. The t_i's must not be interchangable because otherwise the corresponding key–bits would also be interchangable.

The key–scheduling is performed as follows. Before each round the key is cyclically shifted by a defined number of bits and then bitwise added (modulo 2) to a constant string. These values are different for each round and therefore we get a different automaton for each round.

5 Security and Statistics

There are no flaws and no security leaks known to the author. There are also no weak keys because if we get a degenerate automaton in one round (for example too many cycles or too many regularities), we must obtain a "good" automaton in the next round, due to the operations during key–scheduling. It is also impossible to get some isolated states, because the condition of the "plus–one" state being $s_i + 1$ ensures that each state can be reached. However, there should be a sufficient number of rounds performed on the plaintext in order to ensure security and good statistical properties on the ciphertext. No specific tests about security against linear and differential cryptanalysis have been performed.

The statistical tests have been carried out in ECB (Electronic Codebook) and CBC (cipher block chaining) mode. Different tests have been done for different blocksizes and different number of rounds. The statistical properties were measured on:

1. The ciphertext when encrypting normal (ASCII or random) plaintext;
2. The XOR difference of the plaintexts when the original ciphertext was altered one bit (per block);
3. The XOR difference of the ciphertexts when the original plaintext was altered one bit (per block);
4. The XOR difference of the plaintexts when decrypting the ciphertext with a key which differed in one bit from the key used for encryption;
5. The XOR difference of the ciphertexts when encrypting the plaintext with two keys which only differed in one bit; and
6. The (final) ciphertext when encrypting the plaintext 100 times (and taking the ciphertext from the last encryption as the new plaintext).

Changing only one bit of the key (Test 4 and Test 5) could result in a variety of altered automata compared to the original automata according to where the change occurred. That is, whether the "alternative" state part, the variables t_i or the input condition/output bit were changed. In doing the statistical tests, enough random one–bit changes were performed in order to ensure that all the corresponding key–parts were affected. Also observe that the key is shifted cyclically during key–scheduling which means, for example, that in one automaton the "alternative" state part is changed and in the next round the input condition/output bit is altered.

In order to get an idea about the statistical properties, we performed the two–tailed χ^2–test with 255 degrees of freedom. If the value in the cumulative normal distribution was above the 1% and below the 99% mark we recorded a "PASS"; otherwise we recorded a "FAIL". For further details on the χ^2–test, the reader is referred to [8] or any elementary book on statistics.

Table 3 shows some typical values obtained when testing the cryptosystem.

Table 3. Statistical tests in CBC mode

Test	rounds	blksize	avg χ^2	% PASS	% FAIL
1	3	8	253.97	97	3
1	5	8	254.77	97	3
1	10	8	251.86	94	6
1	3	64	252.5	100	0
2	3	8	261.47	90	10
2	5	8	256.23	97	3
2	10	8	250.81	100	0
2	3	64	273.82	94	6
3	3	8	249.99	97	3
3	5	8	250.7	94	6
3	10	8	252.66	100	0
3	3	64	256.1	100	0
4	3	8	330.85	35	65
4	5	8	255.15	100	0
4	10	8	247.43	100	0
4	3	64	249.3	100	0
5	3	8	248.91	97	3
5	5	8	248.39	100	0
5	10	8	251.51	100	0
5	3	64	260.34	100	0
6	5	8	255.48	97	3

The results in CBC mode are all satisfactory except Test 4 with 3 rounds and blocksize 8. Three rounds together with blocksize 8 are obviously too few to gain a satisfactory security level. With the exception of this test all the results are in the same range as statistical tests carried out on DES (in CBC mode). Many of the results are even slightly better.

In the ECB mode, most of the tests turned out to be unsatisfactory. Typical results obtained were: average $\chi^2 \approx 420$, PASS $\approx 25\%$, FAIL $\approx 75\%$. Again these results are in the same range or better than DES (in ECB mode). Tests on blocksize 64 were promising even in ECB mode, but this blocksize failed Test 3 when the plaintext was altered and the XOR difference of the ciphertexts was examined. Values for blocksize 64 were: average $\chi^2 \approx 258$, PASS $\approx 96\%$, FAIL $\approx 4\%$ on tests other than Test 3 and average $\chi^2 \approx 330$, PASS $\approx 70\%$, FAIL $\approx 30\%$ on Test 3.

The encryption and decryption for one round was slightly slower than DES (the actual implementation of DES used was in software on a UNIX–System written by Lawrie Brown, Computer Science, ADFA). The automaton encrypted in one round (with blocksize 8) about 54 KByte per second (on a Sun4). Decryption and CBC mode were just slightly slower than encryption and ECB mode. Increasing the blocksize made encryption and decryption up to 20% faster. When more rounds were used, encryption and decryption times increased almost linearly. These results could be improved by optimizing the program–code.

6   Conclusions

A one–key cryptosystem was presented which the author believes is new. The cryptosystem is based on a nonlinear finite automaton. The automaton can be looked at as an extension of a Mealy machine. Unlike many other cryptosystems based on automata, the key defines the state transition functions and values of internal and output variables. That is, the key defines parts of the automaton itself.

Encryption and decryption is done in a number r of rounds. The output text obtained after one round is reversed in order to get better cryptographic properties of the final ciphertext. The number of rounds and the blocksize are variable and can be specified by the user. The length of the key is also variable, although, in some particular implementations the length has to be even.

Statistical results obtained on the ciphertext and on XOR differences of ciphertexts and plaintexts after carrying out a variety of different tests were satisfactory to very satisfactory.

The way the automaton is defined by the key is simple and there are various other ways of defining an automaton by a key. One can think of many other possibilities how to use this kind of automata in many research areas. Hashing and (en)coding are such possibilities. In this case, the function λ may not need to be reversible anymore.

The author of this article believes that this kind of automata has a great cryptographic potential which has not yet been exploited.

References

1. Solomon W. Golomb, Shift Register Sequences, Aegean Park Press, Laguna Hills, California, 1982.
2. Tao Renji, On finite automaton one–key cryptosystems, Fast Software Encryption, Springer Lectures Notes in Computer Science, **809**, Berlin, 1993.
3. Tao Renji, Shihua Chen, Xuemei Chen, FAPKC3: a new finite automaton public key cryptosystem, to appear.
4. Harry R. Lewis, Christos H. Papadimitriou, Elements of the Theory of Computation, Prentice Hall, New Jersey, 1981.
5. W.M.L. Holcombe, Algebraic Automata Theory, Cambridge studies in advanced mathematics, Cambridge University Press, 1982.
6. P.Guam, Cellular automaton public key cryptosystem, Complex Systems, **1**, 51–56, 1987.
7. Juhani Heino, Finite automata: a layman approach, text posted in sci.crypt newsgroups, October 1994, juhanihe@waltari.helsinki.fi, University of Helsinki, Finland, 1994.
8. David W. Deley, Computer Generated Random Numbers, paper posted in sci.crypt newsgroups, September 1994, deleyd@netcom.com, 1991.

A Cryptanalysis of Clock-Controlled Shift Registers with Multiple Steps

Jovan Dj. Golić[1,2*] and Luke O'Connor[3,1**]

[1] Information Security Research Centre, Queensland University of Technology
GPO Box 2434, Brisbane Q 4001, Australia
[2] School of Electrical Engineering, University of Belgrade
[3] Distributed Systems Technology Centre, Brisbane

Abstract. A clock-controlled shift register that is clocked at least once and at most $d + 1$ times per output symbol is cryptanalyzed using a constrained embedding approach. Upper bounds on the constrained embedding probabilities that are exponentially small in the string length are derived using finite automata theory and generating functions. A known constrained embedding divide-and-conquer attack on a clock-controlled shift-register developed for at most two clocks at a time ($d = 1$) is thus extended to the general case of arbitrary d. The results show that the minimum length of the observed output sequence needed for successful initial state reconstruction is linear in the shift register length, and at least exponential and at most superexponential in d. This proves that by making d large one cannot achieve the theoretical security against the embedding attack. Experimental results obtained by computer simulations indicate that the required output sequence length is only exponential in d, which would mean that the embedding attack is feasible if d is not too large.

1 Introduction

The use of clock-controlled shift registers in keystream generators for stream cipher applications appears to be a good way of achieving sequences with long periods and high linear complexity that are immune to fast correlation attacks [10, 3, 5]. The basic clock-controlled shift register (CCSR) is shown in Fig. 1. A regularly clocked shift register, with not necessarily linear feedback, produces one output bit per clock or step, while an irregularly clocked shift register may be stepped more than once to produce each output bit.

Let $A = \{a_i\}_{i=1}^{\infty}$ be the clocking sequence, let $B = \{b_i\}_{i=1}^{\infty}$ be the bit stream produced by the regularly clocked shift register, and let $C = \{c_i\}_{i=1}^{\infty}$ be the keystream produced by the CCSR. The CCSR operates as follows: when a_i is

* This research was supported in part by the Science Fund of Serbia, grant #0403, through the Institute of Mathematics, Serbian Academy of Arts and Sciences.
** The work reported in this paper has been funded in part by the Cooperative Research Centres program through the Department of the Prime Minister and Cabinet of Australia.

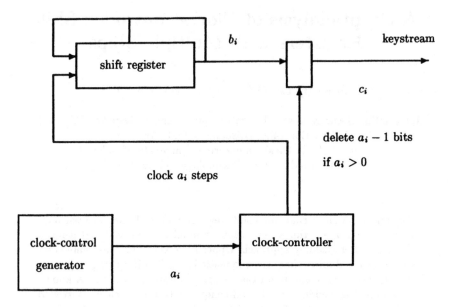

Fig.1. Basic clock-controlled shift register.

generated, the clock-controller advances the shift register a_i steps, discarding the first $a_i - 1$ bits and taking the a_i-th bit as the next bit of output if $a_i > 0$, and repeating the previous bit if $a_i = 0$. It follows that bit c_n of the keystream satisfies

$$c_n = b\left(\sum_{i=0}^{n} a_i\right), \quad n \geq 1.$$

It is clear that C is obtained by a nonuniform decimation of B according to A.

Let us now briefly review the results in the literature on cryptanalyzing certain types of CCSRs. A CCSR is said to be $\{k, m\}$-clocked if $a_i \in \{k, m\}$ for all i, and $\{0, 1\}$-clocked CCSRs are known as stop-and-go shift registers. Cascades of stop-and-go shift registers are cryptanalyzed in [2] based on a specific lock-in effect, and the more general case of $\{k, m\}$-clocked shift registers is treated in [11]. Menicocci has proposed a conditional correlation attack on cascades of stop-and-go linear feedback shift registers of length two in [14] and an unconditional correlation attack on the cascades of an arbitrary length in [15, 16], see also [8]. A conditional correlation attack on these cascades of an arbitrary length has been described and performed in [13]. Živković [19] has devised a divide-and-conquer attack on a $\{1, 2\}$-clocked shift register using a *constrained embedding approach*. The underlying combinatorial problem has been solved in [9]. A more general case of a noised irregularly clocked shift register, where $1 \leq a_i \leq d + 1$, is examined by Golić and Mihaljević [6] using a generalization of the Levenshtein distance. Unconstrained embedding and probabilistic correlation attacks on CCSRs have been proposed and analyzed by Golić in [7]. Recently, Golić [8] has established a linear statistical weakness of arbitrary CCSRs with unconstrained or constrained irregular clocking. Another type of cryptanalytic attacks on CCSRs is a divide and conquer attack on the initial state of a clock-control generator based on the linear consistency test [18] or, in a particular case, on the collision test [1].

In a constrained embedding attack, the cryptanalyst is given a segment of the keystream $C = c_1 c_2 \cdots c_n$ and then attempts to find all the initial contents of the shift register that produce a sequence $B = b_1 b_2 \cdots b_m$ that can be decimated by a_i, $1 \leq a_i \leq d$, to yield C. That is, the cryptanalyst is looking for all the shift register sequences B into which C can be embedded using at most d deletions between any two consecutive bits of C. Assuming that the first bits should match $(b_1 = c_1)$, it suffices to consider $m = n(d+1)$ as the length of B. The attack is successful if upon termination there is only one or very few candidate initial states remaining from all possible initial states. Živković [19] was able to show that for the case of $d = 1$, the number of candidate states is tending to 1 as n, the length of the keystream sample, is increased. He proved this by showing that as n increases, the relative number of all binary strings B that permit an arbitrary binary string C to be embedded is tending to zero.

Our objective is to determine if the constrained embedding attack is still successful if d is allowed to be an arbitrary positive integer. That is, we wish to determine if it is possible to generalize the results of [19] and apply a constrained embedding attack to $[1, d+1]$-clocked shift register, where $a_i \in \{1, 2, \ldots, d+1\}$, for an arbitrary positive integer d. In this case at most d bits are deleted from B before the next bit is selected as a keystream bit. The approach from [19] can not be extended to the general case because it is based on direct counting that makes no sense for arbitrary d.

Our main result is to show that the corresponding minimum length n of the observed sequence C needed for a successful reconstruction of the initial state of a $[1, d+1]$-clocked shift register is approximately

$$n \geq r \cdot \ln 2 \cdot \left(1 - 2^{-d-2(d+1)4^{d+1}+(d+2)2^{d+2}}\right) \cdot 2^{(d+2)(1+2^{d+2})} \tag{1}$$

where r is the length of the shift register. Consequently, we prove that by making d large one cannot achieve theoretical security against the embedding attack, but can significantly improve the practical security of a CCSR. As in [19], we assume that the probability of C embedding into B is well-approximated by the probability that a random string X of length n embeds into a random string Y of length $m = n(d+1)$. The bound in (1) is derived by using regular languages and generating functions.

The rest of the paper is organized as follows. In §2 we describe the constrained embedding problem more formally, and present several related definitions. In §3 we present a theoretical analysis of the general case where $d > 2$, for arbitrary d. Our approach is to consider the embedding probabilities for the constant strings and the alternating strings. An upper bound on the probability that an arbitrary string X will embed can then be derived by factoring X into substrings that are constant and alternating, and combining the respective embedding probabilities. We will use regular languages and generating functions to derive upper bounds on embedding probabilities for the constant and alternating strings. Experimental evidence indicating that the alternating strings maximize the embedding probability is also given in §3. Note that we have presented the results on the upper bound on the embedding probability for the constant strings in [7], but partially and without proofs.

2 Constrained Embedding Problem

Consider two binary strings $X = x_1 x_2 \cdots x_n$ and $Y = y_1 y_2 \cdots y_m$ of lengths n and m, respectively. The string X is said to d-*embed* into Y if there exists a sequence of integers $S = s_1 s_2 \cdots s_n$ such that $x_i = y_{s_i}$ for $1 \leq i \leq n$, with $s_1 = 1$ and $1 \leq s_{i+1} - s_i \leq d+1$ for $1 \leq i < n$. That is, X can be d-embedded into Y if each bit x_i can be matched with a bit y_j such that no consecutive matchings are further than d bits apart. Note that the first bit of X is always matched to the first bit of Y. Equivalently, X can be d-embedded into Y if X can be obtained from a prefix of Y by deleting no more than d consecutive bits at a time, and not deleting the first bit. If it is also the case that $m - s_n \leq d$, then X is said to *strictly* d-embed into Y. To check whether X can be d-embedded into Y, one may use the direct matching algorithm [19] or the constrained Levenshtein distance algorithm [6], where both algorithms have a complexity of $O(nm)$. In the embedding attack, the length n of X should be large enough so that there is one or very few candidates for the initial state to produce a string Y which permits an embedding of X. Note that there will always be at least one candidate since Y is chosen to have maximal length $n(d+1)$.

The embedding attack is successful only if the probability that a given binary string X embeds into a random binary string Y of length $m = n(d+1)$ approaches zero sufficiently fast as n increases. Accordingly, let $P_{d,X}(n)$ denote the probability that a given binary string X of length n can be d-embedded into a random binary string Y of length $m = n(d+1)$, chosen uniformly from all such strings. Also define $P_{d,X}(n,k)$ as the probability that X can be strictly d-embedded into a random string Y of length $n + k$, $0 \leq k \leq nd$. Clearly,

$$P_{d,X}(n) \leq P_{d,X}^*(n) \stackrel{\text{def}}{=} \sum_{k=0}^{nd} P_{d,X}(n,k). \tag{2}$$

Note that $P_{d,X}^*(n)$ may be greater than 1, especially for small values of n (for $n = 1$ it follows that $P_{d,X}^*(1) = (d+1)/2$). The corresponding upper bound that is valid for all X of length n is then

$$P_{d,X}(n) \leq P_d^*(n) \stackrel{\text{def}}{=} \sum_{k=0}^{nd} P_d(n,k) \tag{3}$$

where $P_d(n,k)$ denotes the maximum of $P_{d,X}(n,k)$ taken over all X of length n. The problem of deriving $P_d(n,k)$ for an arbitrary X appears to be very difficult to solve, with similar remarks applying to deriving $P_{d,X}(n,k)$ and $P_{d,X}^*(n)$. The case of $d = 1$ has been resolved in [9]. Our objective is then to obtain a suitable upper bound on $P_{d,X}(n)$. To this end, the following result proves to be very useful.

Lemma 1. Let $X = X_1 X_2$ denote the concatenation of X_1 and X_2 of lengths n_1 and n_2, respectively, where X has length $n = n_1 + n_2$. Then

$$P_{d,X}^*(n) \leq P_{d,X_1}^*(n_1) \cdot P_{d,X_2}^*(n_2). \tag{4}$$

Proof. It is easy to see that for strict d-embeddings we have

$$P_{d,X}(n,k) \leq \sum_{k_1} \sum_{k_2} P_{d,X_1}(n_1,k_1) \cdot P_{d,X_2}(n_2,k_2)$$

where the sum is taken over all $0 \leq k_1 \leq dn_1$ and $0 \leq k_2 \leq dn_2$ such that $k = k_1 + k_2$. Therefore in view of (2),

$$P_{d,X}^*(n) \leq \sum_{k=0}^{nd} \sum_{k_1} \sum_{k_2} P_{d,X_1}(n_1,k_1) \cdot P_{d,X_2}(n_2,k_2) = P_{d,X_1}^*(n_1) \cdot P_{d,X_2}^*(n_2)$$

which completes the proof. □

It is interesting to note, however, that the analogous result does not hold for $P_{d,X}(n)$. So Lemma 1 enables us to combine the upper bounds of the type in (2) into an upper bound of the same type for a string of the form $X = X_1 X_2$.

Unlike the upper bound, a lower bound on $P_{d,X}(n)$ is easily obtained.

Lemma 2. For all X, $P_{d,X}(n) \geq \left(1 - \frac{1}{2^{d+1}}\right)^n$.

Proof. Consider a d-embedding that matches x_i as soon as possible (least index) in Y. □

3 A Solution for Arbitrary d

In this section, a theoretical method is developed for solving the general constrained embedding problem with at most d deletions. Our major concern here is to determine whether the embedding divide-and-conquer attack can be prevented if d is chosen to be sufficiently large. One possibility is to derive an analytical expression for $P_d(n,k)$ and then to use (3). However, this is very difficult to achieve, because it is not clear how to capture the behaviour of $P_{d,X}(n,k)$ for an arbitrary string X.

The approach we propose has two stages, and is based on constant and alternating strings. A string X is constant if it consists of one bit repeated for the length of the string, which will be denoted as 0^n and 1^n. An alternating string X is one in which each bit x_i is the complement of the previous bit x_{i-1}. Let

$P_{d,a}(n, k)$ and $P_{d,c}(n, k)$ denote the embedding probabilities for the constant and alternating strings, respectively. In the first stage we will establish the suitable upper bounds on $P_{d,a}(n, k)$ and $P_{d,c}(n, k)$. Then in the second stage, since an arbitrary string X can be divided into constant and alternating substrings, by virtue of (3) and Lemma 1, we will combine these upper bounds in an exponential upper bound that holds for sufficiently long strings X with probability arbitrarily close to one.

3.1 Constant and alternating strings

In order to obtain the upper bounds on $P_{d,a}(n, k)$ and $P_{d,c}(n, k)$, we begin with the following observations. If the constant string $X = 0^n$ can be strictly d-embedded into a string Y of length $m \geq n$, then Y does not contain the substring 1^{d+1} (with the analogous property holding for 1^n and the substring 0^{d+1}). Also, if an alternating string X of length n can be strictly d-embedded into a string Y of length $m \geq n$, then Y contains neither of the substrings $1^{2(d+1)}$ or $0^{2(d+1)}$. Accordingly, by enumerating all binary strings that possess these properties we can obtain the upper bounds on $P_{d,a}(n, k)$ and $P_{d,c}(n, k)$. We will show that this is indeed possible by using regular expressions from the theory of formal languages [12], and generating functions from combinatorial theory [17].

Consider first the case of a constant string. The set of binary strings that begin with 0 and do not contain $1^l, l = d+1$, as a substring is a regular language [12] for fixed d, which we will denote as L_l^c. Equivalently, there is a deterministic finite automata (DFA) which recognizes (or accepts) the members of L_l^c. The DFA for L_l^c is not unique, but for a given DFA that accepts L_l^c, a regular expression for L_l^c can be determined. Of all such regular expressions consider

$$L_l^c = (0 + 01 + \cdots + 0\underbrace{11\cdots11})^* \tag{5}$$
$$l-1 \text{ times}$$

meaning that each string $X \in L_l^c$ can be obtained by repeated concatenation of strings from the set $\{0\} \cup \{01\} \cup \cdots \cup \{011\cdots11\}$, since the $*$ operator means 'select zero or more times'. The empty string ϵ of length zero is also included. It is crucial to note that each $X \in L_l^c$ can be uniquely decomposed into the substrings $0, 01, 011, \cdots, 011\cdots11$ that define the regular expression (5) for L_l^c. For example, the string 01010010000100 does not contain 11, is thus an element of L_2^c, and is constructed uniquely from (5) as

$$01 \,|\, 01 \,|\, 0 \,|\, 01 \,|\, 0 \,|\, 0 \,|\, 0 \,|\, 01 \,|\, 0 \,|\, 0$$

where $|$ is used to separate the selections from the regular expression. The unique decomposition property allows L_l^c to be enumerated using the generating function $\frac{1}{1-z} = \sum_{i \geq 0} z^i$, and several other basic results for generating functions [17].

Lemma 3. Let $C_l(n)$ denote the number of strings from L_l^c of length $n \geq 0$, $l = d + 1$. Then $C_l(n)$ is equal to the nth coefficient $[z^n]$ of the generating function

$$\mathcal{G}_l^c(z) = \frac{1}{1 - (z + z^2 + \cdots + z^{l-1} + z^l)} = \frac{1}{1 - \sum_{i=1}^{l} z^i}. \qquad (6)$$

Proof. The generating function for $(0 + 01 + \cdots + 0 \underbrace{11 \cdots 11})^i$ with respect to
$l-1$ times

length is $(z + z^2 + \cdots + z^l)^i$. It follows that the number of strings in L_l^c of length n is then the nth coefficient $[z^n]$ of

$$\sum_{i \geq 0} (z + z^2 + \cdots z^l)^i = \frac{1}{1 - \sum_{i=1}^{l} z^i}. \qquad (7)$$

$\square$

Consider now the case of an alternating string. It follows that the set of binary strings that begin with 0 and contain neither 1^l or 0^l, $l = 2(d + 1)$, as substrings is a regular language for fixed d, which we will denote as L_l^a. The language L_l^a is generated by the following regular expression

$$L_l^a = \left((0 + 00 + \cdots + \underbrace{00 \cdots 00})(1 + 11 + \cdots + \underbrace{11 \cdots 11}) \right)^*$$
$$\qquad\qquad\quad l-1 \text{ times} \qquad\qquad\qquad l-1 \text{ times}$$
$$(8)$$

$$(\epsilon + 0 + 00 + \cdots + \underbrace{00 \cdots 00}).$$
$$\qquad\qquad l-1 \text{ times}$$

As was the case for L_l^c, it is easily verified that this regular expression for L_l^a has the unique decomposition property, allowing a direct enumeration.

Lemma 4. Let $A_l(n)$ denote the number of strings from L_l^a of length $n \geq 0$, $l = 2(d + 1)$. Then $A_l(n)$ is equal to the nth coefficient $[z^n]$ of the generating function

$$\mathcal{G}_l^a(z) = \frac{1}{1 - \sum_{i=1}^{l-1} z^i}. \qquad (9)$$

Proof. The generating function with respect to length of the strings defined by

$$\left((0 + \cdots + \underbrace{00 \cdots 00})(1 + \cdots + \underbrace{11 \cdots 11}) \right)^i (\epsilon + 0 + \cdots + \underbrace{00 \cdots 00})$$
$$\qquad\quad l-1 \text{ times} \qquad\qquad l-1 \text{ times} \qquad\qquad\qquad l-1 \text{ times}$$

is

$$\left(\sum_{j=1}^{l} z^j \right)^{2i} \cdot \sum_{j=0}^{l} z^j. \qquad (10)$$

The number of strings in L_l^a of length n is then the nth coefficient $[z^n]$ of

$$\sum_{i\geq 0}\left(\sum_{j=1}^{l}z^j\right)^{2i}\cdot\sum_{j=0}^{l}z^j = \frac{1+\sum_{j=1}^{l}z^j}{1-(\sum_{j=1}^{l}z^j)^2} = \frac{1}{1-\sum_{i=1}^{l-1}z^i}.$$

□

We see that the generating functions for L_l^c and L_l^a have the same form, with the only difference being the degree of the denominator polynomial.

Let us now return to the embedding probabilities $P_{d,a}(n,k)$ and $P_{d,c}(n,k)$. From the previous discussion it follows that

$$P_{d,c}(n,k) \leq 2^{-(n+k)}\cdot C_{d+1}(n+k) \tag{11}$$

$$P_{d,a}(n,k) \leq 2^{-(n+k)}\cdot C_{2d+1}(n+k) \tag{12}$$

where $C_m(n)$ is the nth coefficient $[z^n]$ in the expansion of $G_m(z) = 1/(1 - \sum_{i=1}^{m}z^i)$. It is well-known (see [17]) that one can write an explicit expression for $C_m(n)$ in terms of the roots $\alpha_1, \alpha_2, \cdots, \alpha_m$ of the reciprocal polynomial $P_m(z) = z^m - \sum_{i=0}^{m-1}z^i$ for $G_m(z)$. For the case of $m = 2$, $P_2(z)$ reduces to the reciprocal polynomial corresponding to the generating function of the Fibonacci numbers. However, for $m > 2$, it is not possible to come up with an analytical expression for the roots of $P_m(z)$, and we must therefore use numerical approximations. It is known that the asymptotical behaviour of $C_m(n)$ is dominated by the roots of largest magnitude. This is not satisfactory for us, because we are interested in obtaining an approximate upper bound on $C_m(n)$ which will hold for all values of n. Interestingly enough, it appears that an elegant and sharp bound can indeed be derived.

By ordinary functional analysis, it is easy to show that for $m \geq 2$, the polynomial $P_m(z)$ has a positive real root β_m such that

$$1 < \beta_m < 2 - \frac{1}{2^m}. \tag{13}$$

Hence the sequence $\{c_m\beta_m^n\}_{n=0}^{\infty}$, for any positive constant c_m, is a positive solution to the linear recursion determined by $P_m(z)$, which is $x_n = \sum_{i=1}^{m}x_{n-i}$ for $n \geq m$. The sequence $\{C_m(n)\}_{n=0}^{\infty}$, which is also a positive solution to this recursion, has the initial values $C_m(0) = 1$ and $C_m(n) = 2^{n-1}$, $1 \leq n \leq m-1$, since it enumerates the strings from the language L_m^c. Now, if the constant c_m is chosen large enough to satisfy $C_m(n) \leq c_m\beta_m^n$, $0 \leq n \leq m-1$, then we will have that $C_m(n) \leq c_m\beta_m^n$ is true for all n, because the recursion has positive coefficients. This is satisfied if we pick $c_m = (2/\beta_m)^{m-1}$. Thus we have proved that for all $n \geq 0$ and $m \geq 2$

$$\frac{C_m(n)}{2^n} \leq \left(\frac{\beta_m}{2}\right)^{n-m+1} < \left(1-\frac{1}{2^{m+1}}\right)^{n-m+1} \tag{14}$$

Finally, if we combine (14) with (2) and complete the summation, then we obtain

Theorem 5. Let $P_{d,c}^*(n)$ and $P_{d,a}^*(n)$ denote upper bounds of the type in (2) on the embedding probabilities for a constant and alternating string, respectively. Then for all $n \geq 1$,

$$P_{d,c}^*(n) < 2^{d+2} \left(1 - \frac{1}{2^{d+2}}\right)^{n-d} \tag{15}$$

$$P_{d,a}^*(n) < 4^{d+1} \left(1 - \frac{1}{4^{d+1}}\right)^{n-2d} \tag{16}$$

$\square$

Both of these bounds are greater than one for small values of n, depending on d. For large n however, they tend to zero exponentially fast. For a constant string, the upper bound in (15) is relatively close to the lower bound from Lemma 2, whereas the upper bound (16) for an alternating string is considerably larger than both.

3.2 Random strings

Suppose that we have shown that an alternating string is the most likely string to embed into a random string. Experimental results that we have obtained by direct counting support this conclusion. More precisely, we observed that for $d = 2$ and all $2 \leq n \leq 7$ and $0 \leq k \leq 2n$, the maximum values $P_2(n, k)$ of the strict embedding probabilities $P_{2,X}(n, k)$ are obtained if X is an alternating string (for any given n and k, there are other optimal strings X as well). This leads us to conjecture, which appears to be difficult to prove theoretically for $d \geq 2$ and is proved in [9] for $d = 1$, that the alternating strings maximize the strict embedding probabilities for any d, n, and k. Then, we would have that the upper bound (16) holds for all the strings X, so that by using the criterion

$$2^r \cdot 4^{d+1} \left(1 - \frac{1}{4^{d+1}}\right)^{n-2d} \leq 1 \tag{17}$$

we would then obtain that approximately

$$n \geq (r + 2(d+1)) \cdot \ln 2 \cdot 4^{d+1} \tag{18}$$

consecutive bits of any observed output sequence are sufficient for a successful initial state reconstruction. Compare this with the result following from the lower bound in Lemma 2, by which we have that if the length of any observed sequence satisfies approximately

$$n \leq r \cdot \ln 2 \cdot 2^{d+1} \tag{19}$$

then a successful initial state reconstruction is not possible.

However, it appears very difficult to theoretically prove that the upper bound for alternating strings is also valid for all strings as well. Instead, by noting that every binary string can be divided into constant and alternating substrings, we can still use Theorem 5 in light of Lemma 1. Since the upper bounds in (15) and (16) are greater than one, and hence useless for small n, we are actually interested in finding all the sufficiently long constant and alternating substrings. First, we would like to find the minimum values of n such that (15) and (16) are smaller than one. It is easy to see that if

$$n \geq k_m = m2^m \tag{20}$$

for $m = d + 2$ and $m = 2(d + 1)$, then the bounds in (15) and (16) are smaller than one, respectively.

Consequently, we divide a binary string X of length n into constant runs of ones and zeros, grouping the successive runs of length one into alternating runs. We observe that this division is unique. Only the constant runs of length at least $(d+2)2^{d+2}$ and the alternating runs of length at least $2(d+1)4^{d+1}$ count. If X is arbitrary then all the cases are possible. However, if X is purely random and n is large enough, then with probability arbitrarily close to one (see [4]), there are approximately $n/2^i$ constant runs of length at least i, $i \geq 1$, and $n/2^{i+2}$ alternating runs of length at least i, $i \geq 2$. Therefore it follows that the number of bits contained in constant runs of length at least i, $i \geq 1$, is $(i+1)/2^i$, whereas the number of bits contained in alternating runs of length at least i, $i \geq 2$, is $(i+1)/2^{i+2}$.

Combining Lemma 1 with Theorem 5 it is then simple to prove

Theorem 6. For a purely random string X of length n, for large enough n, with probability arbitrarily close to 1

$$P_{d,X}^*(n) < \left[\left(1 - \frac{1}{2^{d+2}} \right)^{2^{-k_{d+2}}} \cdot \left(1 - \frac{1}{4^{d+1}} \right)^{2^{-k_{2d+2}}} \right]^n \tag{21}$$

where $k_m = m2^m$. $\qquad\qquad\qquad\qquad\qquad\qquad\qquad\qquad\qquad\qquad\qquad \square$

Theorem 6 essentially asserts that given a random string X, the probability that X can be d-embedded into a random string Y exponentially tends to zero with the string length. The result is applicable to the embedding divide-and-conquer attack on a clock-controlled shift register because its output sequence behaves like a random sequence. The corresponding minimum length of the observed sequence needed for a successful reconstruction is approximately

$$n \geq r \cdot \ln 2 \cdot \left(1 - 2^{-d-2(d+1)4^{d+1}+(d+2)2^{d+2}} \right) \cdot 2^{(d+2)(1+2^{d+2})} \tag{22}$$

which is linear in r but superexponential in d. This is a consequence of our theoretical approach, but in practice, the minimum length seems to behave more as for an alternating string, that is, as in (18), which is linear in r and exponential in d.

4 Conclusion

In this paper, a clock-controlled shift register that is clocked at least once and at most $d + 1$ times per output symbol is cryptanalyzed using a constrained embedding approach which originates from [19] and [6]. Upper bounds on the constrained embedding probabilities that are exponentially small in the string length are derived for both the constant and alternating strings by using finite automata theory and generating functions. Starting from these results, an exponential upper bound on the constrained string embedding probability for a random string is also obtained. A constrained embedding divide-and-conquer attack on a clock-controlled shift-register developed in [19] for at most two clocks at a time $(d = 1)$, is thus extended to the general case of an arbitrary d. The results show that the minimum length of the observed output sequence needed for successful initial state reconstruction is linear in the shift register length, and at least exponential and at most superexponential in d. This proves that by making d large one cannot achieve the theoretical security against the embedding attack. Moreover, results obtained by computer simulations indicate that the established upper bound for alternating strings remains valid for arbitrary strings as well. This would then imply that the required minimum length of the observed output sequence is linear in the shift register length and only exponential in d, which would mean that the embedding attack is feasible if d is not too large.

References

1. R. J. Anderson. Solving a class of stream ciphers. *Cryptologia*, 14(3):285–288, 1990.
2. W. G. Chambers and D. Gollmann. Lock-in effect in cascades of clock-controlled shift registers. Advances in Cryptology–EUROCRYPT '88, *Lecture Notes in Computer Science*, vol. 330, C. G. Günther ed., Springer-Verlag, pages 331–342, 1988.
3. C. Ding, G. Xiao, and W. Shan. *The Stability Theory of Stream Ciphers. Lecture Notes in Computer Science*, vol. 561, Springer–Verlag, 1991.
4. W. Feller. *An Introduction to Probability Theory and its Applications.* New York: Wiley, 3rd edition, Volume 1, 1968.
5. J. Dj. Golić and M. V. Živković. On the linear complexity of nonuniformly decimated PN-sequences. *IEEE Transactions on Information Theory*, 34:1077–1079, Sep. 1988.
6. J. Dj. Golić and M. J. Mihaljević. A generalized correlation attack on a class of stream ciphers based on the Levenshtein distance. *Journal of Cryptology*, 3(3):201–212, 1991.
7. J. Dj. Golić and L. O'Connor. Embedding and probabilistic correlation attacks on clock-controlled shift registers. Advances in Cryptology–EUROCRYPT '94, *Lecture Notes in Computer Science*, vol. 950, A. De Santis ed., Springer-Verlag, pages 230–243, 1995.
8. J. Dj. Golić. Intrinsic statistical weakness of keystream generators. Advances in Cryptology–ASIACRYPT '94, *Lecture Notes in Computer Science*, vol. 917, J. Pieprzyk and R. Safavi-Naini eds., Springer-Verlag, pages 91–103, 1995.

9. J. Dj. Golić. Constrained embedding probability for two binary strings. To appear in *SIAM Journal on Discrete Mathematics*.

10. D. Gollmann and W. G. Chambers. Clock controlled shift registers: a review. *IEEE Journal on Selected Areas in Communications*, 7(4):525–533, 1989.

11. D. Gollmann and W. G. Chambers. A cryptanalysis of $step_{k,m}$–cascades. Advances in Cryptology–EUROCRYPT '89, *Lecture Notes in Computer Science*, vol. 434, J.-J. Quisquater, J. Vandewalle eds., Springer-Verlag, pages 680–687, 1990.

12. J. Hopcroft and J. Ullman. *An Introduction to Automata, Languages and Computation*. Reading, MA: Addison Wesley, 1979.

13. S.-J. Lee, S.-J. Park, and S.-C. Goh. On the security of the Gollmann cascades. Advances in Cryptology–CRYPTO '95, *Lecture Notes in Computer Science*, vol. 963, D. Coppersmith ed., Springer-Verlag, pages 148–157, 1995.

14. R. Menicocci. Cryptanalysis of a two-stage Gollmann cascade generator. In *Proceedings of SPRC '93*, Rome, Italy, pages 62–69, 1993.

15. R. Menicocci. Short Gollmann cascade generators may be insecure. In *CODES AND CYPHERS, Cryptography and Coding IV*, P. G. Farrell ed., The Institute of Mathematics and Its Applications, pages 281–297, 1995.

16. R. Menicocci. A systematic attack on clock controlled cascades. Advances in Cryptology–EUROCRYPT '94, *Lecture Notes in Computer Science*, vol. 950, A. De Santis ed., Springer-Verlag, pages 450–455, 1995.

17. F. Roberts. *Applied Combinatorics*. Englewood Cliffs, NJ: Prentice Hall, 1984.

18. K. C. Zeng, C. H. Yang, and T. R. N. Rao. On the linear consistency test (LCT) in cryptanalysis and its applications. Advances in Cryptology–CRYPTO '89, *Lecture Notes in Computer Science*, vol. 435, G. Brassard ed., Springer-Verlag, pages 164–174, 1990.

19. M. V. Živković. An algorithm for the initial state reconstruction of the clock-controlled shift register. *IEEE Transactions on Information Theory*, 37:1488–1490, Sep. 1991.

Discrete Optimisation and Fast Correlation Attacks

Jovan Dj. Golić, Mahmoud Salmasizadeh,
Andrew Clark, Abdollah Khodkar and Ed Dawson

Information Security Research Centre
Queensland University of Technology
GPO Box 2434, Brisbane Qld 4001
Australia
Email: {*golic,salmasi,aclark,khodkar,dawson*} *@fit.qut.edu.au*

Abstract

Modifications to fast correlation attacks on stream ciphers are investigated. Improvements are achieved with both deterministic and random choices for updates of the corresponding posterior probabilities and with fast and partial resetting methods. The use of simulated annealing and appropriate optimisation heuristics is also proposed and successful experimental results are obtained. It is shown that a search for error-free information sets is a necessary technique to be applied after the iterative error-correction algorithms to deal with high noise probabilities.

1 Introduction

Stream ciphers which generate pseudo-random sequences using the output of a number of linear feedback shift registers (LFSRs) combined by some nonlinear function, with or without memory, have long been proposed for use in secure communications. The purpose of nonlinear combiners is to produce a system which can withstand any practical cryptanalytic attack based on low linear complexity of the observed keystream sequence (see [13]) or on high linear correlation to individual LFSR sequences (see [14] and [5]). This paper considers the immunity of these combiners to fast divide and conquer correlation attacks [9]. The problem is to find the conditions under which it is possible to reconstruct the initial contents of individual shift registers using a segment of the keystream generator output sequence. Correlation attacks are based on the statistical dependence between the observed keystream sequence and a set of shift register sequences [5],[14]. If such an attack outperforms an exhaustive search over the initial contents of the corresponding shift registers, it is then called a *fast correlation attack.*

The concept of a fast correlation attack was first introduced in [9] where a bit-by-bit reconstruction procedure based on iterative probabilistic and threshold decoding is proposed (see also [15]). The underlying ideas regarding the probabilistic decoding of low density parity-check linear codes can be found in [4] and [8]. After the pioneering work of Meier and Staffelbach [9] and Zeng and Huang [15], various algorithms of this type have been published and theoretically or experimentally analysed (for example, see [1], [11] and [12]). All of them share two basic features, an iterative error-correction algorithm and a method of obtaining low density parity-checks. Here we present a number of modifications to improve the performance of the fast correlation attacks. These modifications are essentially based on discrete optimisation techniques, including simulated annealing, with appropriate optimisation heuristics.

The basic algorithm and its underlying concepts are described in Section 2. Modified algorithms with fast resetting, deterministic subset selection and simulated annealing are discussed in Sections 3, 4 and 5, respectively. Experimental results obtained by computer simulations are presented in Section 6. LFSR sequence reconstruction is explained in Section 7, and conclusions are given in Section 8. Readers unfamiliar with simulated annealing can find a brief introduction in the Appendix.

2 Basic Algorithm

2.1 Probabilistic Model

The fast correlation attack is based on the model (from [14]) shown in Figure 1. The observed keystream sequence $z = \{z_i\}_{i=1}^N$ is regarded as a noise-corrupted version of the LFSR sequence $a = \{a_i\}_{i=1}^N$, that is, $z_i = a_i + e_i$, where $e = \{e_i\}_{i=1}^N$ is a binary noise sequence. This model is called a Binary Symmetric Memoryless Channel, BSC. The correlation coefficient is defined as $c = 1 - 2p$. Let $f(x)$ be the feedback polynomial of a LFSR of length r. The problem is to reconstruct the LFSR sequence, $a = \{a_i\}_{i=1}^N$, from the observed keystream sequence, $z = \{z_i\}_{i=1}^N$, where $r < N < 2^r - 1$ and the value of N should be as small as possible.

In this paper we consider a decoding algorithm given in [11] which is an iterative procedure employing the parity-checks of possibly different weights and a Bayesian decision rule in error-correction for each bit. Each iteration consists of two main stages. In the first stage the parity-checks are calculated bit-by-bit. In the second stage, the Bayesian bit-by-bit error-correction is made based on the estimation of the relevant posterior probabilities obtained by using posterior probabilities from the previous iteration as the prior probabilities in the current iteration.

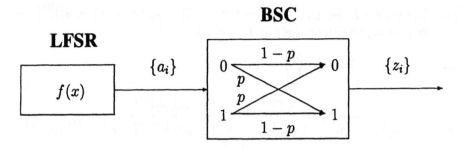

Figure 1: Model of a stream cipher as used in the correlation attack.

2.2 Parity-Checks

A parity-check is any linear relationship satisfied by a LFSR sequence. It is well known that the parity-checks correspond to polynomial multiples of $f(x)$ (see [1]). In [9] a simple algebraic technique is applied to derive a set of parity-checks. This technique involves repeated squaring of $f(x)$ and thus requires a long segment of the observed keystream sequence. To exploit this sequence more efficiently, one may apply a polynomial residue method to generate more parity-checks. Let $h(x)$ be a polynomial multiple of $f(x)$ of degree at most $M \geq r$ such that $h(0) = 1$. In order to find all the polynomial multiples of a certain weight (the number of non-zero terms) W and of degree at most M, the residues of the power polynomials x^m modulo $f(x)$, for all $r < m \leq M$, are first found. If any combination of $W - 1$ of the stored residues sums up to one, then the sum of the corresponding power polynomials plus one is a multiple of $f(x)$.

For larger M, one can use the meet-in-the-middle technique [9] or algorithms for computing discrete logarithms in fields of characteristic two [15]. A set of polynomial multiples such that no power polynomial appears in more than one polynomial multiple is called a set of orthogonal parity-check polynomials. If the exponents of all the power polynomials in all the parity-check polynomials correspond to a full positive difference set, it is possible to use all different phase shifts of the parity-check equations. In this case, for every bit, z_i, where $M \leq i \leq N - M$, the number of parity-checks is W times the number of polynomial multiples. In general, in order to preserve orthogonality some phases of some parity-check polynomials are not used.

2.3 Algorithm

Let $\Pi_i = \{\pi_k(i)\}$, $i = 1, 2, \ldots, N$, $k = 1, 2, \ldots, |\Pi_i|$, ($|\Pi_i|$ denotes the cardinality of Π_i), be a set of orthogonal parity-checks related to the ith bit, that are generated according to polynomial multiples of $f(x)$ as mentioned above. Assume that all the parity-check polynomials have the same weight W and let $w = W - 1$. Let a parity-check value be defined as the modulo 2 sum $c_k(i) = \sum_{l \in \pi_k(i)} z_l$. Assume that in the given model $c_k(i)$ and e_i are realisations of two binary random variables $C_k(i)$ and E_i, for $k = 1, 2, \ldots, |\Pi_i|$, and $i = 1, 2, \ldots, N$. Let $\Pr(E_i, \{C_k(i)\}_{k=1}^{|\Pi_i|})$ be the joint probability of the variables E_i and $C_k(i)$, for $k = 1, 2, \ldots, |\Pi_i|$, and let $\Pr(E_i|\{c_k(i)\}_{k=1}^{|\Pi_i|})$ be the corresponding posterior probability, for $i = 1, 2, \ldots, N$.

$$
p_i = \Pr(E_i = 1|\{C_k(i)\}_{k=1}^{|\Pi_i|} = \{c_k(i)\}_{k=1}^{|\Pi_i|})
$$

$$(1)$$

$$
= \frac{q_i \prod_{l=1}^{|\Pi_i|} q_l(i)^{\overline{c_l}(i)}(1 - q_l(i))^{c_l(i)}}{q_i \prod_{l=1}^{|\Pi_i|} q_l(i)^{\overline{c_l}(i)}(1 - q_l(i))^{c_l(i)} + (1 - q_i) \prod_{l=1}^{|\Pi_i|} (1 - q_l(i))^{\overline{c_l}(i)} q_l(i)^{c_l(i)}}
$$

where p_i and q_i respectively denote the posterior and prior probabilities for the current iteration, $\overline{c_l}(i) = 1 - c_l(i)$, $q_l(i) = (1 - \prod_{t=1}^{w}(1 - 2q_{m_t}))/2$ and $\{m_t\}_{t=1}^{w}$ denotes the set of indices of the bits involved in the parity-check $\pi_l(i)$, for any $l = 1, 2, \ldots, |\Pi_i|$ and $i = 1, 2, \ldots, N$.

In the first iteration the optimal Bayesian decision is made to minimise the symbol error-rate, p_e. In the succeeding iterations the error-rate almost always decreases for two reasons. Firstly, error-correction is introduced to the updated observed keystream sequence by the algorithm, and secondly, recycling (self-composition) of the probability vector in steps 1 – 6 of the algorithm described below cause the probabilities to decrease. Often the algorithm becomes trapped in a local minimum of the error-rate, $p_e \approx 0$, after which the error-correction of the observed keystream sequence ceases. In this case (as was proposed in [9]), when the error-rate is less than ϵ, one can substitute the posterior probability vector for the initial one, and continue from step 1. This is called *resetting* the algorithm. Resetting the algorithm enhances the error-correction capability of the algorithm and increases the number of satisfied parity-checks. The set of iterations between two successive resets is called a *round*. The basic error-correcting algorithm is as follows:

Basic Error-Correction Algorithm

- *Input:* The observed keystream sequence $z = (z_i)_{i=1}^{N}$, p and $\Pi_i = \{\pi_k(i)\}$, $i = 1, 2, \ldots, N$, $k = 1, 2, \ldots, |\Pi_i|$, a set of orthogonal parity-checks.

- *Initialization:* $j = 0$, $k = 0$ and $q_i = p$, $i = 1, \ldots, N$, where j is current iteration index and k is the current round index. Also define the maximum number of rounds k_{max}, the minimum error-rate ϵ and the maximum number of iterations without change in the number of satisfied parity-checks, J.

- *Resetting Criteria:* Probabilities are reset when the average error probability per symbol (error-rate) p_e drops below ϵ or when the number of satisfied parity-checks has not changed for J iterations.

- *Stopping Criteria:* The algorithm stops when the number of rounds reaches k_{max}, or, ideally, when all the parity-checks are satisfied.

- *Step 1:* Calculate the parity-checks, $c_l(i)$, for each bit z_i, $l \in \pi_k(i)$, $i = 1, \ldots, N$, of the observed keystream sequence. If all parity-checks are satisfied, go to step 7. If the number of satisfied parity-checks has not changed for J iterations, go to step 6.

- *Step 2:* Using equation (1) calculate the posterior probabilities $p_i, i = 1, \ldots, N$.

- *Step 3:* If $p_i > 0.5$ set $z_i = z_i \oplus 1$, $p_i = 1 - p_i, i = 1, \ldots, N$.

- *Step 4:* Substitute the posterior probabilities of the current iteration for the prior probabilities of the next iteration: $q_i = p_i$, for $i = 1, 2, \ldots, N$.

- *Step 5:* If $p_e = \frac{1}{N} \sum_{i=1}^{N} p_i > \epsilon$, increment j by 1 and go to step 1.

- *Step 6:* Set $p_i = p, i = 1, \ldots, N$, and increment k by 1. If $k < k_{max}$ go to step 1.

- *Step 7:* Set $a_i = z_i, i = 1, 2, \ldots, N$, and stop the procedure.

- *Output:* The reconstructed LFSR sequence is $\{a_i\}_{i=1}^{N} = \{z_i\}_{i=1}^{N}$.

3   Modified Algorithm with Fast Resetting

The underlying idea of resetting is explained in Section 2. By modifying this idea we have improved the power of resetting which has resulted in increased number of satisfied parity-checks. The resetting used in step 4 of the basic algorithm is called *slow resetting*. We introduce another type of resetting which is called *fast resetting*. Fast resetting is defined in the modified step 5 below.

Reconstruction of the observed keystream sequence is performed in step 3 of the basic algorithm when $z_i = z_i \oplus 1$. However, the complementations are not effective due to the posterior probability transformation $p_i \to 1 - p_i$ (see [16]). They become effective only after resetting, where such a transformation does not take place. Of course, not all of these complementations are correct and the algorithm may introduce new errors to the observed keystream sequence. If the number of correct complementations exceeds the number of incorrect complementations, then the probability of error is reduced in the observed keystream sequence. This may not occur if we wait till the error-rate goes to zero due to the self-composition property of the basic algorithm. Accordingly, one may expect the performance to be improved if the resetting is done before the error-rate falls below a threshold. More precisely, when the cumulative number of complementations in each round reaches a predefined value, C, we substitute $p_i = p, i = 1, \ldots, N$. This is called fast resetting, and significantly improves the performance of the basic algorithm.

- *Step 5:* If $p_e = \frac{1}{n} \sum_{i=1}^{N} p_i > \epsilon$ and the cumulative number of complementations is less than C, increment j by 1 and go to step 1.

In practice, it is possible to optimise C depending on the noise probability, the available keystream sequence or the parity-checks used.

4   Modified Algorithms with Deterministic Subset Selection

The objective of our experiments is to introduce modifications to improve the basic algorithm when it fails to reconstruct the LFSR sequence due to the high probability of noise and limited number of parity-checks. Updating the error probability vector and performing complementations in steps 3 and 4 of the basic algorithm reduce the average error probability. However, when the initial error probability is high, this reduction is mainly due to the self-composition property of the basic algorithm and not to the increase in the number of satisfied parity-checks. In order to avoid this situation and have more reliable decisions we can select a subset containing L significant positions, according to a suitable criterion, from the N positions in the observed keystream sequence. The prior probability vector is then updated only for these L positions. For the remaining $N - L$ positions the prior probability either remains the same (*partial update*) or is set to the initial error probability p (*partial reset*).

The following three criteria are suggested for use in selecting a subset of L significant positions in the observed keystream sequence:

1. Calculate $|p_i - q_i|$, $i = 1, \ldots, N$, in each iteration and select the L positions with the highest absolute difference.

2. Select the L positions with the lowest posterior probabilities, p_i.

3. Select the L positions with the highest number of satisfied parity-checks.

Step 4 of the basic algorithm is replaced by steps 4.1 and 4.2:

- *Step 4.1:* Select L significant positions with respect to a chosen criterion.

- *Step 4.2:* Substitute the posterior probabilities of the current iteration for the prior probabilities of the next iteration for the L significant positions, and the prior probabilities of the remaining $N - L$ positions either remain unchanged or are set to p.

5 Modified Algorithms with Simulated Annealing

An alternative to the deterministic approaches outlined above is to use a discrete optimisation technique such as simulated annealing with an appropriate optimisation heuristic. Techniques like this have proven to be useful tools in the analysis of a number of ciphers (see [2] and [3]). In this section we outline three different approaches for applying simulated annealing in a fast correlation attack on stream ciphers. These approaches are intended to enhance the basic correlation attack as outlined in Section 2. Readers unfamiliar with the simulated annealing algorithm should study the Appendix for a brief description.

First, a cost function (fitness measure or optimisation heuristic) is required. Here, experiments were performed using two different measures. Of course, the ideal cost function would measure the Hamming distance from the LFSR sequence to the proposed solution. To do so would be taking extreme liberties since this assumes knowledge of the answer before the algorithm even begins. Accordingly, the first cost function was chosen to be the error-rate, that is, the average of the probabilities in the error probability vector after complementations. Due to the self-composition property of the basic algorithm the error-rate tends to be minimised, and when the algorithm can correct almost all the errors, it provides an estimate of the relative number of errors in the observed keystream sequence. This gives an error prediction based on the probability that the bit in each position is correct. The second measure used was simply the number of unsatisfied parity-checks in the observed keystream sequence which is iteratively modified by the algorithm. It was found that this too gave a reasonable assessment of the fitness of a solution.

5.1 Random Subset Selection

The basic algorithm described in Section 2 updates every element of the probability vector in each iteration. When the noise is high, the algorithm becomes trapped in an area of local minimum and is unable to escape. For this reason, the first technique which we applied updated only certain elements of the error probability vector. At each temperature a fixed number of elements were picked at random. The probability for each element was updated and the new cost calculated. For each updated element of the updated probability vector, if the Metropolis criterion was satisfied then the change was accepted. Note that when the cost function is the number of unsatisfied parity-checks, the cost will only change if an updated element of the probability vector exceeds 0.5, causing the corresponding bit to be complemented.

5.2 Random Subset Complementation

The second method involves choosing a small number of random positions in the observed keystream sequence and complementing the corresponding bits, without transforming the probability $p_i \rightarrow 1 - p_i$, so that the complementations are effective. After complementing the bits, the parity-checks are recalculated and the probability vector updated. Typically the number of bits complemented is small (less than 10 out of 10,000).

5.3 Random Subset Complementation and Deterministic Subset Selection Combined

As a final option, a combination of random subset complementation and deterministic subset selection can be used. In this algorithm the significant elements of the probability vector are updated as in the modified algorithm with deterministic subset selection and a number of bits are complemented at random. The probability with which bits are complemented depends on the annealing temperature and decreases over time. This is done with the intention of reducing the number of incorrect bit complementations as the algorithm approaches the optimal solution.

6 Experimental Results

In our experiments we are dealing with a 31 bit LFSR with a primitive polynomial of weight three, an observed keystream sequence of length 10, 000 and five orthogonal parity-checks which correspond to a full positive difference set. The parity-checks had a maximum degree of 496 and were all of weight 3. Hence, we are using all three phases of the parity-checks for 90% of the observed keystream sequence.

Experiments regarding the modified algorithm with fast resetting show a considerable improvement over the basic algorithm with slow resetting for initial error probabilities less than 0.42. Figure 2 shows the average minimum Hamming distance for both the basic and modified algorithms. These results were obtained by averaging over 20 different noise samples with the threshold C equal to 1.

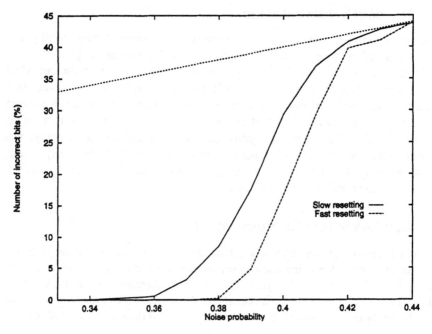

Figure 2: Comparison of basic algorithm with fast and slow resetting.

When the noise probability is high, the experiments have indicated that it is possible to gain an improvement over the modified algorithm with fast resetting by using a combination of the basic algorithm and fast partial resetting with deterministic subset selection of significant elements in the error probability vector. For the subset selection of L significant elements of the probability vector we use the second criterion proposed in Section 4. If the number of complementations reaches a predefined threshold C (for example, 1 or 100), the posterior probabilities of the remaining $N - L$ elements of the probability vector are set to p and the algorithm proceeds to the next iteration (partial resetting). Each round consists of a small number (for example, two) of subrounds of partial resetting. The motivation for this method is twofold. Firstly, it takes advantage of the recycling property of the basic algorithm which involves all the bits in the observed keystream sequence. Secondly, resetting is applied only to bit positions where the probability is not estimated to be significant. It can be seen from Figure 3 that fast partial resetting combined with deterministic subset selection gives an improvement over fast resetting for high noise probabilities, when averaged over 20 different noise samples.

In these experiments the number of significant elements, L, is 5000 and the cumulative number of complementations, C, is 1 or 100.

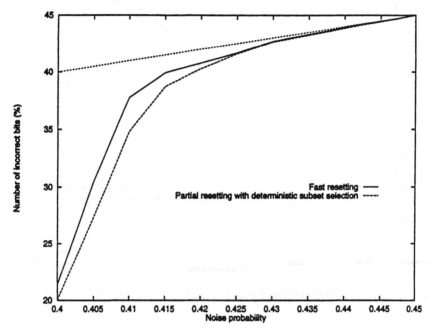

Figure 3: Comparison of fast resetting and partial resetting with deterministic subset selection.

Results for the algorithm adapted for simulated annealing were varied. It was found that a cost function based on the error-rate, rather than the number of satisfied parity-checks was far more effective. This is because the number of unsatisfied parity-checks does not correlate very well with the Hamming distance from the LFSR sequence. Also, the algorithm which updates probabilities at random outperforms the algorithm which complements bits at random.

Figure 4 shows a comparison of the basic algorithm with the one modified with simulated annealing. It can be seen that the improvement is considerable.

The annealing algorithms can also implemented with resetting as described above (i.e., both fast and slow resetting). As can be seen in Figure 5, the fast resetting shows a significant improvement over the slow resetting for the annealing method. The results shown in Figures 4 and 5 were obtained by averaging the minimum Hamming distance found for 25 different noise samples.

7 LFSR Sequence Reconstruction

A fast correlation attack is deemed successful if the entire LFSR sequence can be reconstructed. Applied in the first stage, the iterative error-correction algorithms from previous sections are expected to reduce the initial number of errors in the observed keystream

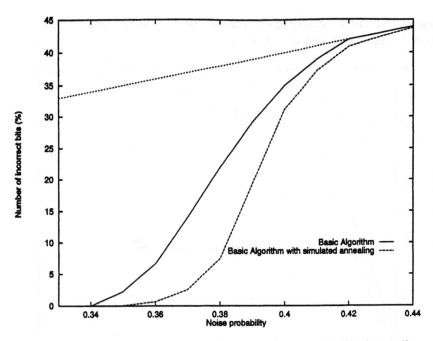

Figure 4: Basic algorithm compared with one modified with simulated annealing.

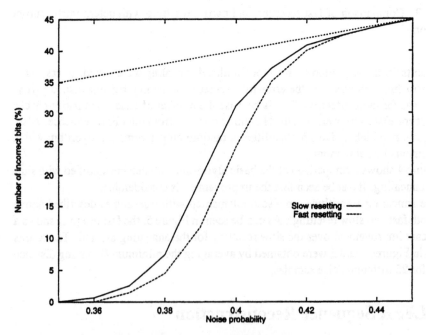

Figure 5: Comparison of fast and slow resetting using simulated annealing.

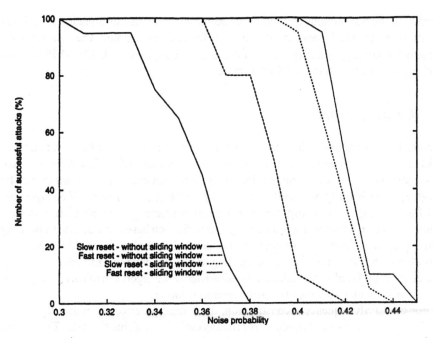

Figure 6: Success rates of LFSR sequence reconstruction.

sequence. In the second stage, the whole LFSR sequence has to be reconstructed. Ideally, if all the parity-checks are satisfied, then the modified observed keystream sequence is already error-free (with high probability). However, this happens only when the noise probability is relatively small. Note that the statistically optimal testing criterion is the Hamming distance between the reconstructed and the observed keystream sequence, see [14].

If some parity-checks are not satisfied, then the modified observed keystream sequence is not error-free. An efficient method to be used then is to search for an error-free *information set*, where an *information set* is any set of r linearly independent bit-positions in the LFSR sequence (r is the LFSR length). For any assumed information set, the whole LFSR sequence is generated and then compared with the observed keystream sequence by the Hamming distance criterion. One can also allow for a small number of random complementations in the modified observed keystream sequence. A sliding window technique without complementations making use of the information sets of r consecutive bits was suggested in [10]. It is important to note that even if the residual number of errors in the modified observed keystream sequence is large, the errors are typically patterned so that such error-free windows exist. In our experiments we used a simplified sliding window technique to search for windows consisting of 31 consecutive bits satisfying the optimal testing criterion mentioned above. For both fast and slow resetting, the search for an error-free sliding window is performed in each round, after the average number of satisfied parity-checks reaches a predefined value.

Figure 6 shows the success rate of the simplified sliding window technique as com- ·

pared with the one without any additional search for error-free information sets. These results were obtained by averaging over 20 different noise samples. The increase in the manageable noise probability is considerable (for example, from 0.30 to 0.39 for slow resetting and from 0.36 to 0.40 for fast resetting).

8   Conclusions

In order to improve the performance of iterative error-correction in fast correlation attacks proposed in [9] and [11], we have introduced a number of modifications with discrete optimisation flavour. The modifications help overcome the problems caused by a high error probability or a limited number of parity-checks. The modified algorithm with fast resetting considerably outperforms the slow resetting of the basic iterative algorithm. A clear improvement is achieved by a modification based on simulated annealing with appropriate optimisation heuristics. Better results are also obtained by a modified algorithm with fast partial resetting according to suitable deterministic selection of significant error probabilities. Such modifications allow the algorithm to move away from regions of local minima towards a better overall solution.

For the LFSR sequence reconstruction, these iterative error-correction algorithms should be combined with procedures for finding error-free information sets. To this end, a simplified sliding window technique is shown to be successful.

References

[1] V. Chepyzhov and B. Smeets, "On a fast correlation attack on stream ciphers," Advances in Cryptology - EUROCRYPT '91, *Lecture Notes in Computer Science*, vol. 547, D. W. Davies ed., Springer-Verlag, pp. 176-185, 1991.

[2] A. J. Clark, "Modern optimisation algorithms for cryptanalysis," in *Proceedings of ANZIIS '94*, pp. 258-262, 1994.

[3] W. S. Forsyth and R. Safavi-Naini, "Automated cryptanalysis of substitution ciphers," *Cryptologia*, vol. 17, no. 4, pp. 407-418, Oct. 1993.

[4] R. G. Gallager, "Low-density parity-check codes," *IRE Trans. Inform. Theory*, vol. IT-8, pp. 21-28, Jan. 1962.

[5] J. Dj. Golić, "Correlation via linear sequential circuit approximation of combiners with memory," Advances in Cryptology - EUROCRYPT '92, *Lecture Notes in Computer Science*, vol. 658, R. A. Rueppel ed., Springer-Verlag, pp. 113-123, 1993.

[6] J. Dj. Golić, "On the security of shift register based keystream generators," Fast Software Encryption - Cambridge '93, *Lecture Notes in Computer Science*, vol. 809, R. J. Anderson ed., pp. 91-101, 1994.

[7] S. Kirkpatrick, C. D. Gelatt, Jr. and M. P. Vecchi, "Optimization by simulated annealing," *Science*, vol. 220, no. 4598, pp. 671-680, 1983.

[8] J. L. Massey, *Threshold Decoding*. Cambridge, MA, MIT Press, 1963.

[9] W. Meier and O. Staffelbach, "Fast correlation attacks on certain stream ciphers," *Journal of Cryptology*, vol. 1, pp. 159-176, 1989.

[10] M. J. Mihaljević and J. Dj. Golić, "A fast iterative algorithm for a shift register initial state reconstruction given the noisy output sequence," Advances in Cryptology - AUSCRYPT '90, *Lecture Notes in Computer Science*, vol. 453, J. Seberry and J. Pieprzyk eds., Springer-Verlag, pp. 165-175, 1990.

[11] M. J. Mihaljević and J. Dj. Golić, "A comparison of cryptanalytic principles based on iterative error-correction," Advances in Cryptology - EUROCRYPT '91, *Lecture Notes in Computer Science*, vol. 547, D. W. Davies ed., Springer-Verlag, pp. 527-531, 1991.

[12] M. J. Mihaljević and J. Dj. Golić, "Convergence of a Bayesian iterative error-correction procedure on a noisy shift register sequence," Advances in Cryptology - EUROCRYPT '92, *Lecture Notes in Computer Science*, vol. 658, R. A. Rueppel ed., Springer-Verlag, pp. 124-137, 1993.

[13] R. A. Rueppel, "Stream Ciphers," in *Contemporary Cryptology: The Science of Information Integrity*, G. Simmons ed., pp. 65-134, New York, IEEE Press, 1991.

[14] T. Siegenthaler, "Decrypting a class of stream ciphers using ciphertext only," *IEEE Trans. Comput.*, vol. C-34, pp. 81-85, Jan. 1985.

[15] K. Zeng and M. Huang, "On the linear syndrome method in cryptanalysis," Advances in Cryptology - CRYPTO '88, *Lecture Notes in Computer Science*, vol. 403, S. Goldwasser ed., Springer-Verlag, pp. 469-478, 1990.

[16] M. Živković, "On two probabilistic decoding algorithms for binary linear codes," *IEEE Trans. Inform. Theory*, vol. IT-37, pp. 1707-1716, Nov. 1991.

Appendix

The Basics of Simulated Annealing

Simulated Annealing is a combinatorial optimisation technique first introduced in 1983 by Kirkpatrick, Gelatt and Vecchi [7]. They used the technique for deciding the optimal placement of components on an integrated circuit (IC) chip. The number of variables in such a problem can be huge, making determination of an optimal solution impossible. Simulated annealing scans a small area of the solution space in the search for the global minimum.

The Metropolis Algorithm

Simulated Annealing utilises a process known as the Metropolis Algorithm which is based on the equations governing the movement of particles in a gas or liquid between different energy states. Equation (2) describes the probability of a particle moving between two energy levels, E_1 and E_2, $\Delta E = E_2 - E_1$,

$$P(E) = e^{\frac{-\Delta E}{kT}} \tag{2}$$

where $\Delta E = E_2 - E_1$, k is Boltzmann's constant and T is the temperature. Equation (2) is used to make a decision as to whether or not a transition between different states should be accepted. The Metropolis Algorithm is based on

$$P(E) = \begin{cases} 1 & \text{if } \Delta E \geq 0 \\ e^{\frac{-\Delta E}{T}} & \text{if } \Delta E < 0 \end{cases} . \tag{3}$$

By allowing a configuration to move to a higher energy level (or higher cost), the search for the global minimum of the cost function is aided, since it is possible to move out of the regions of local minima. This is not the case for the so-called "iterative improvement" techniques which will only update the solution if a better (more optimal) one is found, or, in other words, if $\Delta E > 0$.

The Simulated Annealing Algorithm

1. Generate an initial solution to the problem (usually random).

2. Calculate the *cost* of the initial solution.

3. Set the initial temperature $T = T^{(0)}$.

4. For temperature, T, do many times:
   - Generate a new solution - this involves modifying the current solution in some manner.
   - Calculate the cost of the modified solution.
   - Determine the difference in cost between the current solution and the proposed solution.
   - Consult the Metropolis Algorithm to decide if the proposed solution should be accepted.
   - If the proposed solution is accepted, the required changes are made to the current solution.

5. If the stopping criterion is satisfied the algorithm ceases with the current solution, otherwise the temperature is decreased and the algorithm returns to step 4.

The Cooling Schedule

As mentioned above, the cooling schedule has three main purposes.

1. It defines the Initial Temperature. This temperature is chosen to be high enough so that all proposed transitions are accepted by the Metropolis Algorithm.

2. The cooling schedule also describes how the temperature is reduced. Although there are other methods, two possibilities are presented here.

   (a) An exponential decay in the temperature:

   $$T^{(k+1)} = \alpha \times T^{(k)} = \alpha^k \times T^{(0)}, \text{ where } 0 < \alpha < 1. \tag{4}$$

   Usually $\alpha \approx 0.9$ but can be as high as 0.99.

   (b) Linear decay: here the overall temperature range is divided into a number of intervals, say K.

   $$T^{(k+1)} = \frac{K - k}{K} \times T^{(0)}, \text{ where } k = 1, \ldots, K. \tag{5}$$

3. Finally, the cooling schedule indicates when the annealing process should stop. This is usually referred to as the *stopping criterion*. In the case where a linear decay is used the algorithm can be run for its K iterations, provided K is not too large. For the case where exponential decay is used, the process usually ceases when the number of accepted transitions at a particular temperature is very small (≈ 0).

Keyed Hash Functions

S. Bakhtiari*, R. Safavi-Naini, J. Pieprzyk**

Centre for Computer Security Research
Department of Computer Science
University of Wollongong, Wollongong
NSW 2522, Australia

Abstract. We give a new definition of keyed hash functions and show its relation with strongly universal hash functions and Cartesian authentication codes. We propose an algorithm for a secure keyed hash function and present preliminary result on its performance. The algorithm can be used for fast (about twice the speed of MD5) and secure message authentication.

1 Introduction

Hash functions were introduced in early 1950's [7]. The original aim was to have functions that can uniformly map a large collection of messages into a small set of *message digests* (or *hash values*). A useful application of hash functions is for error detection. Appending message digest to the message allows detection of errors during transmission. In the receiving end, the hash value of the received message is recalculated and compared with the received hash value. If they do not match, an error has occurred. This detection is only for random errors. An active spoofer may intercept a message, modify it as he wishes, and resend it appended with the digest recalculated for the modified message.

With the advent of public key cryptography and digital signature schemes, cryptographic hash functions gained much more prominence. Using hash functions, it is possible to produce a fixed length digital signature that depends on the whole message and ensures authenticity of the message. To produce digital signature for a message x, the hash value of x, given by $H(x)$, is calculated and then encrypted with the secret key of the sender. Encryption may be either by using a public key or a private key algorithm. Encryption of the hash value prevents active intruders from modifying the message and recalculating its checksum accordingly. It effectively divides the universe into two groups: outsiders who do not have access to the key of the encryption algorithm and hence cannot effectively produce a valid checksum, and insiders who do have access to the key and hence can produce valid checksums. We note that in a public key

---

* Support for this project was provided in part by the Ministry of Culture and Higher Education of Islamic Republic of Iran.
** Support for this project was provided in part by the Australian Research Council under the reference number A49530480.

algorithm, the group of insiders consists of only one member (the owner of the private key) and hence the encrypted hash value uniquely identifies the signer. In the case of symmetric key algorithms, both the transmitter and the receiver have access to the secret key and can produce a valid encrypted hash for an arbitrary message and therefore, unique identification based on the encrypted hash is not possible. However, an outsider cannot alter the message or the digest.

Hence, the hash functions can be used for modification detection of both random error and active spoofing and, in this sense, are parallel to error detection codes and authentication codes, respectively. In fact, hashing followed by encryption can be regarded as equivalent to Cartesian authentication codes [13] and hence the security criteria of the two should be equivalent. An important result of this observation is that we can give security criteria for keyed hash functions. In [10], Preneel identified keyed hash functions by relating them to *Message Authentication Codes (MAC)*, and in [4], a formal definition of keyed hash functions was given by Berson, Gong and Lamos. In this paper, we give a new definition of keyed hash functions (K-hash function), or authentication hash functions, which is consistent with authentication codes and Preneel's approach. The important property of this definition is that the security requirements are reduced to those required by authentication codes, which are much less stringent than what were required in [4], and can be obtained by a simple combination of hash functions and encryption functions. This allows development of a much faster algorithm which is equivalent (in terms of the proposed security criteria) to hashing followed by encryption.

In the next section we present our definition of K-hash functions and compare our security criteria with those of Berson, Gong and Lamos [4]. We also study the relation between Cartesian authentication codes, strongly universal classes of hash functions and K-hash functions. In Section 3, the design of a secure K-hash function is presented. Section 4 gives a rudimentary security analysis of the proposed K-hash function. Finally, in Section 5 we conclude the paper.

2   K-hash functions

As noted before, using a secret key in hashing, extends the error detection capability to the detection of active spoofing. That is, a message appended by its keyed hash value allows the owner of the key to validate authenticity of the message and detect modification due to either random noise or active spoofing. Protecting the hash value by a secret key is used by a number of authors. Hiding the initial vector of a hash function, appending a secret key to the message (to be hashed), and hashing followed by encryption are common techniques that are used for this purpose [2, 3]. However, the first attempt to formally define K-hash functions is due to Berson, Gong and Lamos.

Definition 1. (Berson, Gong and Lamos [4]) A function $g()$ that maps a key and a second bit string to a string of a fixed length is a *Secure Keyed One-Way Hash Function (SKOWHF)* if it satisfies five additional properties:

1. Given k and x, it is easy to compute $g(k, x)$.
2. Given k and $g(k, x)$, it is hard to compute x.
3. Given k it is hard to find two values x and y such that $g(k, x) = g(k, y)$, but $x \neq y$.
4. Given (possibly many) pairs x and $g(k, x)$, it is hard to compute k.
5. Without knowledge of k, it is hard to compute $g(k, x)$ for any x.

In our view this definition does not *minimally* capture the essential properties of a K-hash function. That is, because of the existence of a secret key in the hashing process, one may relax the above definition and still retain the main security properties required for a K-hash function.

Let V_n denote the n-dimensional vector space over $GF(2)$.

Definition 2. A K-hash function $H()$ is a class of hash functions $\{h_k : k \in V_n\}$ indexed by a key k such that $h_k() : M \rightarrow V_m$ generates a message digest of length m. $H()$ is a *Secure Keyed One-Way Hash Function (SKOWHF)*, if it satisfies the following properties:

1. The function $h_k()$ is *keyed one-way*. That is,
   (a) Given k and M, it is easy to compute $h_k(M)$.
   (b) Without knowledge of k, it is hard to,
      – find M when $h_k(M)$ is given;
      – find $h_k(M)$ when M is given.
2. The function $h_k()$ is *keyed collision free*. That is, without the knowledge of k it is difficult to find two distinct messages M and M' that collide under $h_k()$.
3. Given a set of t pairs $[M_i, h_k(M_i)]$, $1 \leq i \leq t$, it is difficult to find $h_k(M')$ for any other M', or M' for any other given $h_k(M')$. (These properties imply that it is difficult to find the key.)

We note that the two definitions are markedly different. In our definition, cryptographic properties of the SKOWHF relies on the secrecy of the key, while in Berson *et al.*'s definition, individual hash functions must have collision freeness and one-wayness properties. The relation between the two definition is stated in the following proposition.

Proposition 3. *A SKOWHF satisfying Definition 1 also satisfies Definition 2. (But the inverse is not true.)*

Proof: Let $h_k()$ satisfies Definition 1. Then,

– 1 in Definition 1 implies 1a in Definition 2;
– 2 and 5 in Definition 1 implies 1b in Definition 2;
– 3 in Definition 1 implies 2 in Definition 2;
– 4 in Definition 1 implies 3 in Definition 2. □

One of the main advantages of relaxing Definition 1 is that it motivates design of K-hash functions as a new primitive and not as a combination of other cryptographic primitives (such as hash function followed by encryption), and hence, could result in a more efficient algorithm. The hash function presented in Section 3 justifies this argument.

2.1 Universal Hash Functions and K-Hash Functions

Universal hash functions were defined by Carter and Wegman [5] in an attempt to provide an input independent average linear time algorithm for storage and retrieval of keys in associated memories.

A class H of functions from a set A to a set B is called *universal$_2$*, if for all x and y in A, $\delta_H(x, y) \leq |H|/|B|$, where $|H|$ and $|B|$ are the sizes of H and B, respectively, and $\delta_H(x, y)$ denotes the number of functions $h \in H$ with $h(x) = h(y)$. In [15], they extended their work and defined *strongly universal$_n$* and *almost strongly universal$_2$*, and showed their application to authentication.

A class H of hash functions is strongly universal$_n$ (SU_n) if given any n distinct elements $a_1, \cdots, a_n$ of A and any n (not necessary distinct) elements $b_1, \cdots, b_n$ of B, then $|H|/(|B|^n)$ functions take a_1 to b_1, a_2 to b_2, etc. SU_n hash functions can be used for multiple authentication.

2.2 SU_n, r-fold security, and K-Hash Functions

A *Cartesian authentication code* is a collection $\mathcal{E}$ of mappings from a set S of source states to a set T of tags ($|T| = q$); that is, $e : S \rightarrow T$. A mapping $e \in \mathcal{E}$ is called an encoding rule and determines a subset of $S \times T$ as the valid codewords. The codeword corresponding to the source state $s \in S$ is $s \parallel t$, where $t \in T$ and $\parallel$ denote string concatenation. The transmitter and the receiver share a secret encoding rule and this allows the transmitter to construct valid codewords and the receiver to verify validity of a codeword. The enemy does not know the secret encoding rule. In a *spoofing of order r attack*, the enemy uses his knowledge of r intercepted codewords to construct another valid codeword. The code provides *perfect protection against spoofing of order r attack* and enemy's best chance of success in the attack is $P_r = 1/q$. A Cartesian authentication code provides *r-fold security* if $P_i = 1/q$, $i = 0, \cdots, r$. Cartesian A-codes with r-fold security are equivalent to orthogonal arrays [12].

In [14], van Trung has shown that SU_n's are equivalent to orthogonal arrays and hence we have the following proposition.

Proposition 4. *Cartesian A-codes with n-fold security are equivalent to SU_n's.*

A number of constructions for SU_2 were proposed by Wegman and Carter in [5]. In [15], they also showed a construction of SU_n using polynomials over finite fields. Other constructions of SU_n's can be obtained using constructions of orthogonal arrays. All such constructions are known to be inefficient as they require exponentially many (in terms of the number of tags) encoding rules.

K-hash functions can be seen as parallel to SU_n hash functions in the context of computational security. The first property in Definition 2 ensures efficient forward computation while the second property emphasizes the confusion introduced by the key and finally, the third property is the basic security requirement for multiple authentication.

2.3   Applications of Keyed Hash Functions

K-hash functions are usually used for the applications that provide authentication and require shared secret key between participants. *Message authentication between two or among multiple parties, password checking,* and *software protection* are examples of these applications [2].

Another important application is the construction of other primitives, such as encryption functions, from K-hash functions. In [2], we have presented an encryption algorithm which is based on the scheme that was proposed by Luby and Rackoff in [8]. We have used a K-hash function to provide the one-wayness (based on a secret key).

2.4   Construction of Keyed Hash Functions

Construction of keyed hash functions from encryption functions is the most straightforward approach. However, such construction might not necessarily satisfy the security requirements of K-hash functions. General constructions for keyed hash functions are *construction from existing hash functions, construction from block ciphers,* and *construction from scratch* (cf. [2]). An example of the first construction is given in [3] and it is shown that the resulting algorithm satisfies Definition 1. Different examples of the construction from block ciphers are given in [10] and [16]. This paper gives an example of the construction from scratch (Section 3). The benefit of this approach is the increased speed that is gained by effectively using the secret key in the hashing process.

2.5   Methods of Attack on Keyed Hash Functions

The attacks on hash functions are divided into two major groups: attacks specific to the hashing algorithm, and probabilistic attacks which are applicable to any hashing algorithm. Examples of the methods of attack which are based on the weaknesses of specific hash function, are *Meet in the Middle Attack, Correcting Block Attack, Fixed Point Attack, Differential Cryptanalysis,* and *Linear Cryptanalysis.* Efficiency of probabilistic attacks depend on the message digest length and the key length. *Birthday Attack, Exhaustive Key Search, Random Attack,* and *Padding Attack* are examples of such attacks. [2]

In keyed hash functions, since a secret key contributes to the hashing process, the methods of attack on the secret key should be included. If the cryptanalyst can find a method to extract the secret key, the system is entirely compromised (during the key life time). There might also be some weaknesses in the algorithm that allow the intruder to bypass the secret key. Existence of weak keys is one example of this case, where the algorithm results in the same digest for several keys (cf. weak keys in DES [9]). The intruder can then guess the key, since the key domain is shrunk by some factors. Therefore, design of the keyed hash functions needs careful consideration.

2.6 Requirements for Keyed Hash Functions

We consider the following properties for a SKOWHF.

- *Security Requirements:*
  1. It should satisfy the requirements of Definition 2.
  2. It should use a secret key of at least 128 bits to thwart exhaustive key search.
  3. It should produce a message digest with at least 128 bits to thwart birthday attacks.
  4. It should uniformly distribute the message digest in the message digest space. This thwarts statistical attacks.
- *Design Heuristics:*
  1. It should use the secret key many times in the hashing process — especially at the beginning and the end.
  2. It should use every bit of the message several times in the hashing process (redundancy of the message).
  3. The underlying round function (or step function) should be analyzed very carefully to thwart the attacks that are based on the weaknesses of the round function, such as fixed point attack [10]. It needs special care when the design is based on an existing algorithm, such as an encryption algorithm or a hash function. (Sometimes, it is preferred to construct a keyed hash function from scratch to avoid these problems.)
- *Operational Requirements:*
  1. The secret key must remain secret among the trusted parties. Usage of a symmetric key cannot guarantee the security of the algorithm against attacks from the trusted users.
  2. If the keyed hash function is used in a more general purpose scheme (a protocol), the security of the algorithm should be checked in the new environment. In some cases, other security parameters should be included in the higher level structures. Use of time stamps and serial numbers are examples of such parameters. Use of different keys for different purposes (eg. authentication and/or secrecy) is another example.

Other general requirements for keyed hash functions are *Complementation Freedom, Addition Freedom, Multiplication Freedom, Correlation Freedom*. The reader is referred to [1] for more details about these general requirements.

3 Design of a New Keyed Hash Function (KHF)

In this section, construction of a new keyed hash function *(KHF)* is discussed. Desirable properties of this design are gained by including important features of MD5 [11] and HAVAL [17], nonlinear boolean functions that were proposed in [6], and some additional properties such as variable rotations (cf. Section 3.3). The nonlinear boolean functions, that are used in this design, are (*C language* notations):

$$
f_i(A, B, C, D, E) = \begin{cases}
(A \ \& \ E)\,\hat{}\,(B \ \& \ C)\,\hat{}\,((B\,\hat{}\,C) \ \& \ D), & \text{if } i = 1 \bmod 5. \\
A\,\hat{}\,(B \ \& \ (A\,\hat{}\,D))\,\hat{}\,(((A \ \& \ D)\,\hat{}\,C) \ \& \ E), & \text{if } i = 2 \bmod 5. \\
A\,\hat{}\,(C \ \& \ D \ \& \ E)\,\hat{}\,((A \ \& \ C) \mid (B \ \& \ D)), & \text{if } i = 3 \bmod 5. \\
B\,\hat{}\,((D \ \& \ E) \mid (A \ \& \ C)), & \text{if } i = 4 \bmod 5. \\
D\,\hat{}\,E\,\hat{}\,(((D \ \& \ E)\,\hat{}\,A) \ \& \sim (B \ \& \ C)), & \text{if } i = 0 \bmod 5.
\end{cases}
$$

They are functions of five variables and have the following important properties:

- They are 0-1 balanced.
- They satisfy the Strict Avalanche Criterion (SAC).
- They are highly nonlinear.
- They are pairwise linearly non-equivalent.
- They are far from each other, in terms of the Hamming distance.
- They can be described by short Algebraic Normal Forms (ANF).

KHF takes an arbitrary length message and a secret key (with variable length, between 128 and 384) to compute a secure 128-bit message digest. (In general, we assume that the key length is 384 bits and is presented by three 128-bit buffers $K_X = (X_1, X_2, X_3, X_4)$, $K_Y = (Y_1, Y_2, Y_3, Y_4)$, and $K_Z = (Z_1, Z_2, Z_3, Z_4)$.) The algorithm is software oriented and is designed to be fast on 32-bit register machines, such as SUN SPARC stations. Similar to MD5, this algorithm does not require any large substitution tables (S-boxes).

3.1 Notations

- $\overline{\oplus}$ is an special exclusive-or, where the operands may have different lengths. The XOR will be performed bitwise, starting from the left most bits.
- $\pi_{m,n}$ is a substring from the mth to the nth binary digits of π. For instance, $\pi_{2,5} = 1001$, where $\pi = 11.001001000011\cdots$ (in binary).
- $\parallel$ denotes concatenation.

3.2 Description of the Algorithm

KHF consists of four steps. The first step is to pad the input message by the message length, a fixed block, a part of the key, and a string of a '1' bit followed by some '0' bits; the second step is to initialize the buffers for message and digest chains; the third step is to apply the round function to all message blocks (32 bits each); and the fourth step is to extract the message digest from the digest chain. These steps are described in the following paragraphs. The following three functions are used in this algorithm:

$$add(X, Y, \ldots) = (X + Y + \cdots) \bmod 2^{32},$$

$$rol(X, s) = \text{Rotate left } X \text{ by } s \text{ bits,}$$

$$shr(X, s) = \text{Shift right } X \text{ by } s \text{ bits } (shr(X, s) = X \gg s).$$

Step 1: Adding the Length and Padding Bits Let T denotes the left most 640 binary digits (bits) of π ($T = \pi_{1,640}$) and L denotes the length of an arbitrary input message M. First, $(L \oplus T)$ is pre-pended to the message. This increases the message length by 640 bits. The message is then padded by a single '1' bit followed by enough '0' bits such that the padded message length (in bits) is a multiple of 32. (This padding is at least one bit and at most 32 bits.) Next, the 128-bit key buffer $K_Z = (Z_1, Z_2, Z_3, Z_4)$ is appended to the padded message. Finally, $(L \oplus T)$ is again appended to the previous result. M_i, $1 \leq i \leq n$, denotes the ith block (32 bits) of the final padded message, where n (≥ 45) is the number of blocks. Figure 1 shows the final padded message.

Step 2: Initialization A 4-word (128-bit) buffer $IV_X = (X_1, X_2, X_3, X_4)$, which is used to keep the message digest (in digest chain), is initialized to the 128-bit key K_X. Another 4-word buffer $IV_Y = (Y_1, Y_2, Y_3, Y_4)$, which is used to keep the message block (in message chain), is initialized to the 128-bit key K_Y. A 16-word (512 bits) buffer $B = (B_1, \ldots, B_{16})$, which is used to create message redundancy in the round function, is initialized to $\pi_{641,1152}$.

Step 3: Processing the Message in 32-bit Blocks Processing of the message can be described as follows (T_1, T_2, and T_3 are temporary variables, and '=' denotes assignment):

Do the followings, for $i = 1, \ldots, n$.
- $T_1 = f_i(Y_1, Y_2, Y_3, Y_4, M_i)$.
- $T_2 = shr(T_1, 16) \oplus T_1$.
- $T_3 = rol(\ add(X_4, T_1)\ , (T_2 \bmod 32)\ )$.
- $X_4 = X_1$, $X_1 = X_2$, $X_2 = X_3$, $X_3 = T_3$.
- $T_2 = shr(T_2, 3)$.
- $B_{(T_2 \bmod 16)+1} = B_{(T_2 \bmod 16)+1} \oplus M_i$.
- Do the followings, for $j = 1, \ldots, 4$.
  - $T_2 = shr(T_2, 2)$.
  - $Y_j = rol(\ add(Y_j, B_{((i-1) \bmod 16)+1})\ , (T_2 \bmod 32)\ )$.
- $T_3 = Y_4$, $Y_4 = Y_1$, $Y_1 = Y_2$, $Y_2 = Y_3$, $Y_3 = T_3$.

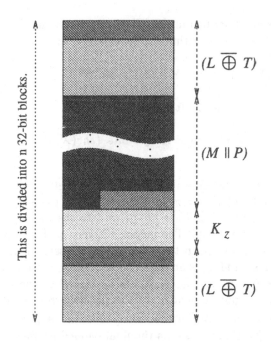

Fig. 1. *Message padding in KHF, where L is the length of the message M, T is a fixed 640-bit block, P is a '1' followed by enough '0's, and K_z is a part of the secret key. The notations '$\|$' and '$\overline{\oplus}$' are, in turn, concatenation and an special XOR operations.*

Step 4: Message Digest The computed message digest is the 4-word (128-bit) output $MD = (X_1, X_2, X_3, X_4)$.

Figure 2 illustrates one round (step) of KHF. In the implementation, the above description can be simplified such that the software becomes more efficient. For instance in the last round $(i = n)$, the following commands can be omitted:

- $T_2 = shr(T_2, 3)$.
- $B_{(T_2 \bmod 16)+1} = B_{(T_2 \bmod 16)+1} \oplus M_n$.
- Do the followings, for $j = 1, \ldots, 4$.
  - $T_2 = shr(T_2, 2)$.
  - $Y_j = rol(\ add(Y_j, B_{((n-1) \bmod 16)+1})\ ,\ (T_2 \bmod 32)\ )$.
- $T_3 = Y_4,\ Y_4 = Y_1,\ Y_1 = Y_2,\ Y_2 = Y_3,\ Y_3 = T_3$.

In our design, we have not used the chaining technique that has been used in most dedicated hash functions. That is, the buffer (X_1, X_2, X_3, X_4) is not added to the result of each round. This thwarts some attacks, when IV_X is initialized to a secret key and a short message is used. In general, five possibilities for the key buffers K_X, K_Y, and K_Z are suggested:

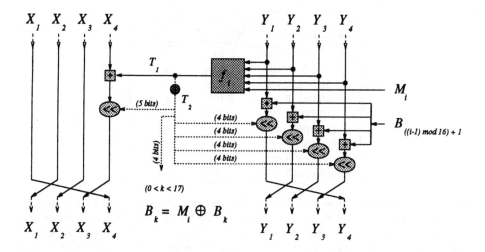

Fig. 2. *One round of KHF. The left chain is digest chain and the right chain is message chain.*

1. K_X, K_Y, and K_Z are all secret keys with $K_X = K_Y = K_Z$ (total key bits $= 128$).
2. K_X is a fixed (public) initial vector and K_Y and K_Z are secret keys with $K_Y = K_Z$ (total key bits $= 128$).
3. K_X is a fixed (public) initial vector and K_Y and K_Z are different secret keys (total key bits $= 256$).
4. K_X, K_Y, and K_Z are all secret keys with $K_X = K_Z$ (total key bits $= 256$).
5. K_X, K_Y, and K_Z are all secret keys with $K_Y = K_Z$ (total key bits $= 256$).
6. K_X, K_Y, and K_Z are different secret keys (total key bits $= 384$).

If the second or the third possibility is chosen, the chaining technique (that was mentioned above) can be added to this design. For instance, the buffer (X_1, X_2, X_3, X_4) can be saved in a temporary buffer to be added (or XORed) to (X_1, X_2, X_3, X_4) after 64 rounds. Again, the result should be saved in a temporary buffer to perform the same procedure for the remaining message blocks.

3.3 Properties of KHF

The following is a brief outline of the properties of this design.

- Block size is only 32 bits long. Therefore, the padded string of '1' followed by '0's will be at most 32 bits.
- Length of the message M is added to the two ends to thwart some attacks that were mentioned in Section 2.5. Pre-pending the length (L) to the message, in particular, thwarts padding attack.
- Rotations are variable. They depend on the result of $f_i()$ functions.

File	Time (seconds)		Rate
Size (bytes)	KHF	MD5	T_{MD5}/T_{KHF}
10	0.100	0.100	1.00
1,000	0.100	0.100	1.00
10,000	0.100	0.100	1.00
50,000	0.100	0.104	1.04
100,000	0.129	0.205	1.59
500,000	0.428	0.700	1.64
1,000,000	0.802	1.311	1.64
5,000,000	3.624	6.535	1.80
10,000,000	7.026	12.998	1.85

Table 1. *Comparison between KHF and MD5, where different file sizes are examined. Each entry is tested 100 times to get the average execution time. The result shows that KHF is faster than MD5 when the file is large. The inclusion of the fixed prefix and suffix makes the algorithm almost the same as MD5 when the file is very short. In this test, the source codes were in C and have been executed on a SUN SPARC station.*

- $f_i()$ functions, themselves, have properties that are very important in this design (especially the nonlinearity).
- Each message block M_i is used several times in the process, to make redundancy. The desirable advantage is the addition of M_i to an unknown step (depending on the result of $f_i()$ functions).
- The digest chain (X_1, X_2, X_3, X_4) does not have a direct (linear) connection to the message chain (Y_1, Y_2, Y_3, Y_4). Therefore, the adversary cannot control the content of the digest chain by choosing or changing the message.
- Different possibilities for the key length are available (128, 256, and 384 bits).
- The structure can be easily extended so that it can be used for 64-bit register machines, and/or it can produce 160-bit message digest. For the earlier one, each block should be 64 bits (instead of 32 bits). For the latter one, a new path should be added to the digest chain (ie. $IV_X = (X_1, X_2, X_3, X_4, X_5)$).
- The 640-bit block added to the beginning and the end of the message thwarts attacks such as correcting block attack (even for the key holders). In particular, the prefix removes any specific format or structure that the secret key values (K_X and K_Y) might have. The suffix makes sure that the last message blocks are used more than once in the evaluation of the digest. (We need at least 16 extra rounds to use all elements of the buffer B which holds the last message blocks.)
- KHF is very fast. Table 1 gives a comparison between KHF and MD5.

We have tried to remove the weaknesses that may allow the key holders to attack the algorithm. However, this protection is not guaranteed, since Definition 2 does not require it.

4 Security Analysis of KHF

In this section applicability of different attacks, including the attacks in Section 2.5, on KHF is examined.

The basic property of a hash function is its random behavior. A hash function should uniformly distribute the message digest in the message digest space (for randomly chosen inputs). We have tested this property in different forms.

1. We have shown that, if a user randomly choose a message M and then form another message M' by flipping a randomly chosen bit of M, the difference between the corresponding digests MD and MD', which is shown by $MD \oplus MD'$, has, on average, the same number of zeros and ones. In other words, flipping one bit of a randomly chosen message results in a completely different message digest, on average.

2. Similarly, one bit change in the secret key results in a completely different message digest. In other words, if K and K' differ only in one bit, the difference between the corresponding digests $(MD \oplus MD')$ has almost the same number of zero and one bits, for a randomly chosen message.

Assuming the above basic properties, and the properties in Section 3.3, different possible attacks are thwarted as follows.

- Because KHF produces a 128-bit message digest, it is secure against Birthday Attack and Meet in the Middle Attack.
- The key length of at least 128 bits thwarts Exhaustive Key Search.
- Random Attack is unsuccessful, because message digest is 128 bits.
- Padding attack is thwarted by pre-pending the message length to the message.
- XORing the message block to the message chaining variable (X_1, X_2, X_3, X_4) of an unknown round provides redundancy in the message block and thwart Correcting Block Attack.
- By both making redundancy in the message blocks (same as above) and adding the message length to the two end of the message, Fixed Point Attack is thwarted.
- $f_i()$ functions provide high nonlinear relationship between the message and the message digest. Therefore, Linear Cryptanalysis is unsuccessful.

Note that, attacks depending on the algorithm are even less successful for the outsiders (who do not know the key). In this case, the intermediate values of the process of keyed hash function are unknown. This prevents the enemy to analyze the components of the algorithm.

5 Conclusion

In this paper, we presented a new definition for keyed hash functions and discussed its relation to authentication codes and universal hash functions. Our definition relaxes Berson *et al.*'s definition and allows design of secure and fast keyed hash functions. The security of the keyed hash functions in our definition heavily relies on the secrecy of the key. This means that the keyed hash function does not need to provide security against cheating insiders. For example, by not requiring collision freeness property for individual hash functions, a transmitter may deny a registered message, and claim another pair of [message , digest] by finding a collision and replacing them. However, by insisting on this property, as in Berson *et al.*'s definition, extra complexity must be provided in the algorithm which results in a slow function. We should emphasize that the main objective in using a keyed hash function is providing security against outsiders and so this extra complexity is not justified. In section 3, we have studied applications, security requirements, and constructions of K-hash functions and proposed a new K-hash function (KHF) that has significant properties and is suitable for fast and secure message authentication.

We gave a brief analysis of the security of KHF against a wide range of commonly used attacks on hash functions. KHF provides flexible security, since the user is free to choose different key sizes: 128, 256, or 384 bits. Its performance is compared with MD5 and is shown to be a very fast and practical keyed hash function. In summary, KHF is an efficient SKOWHF.

References

1. R. Anderson, "The Classification of Hash Functions," in *Codes and Cuphers - Proceedings of Cryptography and Coding IV*, (Essex, UK), pp. 83–93, Institute of Mathematics and its Applications (IMA), 1995.
2. S. Bakhtiari, R. Safavi-Naini, and J. Pieprzyk, "Cryptographic Hash Functions: A Survey," Tech. Rep. 95-09, Department of Computer Science, University of Wollongong, July 1995.
3. S. Bakhtiari, R. Safavi-Naini, and J. Pieprzyk, "Practical and Secure Message Authentication," in *Series of Annual Workshop on Selected Areas in Cryptography (SAC '95)*, (Ottawa, Canada), pp. 55–68, May 1995.
4. T. A. Berson, L. Gong, and T. M. A. Lomas, "Secure, Keyed, and Collisionful Hash Functions," Tech. Rep. (included in) SRI-CSL-94-08, SRI International Laboratory, Menlo Park, California, Dec. 1993. The revised version (September 2, 1994).
5. J. L. Carter and M. N. Wegman, "Universal Class of Hash Functions," *Journal of Computer and System Sciences*, vol. 18, no. 2, pp. 143–154, 1979.
6. C. Charnes and J. Pieprzyk, "Linear Nonequivalence versus Nonlinearity," in *Advances in Cryptology, Proceedings of AUSCRYPT '92*, pp. 156–164, Dec. 1992.
7. D. Knuth, *The Art of Computer Programming: Searching and Sorting*, vol. 3. Addison-Wesley, 1973.
8. M. Luby and C. Rackoff, "How to Construct Pseudorandom Permutations and Pseudorandom Functions," *SIAM Journal on Computing*, vol. 17, pp. 373–386, Apr. 1988.
9. National Bureau of Standard, *Data Encryption Standard*. FIPS publication 46, June 1977. U. S. Department of Commerce.
10. B. Preneel, *Analysis and Design of Cryptographic Hash Functions*. PhD thesis, Katholieke University Leuven, Jan. 1993.

11. R. L. Rivest, "The MD5 Message-Digest Algorithm." RFC 1321, Apr. 1992. Network Working Group, MIT Laboratory for Computer Science and RSA Data Security, Inc.
12. R. Safavi-Naini and L. Tombak, "Combinatorial Characterization of A-Codes with r-Fold Security," in *Advances in Cryptology, Proceedings of ASIACRYPT '94*, vol. 917 of *Lecture Notes in Computer Science (LNCS)*, pp. 211–223, Springer-Verlag, 1995.
13. G. J. Simmons, "Authentication Theory / Coding Theory," in *Advances in Cryptology, Proceedings of CRYPTO '84*, vol. 196 of *Lecture Notes in Computer Science (LNCS)*, pp. 411–431, Springer-Verlag, 1985.
14. T. V. Trung, "Universal Hashing and Unconditional Authentication Codes," in *Proceedings of the IEEE International Symposium on Information Theory*, p. 228, 1993.
15. M. N. Wegman and J. L. Carter, "New Hash Functions and Their Use in Authentication and Set Equality," *Journal of Computer and System Sciences*, vol. 22, pp. 265–279, 1981.
16. Y. Zheng, *Principles for Designing Secure Block Ciphers and One-Way Hash Functions*. PhD thesis, Electrical and Computer Engineering, Yokohama National University, Dec. 1990.
17. Y. Zheng, J. Pieprzyk, and J. Seberry, "HAVAL - A One-Way Hashing Algorithm with Variable Length of Output," in *Advances in Cryptology, Proceedings of AUSCRYPT '92*, vol. 718 of *Lecture Notes in Computer Science (LNCS)*, (Queensland, Australia), pp. 83–104, Springer-Verlag, Dec. 1992.

Some Active Attacks on Fast Server-Aided Secret Computation Protocols for Modular Exponentiation

Hwang, Shin-Jia*; Chang, Chin-Chen**; and Yang, Wei-Pang*

*Institute of Computer and Information Science, National Chiao Tung University,
Hsinchu, Taiwan 300, R. O. C., E-mail: hwangsj@winston.cis.nctu.edu.tw
**Institute of Computer Science and Information Engineering,
National Chung Cheng University, Chiayi, Taiwan 621, R.O.C., E-mail: ccc@cs.ccu.edu.tw

Abstract. Four server-aided secret computation protocols, Protocols 1, 2, 3, and 4, for modular exponentiation were proposed by Kawamura and Shimbo in 1993. By these protocols, the client can easily compute the modular exponentiation M^d mod N with the help of a powerful server, where N is the product of two large primes. To enhance the security, the client was suggested to use a verification scheme and a slight modification on each proposed protocol. In this paper, we propose two new active attacks to break Protocols 3 and 4, respectively. Even if Protocols 3 and 4 have included the slight modification and verification, the untrusted server can still obtain the secret data d. The client cannot detect these attacks by the proposed verification. To adopt these new attacks, the difficulty of finding the value of the secret data d will be decreased drastically.

Keywords: server-aided secret computation protocol, active attack, cryptography, modular exponentiation

1   Introduction

In 1988, Matsumoto et al. first proposed the concepts of server-aided secret computation protocols. In a server-aided secret computation protocol, there are two kinds of members. One is the server with huge computation power, while the other is the clients with small computation power. Through the protocol, a client can easily perform a time consuming function with the help of the server without releasing the secret data of the client.

The server-aided secret computation protocols are important in the public key cryptography. In the public key cryptosystem, the modular exponentiation is the most popular operation. For example, in the RSA public key cryptosystem [Rivest et al. 1978], the decryption operation or signature function is the modular exponentiation M^d mod N, where d is the secret key and N is the product of two large primes. The amount of computations of the modular exponentiation is rather large, so the client can not perform the modular exponentiation with a reasonable speed. Many protocols were proposed to help the client to compute the modular exponentiation M^d mod N. Those protocols are divided into two classes: dependent protocols and independent protocols.

There were many dependent protocols [Matsumoto et al. 1988, Laih et al. 1991, and Laih and Yen 1992]. The characteristic of these protocols is that there exists some relation between the secret data and the transmitted messages. Since the server has chance to derive the secret data d from the transmitted messages, the security of dependent protocols is not clear. Quisquater and Soete [1989] proposed the first independent protocol. In the independent protocol, the secret data and the transmitted messages are independent. In 1993, Kawamura and Shimbo proposed four fast protocols which supply more effect server-aided secret computation than that of Quisquater and Soete's protocol. Because the transmitted messages and the secret data of the client are independent, the independent protocols are more secure than the dependent protocols.

In [Kawamura and Shimbo 1993], the four protocols are announced to be secure. Including a slight modification and a verification mechanism, the four protocols were against the proposed active attacks [Shimbo and Kawamura 1990, Pfitzmann and Waidner 1992]. Unfortunately, we will propose two active attacks on Protocols 3 and 4, respectively, when N is the product of two large primes. In Section 2, Protocols 3 and 4 are stated. In the same section, the verification mechanism and the slight modification are introduced. Then the two active attacks are proposed in Section 3. Finally, the conclusions are given in Section 4.

2   Protocols Descriptions

The descriptions of Protocols 3 and 4 proposed by Kawamura and Shimbo [1993] are given. Since Protocol 3 is based on Yao's exponentiation evaluation scheme [1976], the Yao's scheme is described first. Then Protocol 3 is given next. Since Protocol 4 is based on Knuth's exponentiation evaluation scheme [1981], the description of Knuth's scheme is stated prior to Protocol 4.

Assume that M^d mod N is the modular exponentiation and k is the bit size of a window, where N is a large prime number or the product of two large primes. Let $n = \lceil \log_2 N \rceil$ and $b = 2^k$. In Yao's scheme, the exponent d is divided into many small k-bit windows $d_0, d_1, ..., d_{l-1}$, from the least signification bit (LSB) to the most signification bit (MSB), where $l = \lceil n/k \rceil$. Then $d = \sum_{i=0}^{l-1} d_i (b)^i = \sum_{j=1}^{b-1} [j(\sum_{d_i=j} b^i)]$. So

$M^d \bmod N$ is computed as $M^d \equiv \prod\limits_{j=1}^{b-1} [\prod\limits_{d_i=j} M^{(b^i)}]^j \equiv \prod\limits_{j=1}^{b-1} (z_j)^j \pmod{N}$, where $z_j =$

$[\prod\limits_{d_i=j} M^{(b^i)}]$ for $j=1, 2, ..., b-1$. Based on this approach, Protocol 3 was proposed.

Protocol 3

Step 0: [Precomputation of the client]

 Step 0-1: Generate $b-1$ random positive integers $r_1, r_2, ..., r_{b-1}$, where $r_j \in Z_N$, for $j=1, 2, ..., b-1$. Here Z_N denotes the set $\{1, 2, ..., N-1\}$.

 Step 0-2: Compute $R = \prod\limits_{j=1}^{b-1} (r_j)^j \bmod N$.

 Step 0-3: Compute R^{-1} such that $R(R^{-1}) \bmod N = 1$.

Step 1: [Client] Send the message $\{M, N\}$ to the server.

Step 2: [Server] After receiving $\{M, N\}$, execute the following substeps:

 Step 2-1: $y_0 = M$.

 Step 2-2: For $i= 1, 2, ..., l-1$, compute $y_i = (y_{i-1})^b \bmod N$, where $y_0 = M$.

 Step 2-3: Send $y_1, y_2, ..., y_{l-1}$ to the client.

Step 3: [Client] Accumulate y_i as follows:

 Step 3-1: For $j= 1, 2, ..., b-1$, $z_j = r_j$.

 Step 3-2: For $i=0, 1, 2, ..., l-1$, if $d_i \neq 0$, then execute $z_{d_i} = (z_{d_i} \times y_i) \bmod N$.

 Step 3-3: Send $z_1, z_2, ..., z_{b-1}$ to the server.

Step 4: [Server] Help the client to merge all z_j's as below.

 Step 4-1: Execute $z_j = (z_j \times z_{j+1}) \bmod N$ from $j= b-2$ to $j=1$.

 Step 4-2: Execute Step 4-1 again.

 Step 4-3: Send z_1 to the client.

Step 5: [Client] Obtain $S = z_1 \times R^{-1} \bmod N$ as the final result.

In Knuth's scheme, the exponent d is also scanned from LSB to MSB. The major difference between Knuth's and Yao's schemes is that the LSB of each window must be '1' in Knuth's scheme. If the scanned bit is '1,' then this bit and the successive $k-1$ bits are collected to form a new window u_i and the position of the

LSB of the new window is stored as w_i; otherwise, to scan the next bit. Here the position of the LSB is 0 and the position of MSB is n-1. Suppose that there are l

small windows $u_0, u_1, ..., u_{l-1}$, then $d = \sum_{i=0}^{l-1} u_i (2)^{w_i} = \sum_{j=1}^{b/2} [(2j-1)(\sum_{u_i=2j-1} (2)^{w_i})]$,

where w_i is the position of the LSB of the window u_i, for i= 0, 1, ..., l-1. Therefore,

the exponentiation M^d mod N is computed as $M^d \equiv \prod_{j=1}^{b/2} [\prod_{u_i=2j-1} M^{(2^{w_i})}]^{2j-1} \equiv$

$\prod_{j=1}^{b/2} [z_j]^{2j-1}$ (mod N), where $z_j = [\prod_{u_i=2j-1} M^{(2^{w_i})}]$ for j=1, 2, ..., b/2. In the following, Protocol 4 is stated.

Protocol 4

Step 0: [Precomputation of the client]
  Step 0-1: Generate b/2 random integers $r_1, r_2, ..., r_{b/2}$, where $r_j \in Z_N$, for j=1, 2, ..., b/2.

  Step 0-2: Compute R= $\prod_{j=1}^{b/2} (r_j)^{2j-1}$ mod N.

  Step 0-3: Compute R^{-1} such that $R \times R^{-1}$ mod N $\equiv$ 1.
Step 1: [Client] Send the message {M, N} to the server.
Step 2: [Server] After receiving {M, N}, execute the following substeps:
  Step 2-1: y_0= M.

  Step 2-2: For i= 1, 2, ..., n-1, compute $y_i = (y_{i-1})^2$ mod N.

  Step 2-3: Send $y_1, y_2, ..., y_{n-1}$ to the client.
Step 3: [Client] Accumulate y_i as follows:
  Step 3-1: For j= 1, 2, ..., b/2, $z_j = r_j$.
  Step 3-2: For i=0, 1, 2, ..., l-1, execute $z_{(u_i+1)/2} = (z_{(u_i+1)/2} \times y_{w_i})$ mod N, where y_0= M.

  Step 3-3: Send $z_1, z_2, ..., z_{b/2}$ to the server.

Step 4: [Server] Help the client to merge all z_j's as below.
  Step 4-1: Execute $z_j = (z_j \times z_{j+1})$ mod N from j= (b/2)- 1 to 2.

Step 4-2: Let h= z_2.

Step 4-3: Execute Step 4-1 again.

Step 4-4: Compute S= $z_1 \times h \times (z_2)^2$ mod N and send S to the client.

Step 5: [Client] Obtain the final result as S= SR^{-1} mod N.

The major differences between Protocols 3 and 1 (or 4 and 2) are Steps 0 and 4. In order to improve the efficiency, in Protocols 3 and 4, the server helps the client to merge all z_j's at Step 4 while, in Protocols 1 and 2, the client merges all z_j's by himself at Step 4. Since the server is powerful, Protocols 3 and 4 are more efficient than Protocols 1 and 2, respectively. To protect all z_j's by the random integers r_j's, the client must perform the precomputation in Step 0; otherwise, the server can derive the secret key d from all z_j's. The random integers r_j's should be used for many message blocks; otherwise, Protocols 3 and 4 are useless to improve the efficiency of Protocols 1 and 2. Thus, the client uses the same random integers R and R^{-1} for the message blocks of the same stage.

To be secure against the active attacks, Kawamura and Shimbo proposed a slight modification and a verification mechanism for Protocols 1, 2, 3, and 4. In the modification part, the client sends all final results S out at the final stage of the protocol. To check the correctness of S, a verification mechanism $S^e \equiv M \pmod{N}$ with sufficiently small e is proposed, where N is the product of two large primes, (e, N) is the public key of RSA cryptosystem, and d is the secret key of RSA cryptosystem [Rivest et al. 1978]. Since the computation power of the client is very small, e must be small. If the final result is not correct, the client does not send the result to anyone. By using the slight modification and the verification mechanism, the server cannot obtain the influenced result S' after the protocol, since $(S')^e$ mod N $\neq$ M. So, Protocols 3 and 4 were inferred to be secure against the active attacks [Kawamura and Shimbo 1993]. However, two new active attacks will be proposed to break Protocols 3 and 4.

3   Active Attacks

Before stating the new undetectable attacks, two assumptions are given. The first assumption is that the server obtains the correct result S of some message M and finds R and R^{-1} at the beginning of each stage. The server has many ways to get the correct result S, so this assumption is reasonable. For example, the server intercepts S from the public communication channel between the client and the receiver. Since the message blocks belonging to the same stage sharing the same random integers R and R^{-1}, the server obtains R by R= $z_1 \times S^{-1}$ mod N and R^{-1} by $R \times R^{-1} \equiv 1 \pmod{N}$. The next assumption is that the server can execute S^e mod N quickly. Since the client uses the public e to verify the correctness of the final result S, the public e should be small. The server can execute S^e mod N quickly, because the server has huge computation power. Based on these assumptions, two active attacks are proposed. The active attack on Protocol 3 is described in Section 3.1 and the active attack on Protocol 4 is described in Section 3.2.

3.1   Active Attack on Protocol 3

The basic attack on finding the Lth window d_L is described first, where $1 \leq L < l$ and $l = \lceil n/k \rceil$.

Basic Attack on Protocol 3

Step 1' are the same as *Step 1* in Protocol 3.
Step 2': [Server] After receiving {M, N}, execute the following substeps:

> *Step 2'-1*:   Randomly choose a positive integer x and compute x^{-1} such that $x^{-1} \times x \equiv 1$ (mod N).
>
> *Step 2'-2*: $y_0 = M$.
>
> *Step 2'-3*:   For i= 1, 2, ..., l-1, compute $y_i = (y_{i-1})^b$ mod N. Then compute $y_L = x \times y_L$ mod N.
>
> Step 2'-4:   Send $y_1, y_2, ..., y_{l-1}$ to the client.

Step 3' is the same as *Step 3* in Protocol 3.
Step 4': [Server] Help the client to merge all z_j's. Then obtain d_L.

> *Step 4'-1*:   Execute $z_j = z_j \times z_{j+1}$ mod N from j= b-2 to j=1.
>
> *Step 4'-2*:   Execute *Step 4'-1* again.
>
> *Step 4'-3*:   Remove R from z_1 by $z_1 = z_1 R^{-1}$ mod N.
>
> *Step 4'-4*:   Obtain d_L by performing the following:
>
>> $d_L = 0$.
>>
>> While (($z_1)^e \equiv M$ (mod N) does not hold) do $d_L = d_L + 1$ and $z_1 = z_1 \times x^{-1}$ mod N.
>
> *Step 4'-5*:   Compute $z_1 = Rz_1$ mod N.
>
> *Step 4'-6*:   Send z_1 to the client.

Step 5' is the same as *Step 5* in Protocol 3.

The reason why the server obtains d_L is stated below. After *Step 4'-2*, $z_1 \equiv Rx$ $x^{d_L} \times S \equiv Rx \times x^{d_L} \times M^d$ (mod N). After removing R from z_1 at *Step 4'-3*, $z_1 \equiv x^{d_L} \times S \equiv$

$x^{d_L} \times M^d$ (mod N). Thus, the repeating times of the while loop at *Step 4'-4* is equal to d_L.

In order to pass the verification of the client, the server constructs z_1 at *Step 4'-5*. After *Step 4'-4*, $z_1 \equiv S \equiv (M^d)$ (mod N), then the server can construct z_1 such that $z_1 \equiv R \times S \equiv R \times M^d$ (mod N). So the new z_1 can pass the verification of the client after removing R from the new z_1. In other words, the client cannot find that the server is trying to guess the secret value d. Since this is an undetectable attack, the server can repeat this basic attack many times. Finally, the server can find l-1 windows d_1, d_2, ..., d_{l-1}. Although the server has no chance to find d_0, he can conduct exhaustive search to obtain d_0.

Step 4'-4 is the time consuming part of this attack, when k is large. Here a parallel method is proposed to speed up *Step 4'-4*. The server can compute x^{-1}, x^{-2}, ..., $x^{-(2^k-1)}$ in advance. Then uniformly distribute these values to 2^k-1 processors. The ith processor is responsible for verifying $(z_1 \times x^{-i})^e \equiv M$ (mod N) for the value x^{-i} received from the server, where i= 1, 2, ..., 2^k-1. One of these processors will tell the server the final result S and the value i. So *Step 4'-4* can then be speeded up.

Even if the client sends the final result at final stage, the server can derive incorrect result S' directly from z_1 at *Step 4'-3*. So the slight modification idea of Protocol 3 is useless for this active attack.

Through this active attack, the number of possible values of d is reduced from 2^{n-1} to $\lceil n/k \rceil 2^k$. For example, when k= 8 and n=512, the possible largest value of d is reduced from 2^{511} to 64×2^8. Then the exhaustively searching space of d is reduced drastically for k << n. The following example is used to illustrate this active attack.

Example 1

Suppose that n=10, k=2 and d= 611_{10}= $(1001100011)_2$, then b=4, d_0= 3, d_1= 0, d_2= 2, d_3= 1, d_4= 2, and l= 5. Assume that the server wants to obtain d_2 , that is L=2. The details of this attack are shown step by step.

Step 1': [Client] Send the message {M, N} to the server.
Step 2': [Server] After receiving {M, N}, execute the following substeps:

Step 2'-1: Choose a random positive integer x and compute x^{-1} such that $x(x^{-1}) \equiv 1$ (mod N).

Step 2'-2: y_0= M.

Step 2'-3: y_1= M^4 mod N, y_2= xM^{4^2} mod N= xM^{16} mod N, y_3= M^{64} mod N, and y_4= M^{4^4} mod N= M^{256} mod N.

Step 2'-4: Send $\{y_1, y_2, y_3, y_4\}$ to the client.

Step 3': [Client] Accumulate all y_i as follows:

Step 3'-1: $z_1 = r_1$, $z_2 = r_2$, and $z_3 = r_3$.

Step 3'-2: $z_1 = r_1 \times y_3 \bmod N = r_1 \times M^{64} \bmod N$, $z_2 = r_2 \times (y_4 \times y_2) \bmod N = x \times r_2 \times M^{272} \bmod N$, and $z_3 = r_3 \times y_0 \bmod N = r_3 \times M \bmod N$.

Step 3'-3: Send $\{z_1, z_2, z_3\}$ to the server.

Step 4': [Server] Help the client to merge all z_j's as below.

Step 4'-1: $z_1 = z_1 \times z_2 \times z_3 = x \times r_1 \times r_2 \times r_3 \times M^{64+272+1} \bmod N = x \times r_1 \times r_2 \times r_3 \times M^{337} \bmod N$, $z_2 = z_2 \times z_3 = x \times r_2 \times r_3 \times M^{272+1} \bmod N = x \times r_2 \times r_3 \times M^{273} \bmod N$, and $z_3 = r_3 \times y_0 \bmod N = r_3 \times M \bmod N$.

Step 4'-2: $z_1 = z_1 \times (z_2)^2 \times (z_3)^3 = (x)^2 \times r_1 \times (r_2)^2 \times (r_3)^3 \times M^{611} \bmod N = (x)^2 \times R \times M^{611} \bmod N$, $z_2 = z_2 \times (z_3)^2 = x \times r_2 \times (r_3)^2 \times M^{273} \bmod N$, and $z_3 = r_3 \times y_0 \bmod N = r_3 \times M \bmod N$.

Step 4'-3: Remove R from z_1 by $z_1 = z_1 R^{-1} \bmod N = (x)^2 \times R \times R^{-1} \times M^{611} \bmod N = (x)^2 \times M^{611} \bmod N$.

Step 4'-4: Since the repeating times of the while loop is 2, $d_L = 2$. Then $z_1 = M^{611} \bmod N$.

Step 4'-5: Compute $z_1 = R \times z_1 \bmod N = R \times M^{611} \bmod N$.

Step 4'-6: Send z_1 to the client.

Step 5: [Client] Obtain the final result as $S = (z_1)R^{-1} \bmod N = M^{611} \bmod N$.

It is easy to see that the final result is correct, the result can pass the verification of the client. So the client is not able to detect this basic active attack.

3.2 Active Attack on Protocol 4

Inspired by the active attack on Protocol 3, an active attack on Protocol 4 is also proposed. The basic active attack is proposed first. Through this basic attack, the server gets one bit or one window. Since this attack is undetectable, the server repeats this basic attack many times to derive the secret d. Since d is an odd number, then the 0th window is consisted of the k least signification bits of d. So the server starts from $L = k$ to $n-1$ to obtain the other $n-k$ bits of the secret d. Let L denote the bit position. The basic attack is stated below.

Basic Attack on Protocol 4

Step 1' is the same as *Step 1* in Protocol 4.

Step 2': [Server] After receiving $\{M, N\}$, execute the following steps:

    Step 2'-1: Choose a random positive integer x, compute x^{-1} such that $x(x^{-1}) \equiv 1$ (mod N), and compute $x^{-2} = (x^{-1})^2 \bmod N$.

    Step 2'-2: $y_0 = M$.

    Step 2'-3: For i= 1, 2, ..., n-1, compute $y_i = (y_{i-1})^2 \bmod N$. Compute $y_L = x \times y_L$ mod N.

    Step 2'-3: Send $y_1, y_2, ..., y_{n-1}$ to the client.

Step 3' is the same as *Step 3* in Protocol 4.

Step 4': [Server] During merging all z_j's, the server also derives the Lth bit or some window at the same time.

    Step 4'-1: Execute $z_j = (z_j) \times (z_{j+1}) \bmod N$ from j= (b/2)- 1 to 2.

    Step 4'-2: Let $h = z_2$.

    Step 4'-3: Execute Step 4-1 again.

    Step 4'-4: Compute $S = z_1 \times h \times (z_2)^2 \bmod N$.

    Step 4'-5: Remove R from S by computing $S = R^{-1} \times S \bmod N$.

    Step 4'-6: Compute the Lth bit or the window U from the Lth to the (L+k-1)th bits as follows.

        If $((S)^e \equiv M \pmod N$ holds) then T= 0; else T=1 and $S = S \times x^{-1} \bmod N$.

        While $((S)^e \equiv M \pmod N$ does not hold) do T= T +2 and $S = S \times x^{-2}$ mod N.

        If (T= 0) then the Lth bit is '0' and L=L+1; otherwise, U= T and L= L+k.

    Step 4'-7: Reconstruct S by computing $S = R \times S \bmod N$.

    Step 4'-8: Send S to the client.

Step 5' is the same as *Step 5* in Protocol.

After this basic attack, the server has found the Lth bit or the window U from the Lth to the (L+k-1)th bits. The inference is shown as below:

    After *Step 4'-4*, if the Lth bit is '0', $S = R \times M^d \bmod N$; otherwise, $S = R \times x^U \times M^d$ mod N.

    After *Step 4'-5*, if the Lth bit is '0', $S = M^d \bmod N$; otherwise, $S = x^U \times M^d \bmod N$.

    After *Step 4'-6*, if $S = M^d \bmod N$, then T=0; otherwise, T= U for $S = x^U \times M^d \bmod N$. Then $S = M^d \bmod N$.

    After *Step 4'-7*, $S = R \times M^d \bmod N$.

At *Step 4'-5*, if the incorrect y_L is accumulated into some z_j, then the server obtains the incorrect result $S = x^U \times M^d \bmod N$, where U= 2j-1. The server can derive incorrect final result S by himself, even if the client sends the final result S at final stage. So the slight modification of Protocol 4 is useless for this active attack.

At *Step 5'*, the client gets the correct result $S = M^d \bmod N$. So the client cannot use the verification $S^e \equiv M \pmod{N}$ to detect this attack. *Step 4'-6* is also very time consuming, when k is large. This step can be parallelized in the same manner of the attack on Protocol 3. Since the server does not help the client to compute M, the server has no chance to derive u_0. Eventually, the server can collect the other bits and exhaustively search u_0, since $1 \le u_0 < 2^k$, then all bits of d can thus be obtained.

By the above active attack, the number of possible values of the secret d is reduced from 2^{n-1} to at most $\lceil n/k \rceil 2^{k-1}$. In average case, the number of possible values of the secret d is reduced to $\lceil n/(k+1) \rceil 2^{k-1}$. When k= 8 and n=512, the possible values of d are reduced from 2^{511} to 64×2^7, in worse case, and to 57×2^7, in average case. Since k<< n, the exhaustively searching space of d is reduced drastically. Example 2 illustrates this active attack.

Example 2

Suppose that n=10, k= 2 and d= 619_{10}= $(1001101011)_2$, then (u_0, w_0)= (3, 0), (u_1, w_1)= (1, 3), (u_2, w_2)= (3, 5), (u_3, w_3)= (1, 9), and l= 3. Assume that the server starts at L= 2. This example shows how to find the second bit and u_1.

Step 1': [Client] Send the message {M, N} to the server.
Step 2': [Server] After receiving {M, N}, execute the following steps:

    Step 2'-1: Choose a random positive integer x, compute x^{-1} such that $x(x^{-1}) \equiv 1 \pmod{N}$, and compute $x^{-2} = (x^{-1})^2 \bmod N$.

    Step 2'-2: $y_0 = M$.

    Step 2'-3: For i= 1, 2, ..., 9, compute $y_i = (y_{i-1})^2 \bmod N$. Compute $y_2 = x \times y_2 \bmod N$.

    Step 2'-3: Send $y_1, y_2, ..., y_9$ to the client.

Step 3': [Client] Accumulate y_i as follows:

    Step 3'-1: $z_1 = r_1$ and $z_2 = r_2$.

    Step 3'-2: $z_1 = r_1 \times y_3 \times y_9 \bmod N = r_1 \times M^{2^3+2^9} \bmod N = r_1 \times M^{520} \bmod N$, and

        $z_2 = r_2 \times y_0 \times y_5 \bmod N = r_2 \times M^{2^0+2^5} \bmod N = r_2 \times M^{33} \bmod N$.

    Step 3'-3: Send z_1 and z_2 to the server.

Step 4': [Server] After merge all z_j's, the server derives the second bit.

    Step 4'-1: $z_1 = r_1 \times M^{520} \bmod N$, and $z_2 = r_2 \times M^{33} \bmod N$.

Step 4'-2: Let $h = z_2 = r_2 \times M^{33}$ mod N.

Step 4'-3: $z_1 = r_1 \times M^{520}$ mod N, and $z_2 = r_2 \times M^{33}$ mod N.

Step 4'-4: $S = r_1 \times (r_2)^3 \times M^{619}$ mod N $= R \times M^{619}$ mod N.

Step 4'-5: Remove R from S by computing $S = R^{-1} \times S$ mod N $= M^{619}$ mod N.

Step 4'-6: Since $S^e \equiv M$ (mod N) holds, T=0, L= 2+1= 3 and S= M^{619} mod N.

Step 4'-7: Reconstruct S by computing $S = R \times S$ mod N $= R \times M^{619}$ mod N.

Step 4'-8: Send S to the client.

Step 5': [Client] Obtain the final result as $S = S \times R^{-1}$ mod N $= R^{-1} \times R \times M^{619}$ mod N $= M^{619}$ mod N.

Now the server finds that the second bit is '0'. He wants to find u_1 as below.

Step 1': [Client] Send the message {M', N} to the server.

Step 2': [Server] After receiving {M', N}, execute the following steps:

Step 2'-1: Choose a random positive integer x, compute x^{-1} such $x(x^{-1}) \equiv 1$ (mod N), and then compute $x^{-2} = (x^{-1})^2$ mod N.

Step 2'-2: $y_0 = M'$.

Step 2'-3: For i= 1, 2, ..., 9, compute $y_i = (y_{i-1})^2$ mod N. Compute $y_3 = x \times y_3$ mod N.

Step 2'-3: Send $y_1, y_2, ..., y_9$ to the client.

Step 3': [Client] Accumulate y_i as follows:

Step 3'-1: $z_1 = r_1$ and $z_2 = r_2$.

Step 3'-2: $z_1 = r_1 \times y_3 \times y_9$ mod N $= r_1 \times x \times M'^{2^3+2^9}$ mod N $= r_1 \times x \times M'^{520}$ mod N, and $z_2 = r_2 \times y_0 \times y_5$ mod N $= r_2 \times M'^{2^0+2^5}$ mod N $= r_2 \times M'^{33}$ mod N.

Step 3'-3: Send z_1 and z_2 to the server.

Step 4': [Server] After merging all z_j's, the server derives u_1.

Step 4'-1: $z_1 = r_1 \times x \times M'^{520}$ mod N, and $z_2 = r_2 \times M'^{33}$ mod N.

Step 4'-2: Let $h = z_2 = r_2 \times M'^{33}$ mod N.

Step 4'-3: $z_1 = r_1 \times x \times M'^{520}$ mod N, and $z_2 = r_2 \times M'^{33}$ mod N.

Step 4'-4: $S = r_1 \times x \times (r_2)^3 \times M'^{619}$ mod N $= R \times x \times M'^{619}$ mod N.

Step 4'-5: Remove R from S by computing $S = (R^{-1}) \times S$ mod N $= x \times M'^{619}$ mod N.

Step 4'-6: Since $(S \times x^{-1})^e \equiv M'$ (mod N) holds, $u_1 = T = 1$, $L = 3+2 = 5$ and $S = M'^{619}$ mod N.

Step 4'-7: Reconstruct S by computing $S = R \times S$ mod $N = R \times M'^{619}$ mod N.

Step 4'-8: Send S to the client.

Step 5: [Client] Obtain the final result as $S = SR^{-1}$ mod $N = R^{-1} \times R \times M'^{619}$ mod $N = M'^{619}$ mod N.

So the server finds that $u_1 = 1$ and the server repeats this basic attack to find the other bits and windows, except u_0.

4   Conclusions

Two new active attacks are proposed to break Protocols 3 and 4 in [Kawamura and Shimbo 1993], respectively, when N is the product of two large primes. The major assumption is that the server obtains the correct result M^d mod N and the random integers shared by the message blocks of the same stage. By these attacks, the server finds the incorrect final result by himself, uses the incorrect result to derive the partial bits of the secret d, then constructs a correct result for the client. So the client cannot detect this attack. To adopt these active attacks, the number of the possible values of the secret data d is reduced from 2^{n-1} to $\lceil n/k \rceil 2^k$ or $\lceil n/k \rceil 2^{k-1}$, in worst case, where n is the bit length of d and k is the size of the partition window of d. Since $n \gg k$, the number of possible values of the secret data d is greatly reduced.

References

1. Kawamura, S. and Shimbo, A. (1993): "Fast Sever-Aided Secret Computation Protocols for Modular Exponentiation," *IEEE Journal on Selected Areas in Communications*, Vol. 11, No. 5, 1993, pp. 778- 784.
2. Knuth, D. E. (1981): "*The Art of Computer Programming Vol 2: Seminumerical Algorithms*," 2nd Ed., Addition-Wesley, Reading, MA, 1981, pp. 451.
3. Laih, C. S. and Yen, S. M. (1992): "Secure Addition Sequence and Its Application on the Server Aided Secret Computation Protocols," *AUSCRYPT '92*, Gold Coast, Australia, Dec. 1992, pp. 6.1- 6.7.
4. Laih, C. S., Yen, S. M. and Harn, L. (1991): "Two Efficient Server-Aided Secret Computation Protocols based on the Addition Sequence," *ASIACRYPT '91*, Fuijyoshida, Japan, Nov. 1991, pp. 270- 274.
5. Matsumoto, T., Kato, K. and Imai, H. (1988): "Speed up Secret Computations with Insecure Auxiliary Devices," *Advances in Crytpology- CRYPTO '88*, Springer Verlag, New York, 1990, pp. 497- 506.
6. Pfitzmann, B. and Waidner, M. (1992): "Attacks on Protocols for Server-Aided RSA Computation," *EUROCRYPT '92*, Balatonfured, Hugary, 1992, pp. 139 -146.

7. Quisquater, J.-J., and Soete, M. De (1989): "Speeding up Smart Card RSA Computation with Insecure Coprocessors," *Proc. SMART CARD 2000*, Amsterdam, North-Holland, Oct. 1989, pp. 191- 197.
8. Rivest, R. L., Shamir, A. and Adleman, L. (1978): "A Method for Obtaining Digital Signatures and Public Key Cryptosystems," *Communications of ACM*, Vol. 21, No. 2, 1978, pp. 120-126.
9. Shimbo, A. and Kawamura, S. (1990): "A Factorization Attack on Certain Server-aided Computation Protocols fro RSA Secret Transformation," *Electronic Letters*, Vol. 26, No. 17, 1990, pp. 1387- 1388.
10. Yao, A. C. (1976): "On the Evaluation of Powers," *SIAM J. Comput.*, Vol. 5, No. 1, 1976, pp. 100-103.

Cryptanalysis of the Enhanced ElGamal's Signature Scheme

Chi-Sung Laih and Wen-Chung Kuo

Department of Electrical Engineering
National Cheng Kung University,
Tainan,Taiwan,Republic of China
E-mail:laihcs@eembox.ncku.edu.tw

Abstract. Recently, He and Kiesler proposed a new signature scheme whose security is based on both factorization and discrete logarithms to enhance the security of ElGamal's scheme. In this paper, we show that He-Kiesler signature scheme is insecure under the Known-Signature attack.

1 Introduction

ElGamal's signature scheme is one of the most important digital signature schemes whose security is based on the difficulty of computing discrete logarithms. In [1], He and Kiesler proposed two modified versions to enhance the security of the original ElGamal signature scheme. The security of their first version is the same as that of the original ElGamal scheme. He and Kiesler claimed that the security of their second version is based on the two hard problems, i.e., the discrete logarithm problem and the integer factorization problem. In [2], Harn pointed out that the security of their second version is not as secure as they have claimed. He showed that the security of the modified version 2 only relies on the integer factorization problem when a pair of message and signature is given. In this paper, we further show that the second version of He-Kiesler signature scheme is insecure under the Known-Signature attack. That is, given two pairs of messages and signatures, an attacker can derive one of the signer's secret keys and then can produce a valid signature for any message with significant probability.

2 Modified version 2 of ElGamal's scheme

In this section, we give a brief description on the second version of modified ElGamal's scheme[1]. The following parameters are commonly used in the system:

- p: a large prime such that $p - 1$ has two large prime factors p_1 and q_1.
- n: the product of p_1 and q_1.
- g: a primitive element of $GF(p)$.

Note that the two factors of n, i.e., p_1 and q_1 are kept secret from every user. Any user A has the following keys:

- secret key: x_1 which is coprime with $p - 1$.
- public key: $z \equiv g^{x^2}$ (mod p), where $x \equiv x_1^2$ (mod $(p-1)$) is an immediate secret integer used for signing the signature.

Signature generation:
To sign a message m, A does the following steps:

1. Chooses a random integer t_1 between 0 and n such that $\gcd(t_1, p - 1) = 1$ and compute $t \equiv t_1^2$ (mod $(p - 1)$).
2. Calculates $k \equiv t^2$ (mod $(p - 1)$).
3. Calculates $r \equiv g^k$ (mod p) with $r \neq 1$.
4. Computes $s \equiv t^{-1}(m - xr)$ (mod $(p - 1)$).
   Note that the following equation is hold.

$$m \equiv xr + ts \quad (\text{mod } (p - 1)). \tag{1}$$

5. Compute $c \equiv x_1 t_1$ (mod $(p - 1)$).

A then sends (r, s, c) to a verifier as the signature on m. To verify whether (r, s, c) is a valid signature of m or not, the verifier checks the identity

$$g^{m^2} \equiv z^{r^2} r^{s^2} g^{2rsc^2} \quad (\text{mod } p). \tag{2}$$

If Eq.(2) holds, (r, s, c) is a valid signature of m signed by A. Otherwise, it is not a valid signature.

3   The Attack

In [1], the authors claimed that the security of their scheme is based on both two hard problems: integer factorization and discrete logarithms. In [2], Harn showed that if an attacker can solve the factorization problem, then the immediate secret integer x can be recovered under the Known-Signature attack. Harn's attack can be described as follows. Given a valid signature (r, s, c) on m, multiplying x on both sides of Eq.(1) the attacker has

$$mx \equiv x^2 r + stx \quad (\text{mod } (p - 1)). \tag{3}$$

Since $c^2 \equiv tx$ (mod $(p - 1)$), the attackers obtains the following quadratic congruence

$$rx^2 - mx + sc^2 \equiv 0 \quad (\text{mod } (p - 1)). \tag{4}$$

The quadratic congruence can be solved if the factors of $p - 1$ can be factorized. Once x has been derived, the attacker can obtain the secret key x_1 by finding the square root of x (mod $(p - 1)$). Note that solving Eq.(4) and computing $x_1 \equiv \sqrt{x}$ (mod $(p - 1)$) need the factorization of $p - 1$.

In this paper, we show that even if the attacker cannot factor $p - 1$, he can still forge a valid signature for any message. Given two signatures (r, s, c) and

(r', s', c') on messages m and m', respectively, the attacker obtains the following two equations

$$rx^2 - mx + sc^2 \equiv 0 \pmod{(p-1)} \tag{5}$$

and

$$r'x^2 - m'x + s'(c')^2 \equiv 0 \pmod{(p-1)}. \tag{6}$$

Rewriting the equations the attacker finds

$$\begin{bmatrix} r & -m \\ r' & -m' \end{bmatrix} \begin{bmatrix} y \\ x \end{bmatrix} \equiv \begin{bmatrix} -sc^2 \\ -s'(c')^2 \end{bmatrix} \pmod{(p-1)} \tag{7}$$

where $x^2 \equiv y \pmod{(p-1)}$. Let $\Delta \equiv -rm' + r'm \pmod{(p-1)}$. If $\Delta \neq 0$ and $\gcd(\Delta, p-1) = 1$, then

$$\begin{bmatrix} y \\ x \end{bmatrix} \equiv \begin{bmatrix} r & -m \\ r' & -m' \end{bmatrix}^{-1} \begin{bmatrix} -sc^2 \\ -s'(c')^2 \end{bmatrix} \pmod{(p-1)}. \tag{8}$$

Thus, x can be found by the attacker without factoring $p-1$. The probability that $\Delta \neq 0$ and $\gcd(\Delta, p-1) = 1$ is $\frac{\phi(p-1)}{p-1}$, where ϕ is Euler's totient function. Let $p - 1 = q_1 p_1 p_2^{e_2} \cdots p_k^{e_k}$, then the probability that Δ satisfies the conditions as above is $(1 - \frac{1}{q_1})(1 - \frac{1}{p_1})(1 - \frac{1}{p_2}) \cdots (1 - \frac{1}{p_k})$, which cannot be neglected. For example, if $p - 1 = 2p_1 q_1$, then the probability is $(1 - \frac{1}{q_1})(1 - \frac{1}{p_1})(1 - \frac{1}{2}) \approx \frac{1}{2}$, since q_1 and p_1 are large primes.

We now show that the attacker can forge a valid signature for any message even if he cannot find the secret key x_1, i.e., the square root of $x \pmod{(p-1)}$. Note that once x is revealed the attacker can obtain t from Eq.(1). Let (s, r, c) be a valid signature of m and let m_a be the message to be forged. The attacker does the following steps:

1. Randomly chooses an integer b and calculate $t_a \equiv b^2 t \pmod{(p-1)}$ and makes sure that $\gcd(t_a, p-1) = 1$.
2. Computes $k_a \equiv t_a^2 \pmod{(p-1)}$.
3. Computes $r_a \equiv g^{k_a} \pmod{p}$, with $r_a \neq 1$.
4. Finds s_a such that

$$m_a \equiv xr_a + t_a s_a \pmod{(p-1)} \tag{9}$$

i.e.,

$$s_a \equiv (m_a - xr_a) t_a^{-1} \pmod{(p-1)}. \tag{10}$$

5. Calculates $c_a \equiv bc \pmod{(p-1)}$.

Then we claim that (r_a, s_a, c_a) is a valid signature of m_a. From Eq.(9), we have

$$m_a^2 \equiv x^2 r_a^2 + 2x r_a t_a s_a + t_a^2 s_a^2 \pmod{(p-1)}.$$

Since $x t_a \equiv x b^2 t \equiv b^2 c^2 \equiv c_a^2 \pmod{(p-1)}$, we have

$$g^{m_a^2} \equiv \left(g^{x^2}\right)^{r_a^2} g^{2r_a s_a c_a^2} \left(g^{t_a^2}\right)^{s_a^2} \pmod{p}$$

or

$$g^{m_a^2} \equiv z^{r_a^2} r_a^{s_a^2} g^{2r_a s_a c_a^2} \pmod{p}.$$

Hence (r_a, s_a, c_a) is a valid signature of m_a which can be forged by the attacker who does not know the signer's secret key x_1. Concluding the above descriptions, we have the following theorem.

Theorem 1. The modified version 2 of ElGamal's signature scheme proposed by He and Kiesler is totally insecure under the Known-Signature attack.

Acknowledgement: The authors would like to thank Prof. L. Harn with whom they had many useful discussions.

References

1. J. He and T. Kiesler " Enhancing The Security of ElGamal's Signature Scheme", *IEE Proc.-Comput.Digit.Tech.*, Vol.141, No.4, pp.249-252, July 1994.
2. L. Harn " Comment Enhancing The Security of ElGamal's Signature Scheme", to be appeared *IEE Proc.-Comput.Digit.Tech.*.

Access with Pseudonyms

Lidong Chen

Texas A&M University, U.S.A.
(lchen@math.tamu.edu)

Abstract. In some systems, users might want to identify themselves by their pseudonyms. If access control is necessary, then a certificate of authorized access employing pseudonyms must be unforgeable. We call a certificate of authorized access a credential. If different pseudonyms which identify a user are to be unlinkable, the user must be able to choose his pseudonyms at random and to transfer the credential issued on one pseudonym to another pseudonym untraceably. However, in order to prevent forgery of the certificate, pseudonyms must be formed in a specific way. This work presents a pseudonym validation process based on discrete logarithms without using cut-and-choose. The certificates issued with pseudonyms are unforgeable. The privacy of users is protected unconditionally. This pseudonym system has the novel feature that each user has a validated public key relevant to each pseudonym, so that signatures can be made by pseudonyms.

In order to protect the privacy of individuals, a system might grant users access based only on their pseudonyms. In order to control access, a credential, such as a certificate of authorized access, can be assigned to a pseudonym. The same person may use different pseudonyms in dealing with different organizations. In such a case, an individual should be able to transfer a credential issued to one pseudonym to another pseudonym in order to show it to some organization. The phrase credential mechanism is used in the literature to describe the whole process including establishing pseudonyms, issuing and transferring credentials (see [CE86]).

1 Main idea and basic protocol

The basic requirement for credentials is *unforgeability, i.e.* an individual cannot show a credential to an organization unless it has been properly issued to him. If the credentials are based on pseudonyms, then they must guarantee *unlinkability* of pseudonyms. This includes two aspects: no pseudonym can be linked to the identity of the individual, and two different pseudonyms cannot be found corresponding to one individual.

In order to protect the privacy of individuals, the pseudonyms must have some randomness. However, a credential is a kind of certificate based on a pseudonym. So, to prevent individuals from abusing the system, all pseudonyms must be formed according to some rules. For example, if credentials are RSA signatures (see [RSA]) on pseudonyms with public key e and secret key d, then the credential is easy to forge with any pseudonym formed as r^e for a random number r known by the individual who uses the pseudonym. However, suppose that a pseudonym is constructed as ur^e. If an individual using the pseudonym does not know u^d but knows r, then it seems more difficult to forge the credential (see [CE86]). So the system must include a process which forces individuals to form their pseudonyms in a proper manner, but does not require individuals to reveal anything about which random factors they have actually used to construct their pseudonyms. This process is called validating pseudonyms.

Chaum and Evertse (see [CE86]) described an RSA based credential mechanism, where the credentials are RSA signatures on pseudonyms which are validated by cut-and-choose. Damgård [Da88] published another construction of a credential mechanism based on a multiparty computing protocol with secret inputs and outputs. The credential system proposed here uses a discrete logarithm setting. The whole process, from validating pseudonyms to transferring credentials, is based on a three move divertible proof of knowledge.

As in [CE86], [Da88], individuals are identified by pseudonyms. The participants are: a trusted center C; a set of organizations $\{O_j \mid j \in \mathcal{J}\}$; and a set of individuals $\{I_k \mid k \in \mathcal{K}\}$.

The center C will work as a notary office to validate pseudonyms. For any k, I_k will be identified with pseudonym U_{kj} in organization O_j, $j \in \mathcal{J}$. After pseudonym validation, each individual I_k has different pseudonyms in different organizations $\{U_{kj}, j \in \mathcal{J}\}$. And each organization O_j has a group of individuals who are identified by pseudonyms $\{U_{kj}, k \in \mathcal{K}\}$. Thereafter, C will no longer be needed.

Each organization O_j is supposed to issue some kinds of credentials. For example, a police station may issue drivers' licenses and a hospital may issue health certificates. When organization O_j issues a credential to an individual identified as U_{kj}, it may need to show the credential to another organization $O_{j'}$ where the individual is identified as $U_{kj'}$. In this case, it is necessary to perform an untraceable transfer of the credential from U_{kj} to $U_{kj'}$ by the individual who uses the synonym so that others cannot perceive that U_{kj} and $U_{kj'}$ represent a same person.

Suppose that p is a prime, that q is the largest prime factor of $p-1$, and that g is a generator of the multiplicative group G_q of order q. An individual with the physical identity ID_k (name, address, birth date, photo, etc.) will be represented by I_k. C issues a personal identification number u_k to I_k. The pseudonym of I_k in O_j will then have the form

$$U_{kj} = u_k g^{s_{kj}},$$

where s_{kj} is chosen by I_k from $\mathbb{Z}_q^*$ randomly and secretly. The pseudonym U_{kj} is independent of the physical identity ID_k. If $U_{kj'}$ is the pseudonym of I_k in $O_{j'}$, then

$$U_{kj'} = u_k g^{s_{kj'}} = U_{kj} g^{s_{kj'} - s_{kj}}$$

for some $s_{kj'}$ which only I_k knows. So different pseudonyms are also independent of each other.

In the credential system, suppose the organization O_j issues a credential with public key (g, h) and corresponding secret key $\log_g h = x$. The credential of the pseudonym U_{kj} is designated as

$$Z_{kj} = U_{kj}^x$$

together with a signature based on a proof that

$$\log_g h = \log_{U_{kj}} Z_{kj},$$

which is shown in Section 1.1 and Figure 1 as the basic protocol, where U_{kj} and Z_{kj} are simply written as U and Z.

1.1   The basic protocol

The prover P and verifier V execute the following protocol with common input (g, h, U, Z), given prime p, q, the largest prime factor of $p-1$, and a generator g of the group G_q.

1. P chooses t randomly from $\mathbb{Z}_q^*$, computes $a = g^t$ and $b = U^t$, and then sends (a, b) to V.
2. V chooses c randomly from $\mathbb{Z}_q^*$, and sends it to P as a challenge.
3. P computes $r = t + cx$, and sends it to V as a reply to challenge c.
4. V determines whether it is true that $g^r = ah^c$ and $U^r = bZ^c$. If it is true, V will accept the proof; otherwise, V will reject.

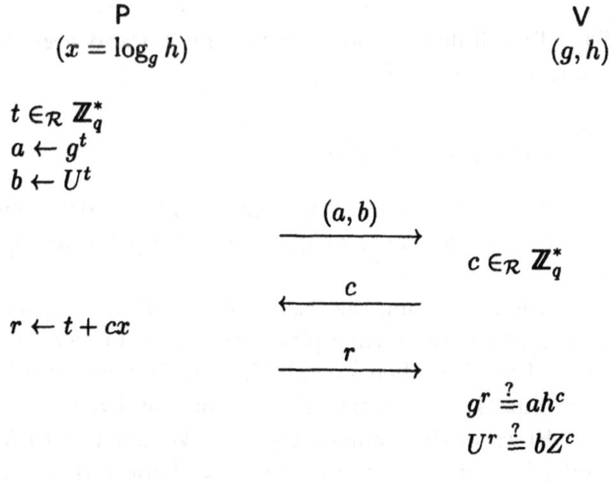

$$
\begin{array}{ll}
\mathsf{P} & \mathsf{V} \\
(x = \log_g h) & (g, h)
\end{array}
$$

$$
\begin{aligned}
& t \in_{\mathcal{R}} \mathbb{Z}_q^* \\
& a \leftarrow g^t \\
& b \leftarrow U^t
\end{aligned}
$$

$$\xrightarrow{\quad (a, b) \quad}$$

$$c \in_{\mathcal{R}} \mathbb{Z}_q^*$$

$$\xleftarrow{\quad c \quad}$$

$$r \leftarrow t + cx$$

$$\xrightarrow{\quad r \quad}$$

$$
\begin{aligned}
g^r & \overset{?}{=} ah^c \\
U^r & \overset{?}{=} bZ^c
\end{aligned}
$$

Fig. 1. Basic protocol: proof that $\log_g h = \log_U Z$

This protocol was originally introduced by Chaum and Pedersen in [ChaP92]. It is a proof of knowledge of secret key x and of the equality $\log_g h = \log_U Z$. It can be proved to be witness hiding only when the challenge set is a subset A in $\mathbb{Z}_q$ and $|A|$ is a polynomial in $|q|$. In [ChaP92], Chaum and Pedersen conjectured that no matter which $c \in \mathbb{Z}_q$ is chosen as a challenge, the prover reveals no information other than the fact that $\log_g h = \log_U Z$. The signature scheme based on the basic protocol is defined in Section 1.2 immediately below.

1.2 The signatures based on basic protocol

Let $\mathcal{H}$ be a hash function (as in the Fiat-Shamir signature scheme, see [FS87]). Given this function, and the basic protocol, the signature on U with secret key

$$x = \log_g h$$

is

$$\sigma_x(U) = (Z, a, b, r).$$

It is correct if $c = \mathcal{H}(U, Z, a, b)$ and

$$g^r = ah^c \qquad \text{and} \qquad U^r = bZ^c.$$

The security of this signature scheme depends on the properties of the hash function $\mathcal{H}$ to a great extent. The following assumption is used.

Assumption 1. $\mathcal{H}$ has the property that if the basic three-move protocol is a proof of knowledge, then it is as difficult to convince a verifier who chooses $c = \mathcal{H}(U, Z, a, b)$ as to convince a verifier who chooses c at random.

After getting U_{kj}^x, I_k can easily compute $Z_{kj'} = U_{kj'}^x = U_{kj}^x h^s$ for his pseudonym in $O_{j'}$, since he knows $s = s_{kj'} - s_{kj}$. In order to show the credential with pseudonym $U_{kj'}$ to organization $O_{j'}$, the basic protocol is diverted with input $(g, h, U_{kj'}, Z_{kj'})$. The divertibility is shown in Section 1.3 and Figure 2 where $U_{kj'}$ and $Z_{kj'}$ are simply written as U_1 and Z_1.

1.3 Divertibility of the basic protocol

The basic protocol can be diverted to a third party. The readers who are interested in divertibility of protocols are suggested to refer [CheDaP94], [OO89], [Sim84].

If a middle party, when executing the basic protocol with a prover P, plays the role of verifier, and at the same time plays the role of prover with a third party V, which is called verifier, then we call the middle party warden W for historical reasons (see [Sim84]). The common input for P and W is (g, h, U, Z) as in the basic protocol. However the common input for W and V is (g, h, U_1, Z_1), where $U_1 = Ug^s$ and s is a random number which W knows. By diverting the protocol online, W can prove to V that $\log_g h = \log_{U_1} Z_1$ even though he is not supposed to hold the secret key $x = \log_g h$. W will behave as follows.

1. When getting (a, b) from P, W chooses t_1 and e randomly from $\mathbb{Z}_q^*$, computes $a_1 = ag^{t_1}h^e$ and $b_1 = bU_1^{t_1}Z_1^e a^s$, and then sends (a_1, b_1) to V.

2. When receiving the challenge c_1 from V, W forms $c = c_1 + e$ and sends c to P as a challenge.

3. When getting a reply r of P to challenge c, W computes $r_1 = r + t_1$ and sends it to V as the reply to challenge c_1.

It is easy to verify that $g^{r_1} = a_1 h^{c_1}$ and $U_1^{r_1} = b_1 Z_1^{c_1}$, if the reply r satisfies $g^r = ah^c$ and $U^r = bZ^c$. So V will accept, even though what he gets is not from the prover P directly. The divertibility of the basic protocol is an online property.

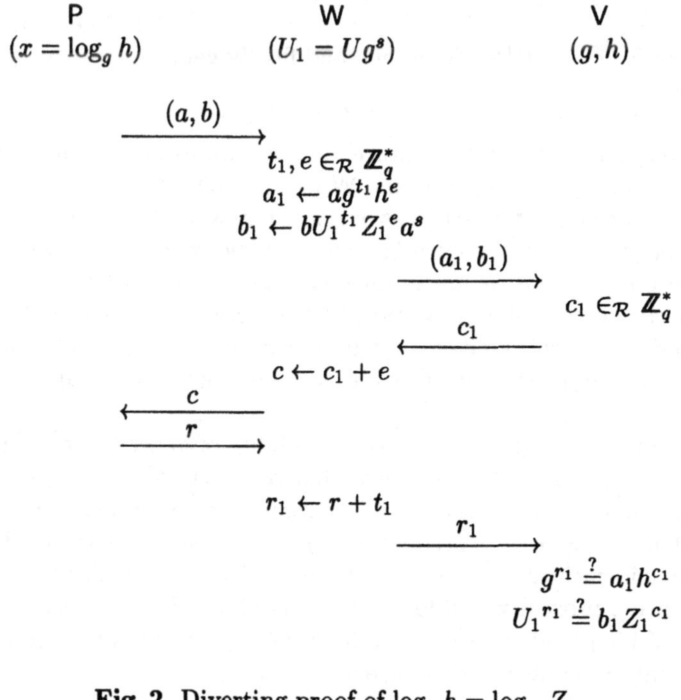

$$
\begin{array}{ccc}
\text{P} & \text{W} & \text{V} \\
(x = \log_g h) & (U_1 = Ug^s) & (g, h)
\end{array}
$$

$$\xrightarrow{\quad (a, b) \quad}$$

$$
\begin{aligned}
t_1, e &\in_{\mathcal{R}} \mathbb{Z}_q^* \\
a_1 &\leftarrow ag^{t_1}h^e \\
b_1 &\leftarrow bU_1^{t_1}Z_1^e a^s
\end{aligned}
$$

$$\xrightarrow{\quad (a_1, b_1) \quad}$$

$$c_1 \in_{\mathcal{R}} \mathbb{Z}_q^*$$

$$\xleftarrow{\quad c_1 \quad}$$

$$c \leftarrow c_1 + e$$

$$\xleftarrow{\quad c \quad}$$

$$\xrightarrow{\quad r \quad}$$

$$r_1 \leftarrow r + t_1$$

$$\xrightarrow{\quad r_1 \quad}$$

$$g^{r_1} \overset{?}{=} a_1 h^{c_1}$$
$$U_1^{r_1} \overset{?}{=} b_1 Z_1^{c_1}$$

Fig. 2. Diverting proof of $\log_g h = \log_U Z$

However, in most cases, a credential needs to be shown to some organization afterwards. The blind signature scheme based on the basic protocol introduced in Section 1.4 and Figure 3 will demonstrate the basic idea of how to transfer the credential offline.

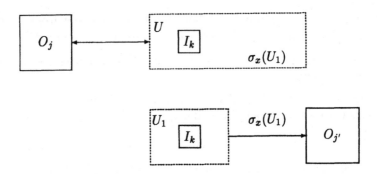

Fig. 3. Transferring a credential

1.4 The blind signatures

In order to get a signature on $U_1 = Ug^s$, the warden W computes U_1^x from U^x. When executing the protocol with input (g, h, U, Z), he diverts the protocol with input (g, h, U_1, Z_1). But instead of getting a random challenge c_1 from verifier V, he computes $c_1 = \mathcal{H}(U_1, Z_1, a_1, b_1)$, and finally gets

$$\sigma_x(U_1) = (Z_1, a_1, b_1, r_1).$$

It is clear that the blind signature is information-theoretically independent of the communication between P and W (see [OO89]).

By all the protocols and schemes presented above, we will establish a pseudonym system such that the credential can be issued with pseudonyms and transfered from one pseudonym to another pseudonym. Under Assumption 1, the signature scheme presented in Section 1.2 is unforgeable ([ChaP92]). So the credentials designed as this kind of signatures on pseudonyms should be unforgeable too. However, the divertibility provides some opportunity for abuse of the system.

The example here will point up the possibility of abuse by colluding individuals. If I_k and $I_{k'}$ collude in such a way that I_k tries to help $I_{k'}$ to pass a driver's license examination, then they must find a way to transfer a credential issued to I_k's pseudonym U_{kj} to $I_{k'}$'s pseudonym $U_{k'j}$. A possible way is that $I_{k'}$ borrows I_k's personal identification number u_k to form his pseudonym as $U_{k'j} = u_k g^s$. So we must have a process which forces each individual I_k to use u_k as a factor of every one of his pseudonyms, and which prevents an individual from swapping his identification number with someone else's.

2 Validating pseudonyms and public keys

We assume the center C is trusted not to produce illegal pseudonyms for itself, just as a bank is trusted not to produce illegal money. The privacy of individuals does not depend on C.

For each individual I_k, the center C chooses $u_k \in G_q$ as the personal identification number of I_k. It will be used as a factor of all his pseudonyms. It is reasonable to suppose that I_k does not know $\log_g u_k$, since u_k is chosen by C. We regard u_k as uniquely corresponding to I_k.

In this section, we omit the index j of I_k's pseudonym in O_j, when we state how to validate pseudonyms of individuals in O_j, for a fixed j.

Before validating the pseudonyms for some organization O_j, C chooses $\alpha \in \mathbb{Z}_q^*$, as the secret key that will be used in the validating process, and publishes the corresponding public key (g, g^α). The process of validating pseudonyms goes as follows.

1. I_k chooses $s_k \in \mathbb{Z}_q^*$ at random. He forms his pseudonym $U_k = u_k g^{s_k}$. Here U_k is represented as a number in $\mathbb{Z}_p$.

2. C computes

$$v_k = u_k^\alpha,$$

and sends it to I_k.

3. C proves to I_k that v_k is correct by executing the basic protocol with input (g, g^α, u_k, v_k). I_k gets the following blind signature as the validator for his pseudonym,

$$V_k = \sigma_\alpha(U_k).$$

4. I_k sends U_k and V_k to O_j. If V_k is a correct validator of U_k, then U_k will be registered in O_j as a valid pseudonym.

If necessary, I_k will choose m_k as his secret key under the pseudonym U_k, and register $Q_k = U_k^{m_k}$ as his public key. Q_k can be validated in a fashion similar to the way U_k is validated, since the basic protocol with common input (g, h, U, Z) can be diverted with input (g, h, U', Z') where $U' = U^y g^s$, with $y \neq 0$ (see **Appendix A**).

Sometimes (U_k, Q_k) will be used as a validated pseudonym and a public key together. The individual with pseudonym U_k will sign messages by use of the secret key m_k in the signature scheme aforementioned. The signature can be verified by anybody as a legal signature of the individual with pseudonym U_k.

In the validating process, the secret key α could be chosen by both C and O_j jointly, depending on how much trust is placed in O_j. In some cases, O_j will not be trusted. For example, in a voting scheme, the voting organization is not trusted, *i.e.* is presumed to be willing to forge votes. If it can produce valid pseudonyms, then it can forge votes. In such a case, we will prevent O_j from knowing α.

In order to validate the pseudonyms for another organization $O_{j'}$, the same process will be repeated using a different key β.

After this process, each organization identifies a group of individuals by a list of valid pseudonyms, and each individual has different pseudonyms in several relevant organizations.

3   Issuing and transferring credentials

Each organization O_j is authorized to issue a particular type of credential. For example, a police station may issue drivers' licenses.

Suppose that O_j issues the credential based on the public key (g, h) and corresponding secret key $\log_g h = x$.

The individual I_k is authenticated to get the credential by O_j, where I_k is identified as U_{kj}. The credential of U_{kj} is the signature

$$\sigma_x(U_{kj}) = (Z_{kj}, a, b, r),$$

where $Z_{kj} = U_{kj}^x$. It is correct if $c = \mathcal{H}(U_k, Z_k, a, b)$ and

$$g^r = ah^c \qquad \text{and} \qquad U_{kj}^r = bZ_{kj}^c.$$

In order to show this credential to organization $O_{j'}$, I_k must transfer it to his pseudonym $U_{kj'}$ which identifies him in $O_{j'}$. Here

$$U_{kj'} = U_{kj}g^s$$

for some s which I_k knows. One way to transfer the credential is the following. Suppose that I_k needs to show it to $O_{j'}$. He can execute the basic protocol with O_j, playing the role of warden, in order to get the credential with pseudonym $U_{kj'}$. But this way is inconvenient, since every time I_k needs to show the credential to some organization, he has to go to O_j and execute the basic protocol. The discussion immediately below gives a more convenient way to get several versions of the credential when it is issued, in order to be able to show it to several organizations independently.

When I_k gets the credential, he may have no idea which organizations he may have to show it to afterwards. So versions shouldn't be customized for any predetermined organizations. One version of the credential of I_k is the signature

$$\sigma_x(U_k^*),$$

where $U_k^* = u_k g^t$, and t is chosen randomly by I_k. U_k^* is not necessarily any of I_k's pseudonyms. I_k can get the credential on U_k^* by executing the basic protocol with O_j.

When I_k needs to show the credential to $O_{j'}$, he shows one of the versions $\sigma_x(U_k^*)$ and proves to $O_{j'}$ that he knows $\log_g(\frac{U_{kj'}}{U_k^*})$ by executing the basic protocol with input (g, g^d, h, h^d) where $U_{kj'} = U_k^* g^d$ is the pseudonym of I_k in $O_{j'}$.

Remark. Different versions of the credential are independent. But any given version $\sigma_x(U_k^*)$ can only be used once. Otherwise it links different pseudonyms.

For communication between individuals and organizations, an anonymous channel is not assumed, because it is rather restrictive. Some transactions must be face to face, or I_k may find a "representative" as in [CE86]. Some documents, such as a driver's license or a health certificate, can be issued only after physical identification.

For practical credential mechanisms, unlinkability cannot preclude revelation of "strictly necessary" information (see [CE86]). If O_j is supposed to issue the credential with secret key x, we define some sets for the organizations corresponding to this credential:

- for O_j, the set of the individuals who have received the credential is denoted by $O(x)$;
- for $O_{j'}$, the set of individuals who have shown the credential with public key (g, h) is denoted by $O'(g, h)$.

The strictly necessary information is precisely described as the fact that any individual in $O'(g, h)$ is in $O(x)$. It is required that he is identified with a special individual in $O(x)$ with the probability $\frac{1}{|O(x)|}$. If $O(x)$ consists of just a single individual, then it is impossible to avoid linking the pseudonyms as a result of showing the credential to some other organization $O_{j'}$.

In our system, unlinkability holds in the following sense.

Definition 2. (unlinkability) A credential mechanism satisfies unlinkability in following sense. Consider an individual I_k. The amount of Shannon information revealed about either the connection between his identity and any of his pseudonyms, or the connection between any of his two different pseudonyms, is the information-theoretic minimum.

Theorem 3. *The credential mechanism proposed above satisfies unlinkability, even if the center and some organizations collude.*

Proof. The validator of a pseudonym, which is a blind signature on the pseudonym, is independent of u_k. So it does not reveal any connection between u_k and the pseudonym. Transferring a credential from one pseudonym to another by blind signature will not link those pseudonyms either. □

4 Unforgeability of credentials

Consider the credential with public key (g, h) and secret key $x = \log_g h$. For any individual I_k, forging one version of this credential is equivalent to forging the credential with one of his pseudonyms. So we treat only forging credentials of pseudonyms in this section. We also omit the index j of the pseudonyms here, since we will discuss unforgeability of the credential issued by O_j for any fixed $j \in \mathcal{J}$.

Definition 4. (unforgeability) The credential mechanism satisfies unforgeability, if no individual I_k can forge the credential with any of his valid pseudonym U_k after a cooperating group of individuals obtain these credentials.

The unforgeability of our scheme depends on the discrete logarithm assumption and the following conjecture.

Conjecture 5. For any polynomial time warden W, if the basic protocol with input (g, h, U, Z) can be diverted by W with input (g, h, U', Z'), then either W knows the secret key $x = \log_g h$, or W knows some y, s such that $U' = U^y g^s$ and $y \neq 0$.

This conjecture is assumed to be true in [Bran94a]. Some arguments have been presented in [Bran94b], which suggest that finding a counterexample to it requires breaking either the Schnorr scheme (see [Sch90]) or the Diffie-Hellman assumption (see [DH76]).

Theorem 6. *The credential described above satisfies unforgeability under the discrete logarithm assumption, if the conjecture is true and the hash function in the signature scheme has the property stated in Assumption 1.*

Proof. By Assumption 1, it is infeasible for any I_k to compute the credential by himself without knowing the secret key x, after sufficiently many cooperating individuals have obtained the credential, no matter how U_k is formed.

If I_k can transfer a credential issued to somebody else, say $I_{k'}$, $k \neq k'$, with pseudonym $U_{k'}$, to his own pseudonym U_k, then by the conjecture,

$$U_k = U_{k'}^i g^s$$

for some i, s, $i \neq 0$, known by I_k. We will prove that this is infeasible even if $I_{k'}$ collusion with I_k.

By the conjecture, we can establish that the pseudonym U_k of I_k with personal identification number u_k can be validated, if and only if

$$U_k = u_k^{\lambda_1} g^{\lambda_2},$$

where I_k knows λ_1 and λ_2, $\lambda_1 \neq 0$. Similarly, if $U_{k'}$ is $I_{k'}$'s valid pseudonym then

$$U_{k'} = u_{k'}^{\tau_1} g^{\tau_2},$$

and I_k knows τ_1 and τ_2, $\tau_1 \neq 0$.

If for some i, s, the equality $U_k = U_{k'}^i g^s$ holds, then I_k can compute ν_1 and ν_2 such that

$$u_k = u_{k'}^{\nu_1} g^{\nu_2},$$

which contradicts the discrete logarithm assumption. $\square$

In our credential mechanism, we suppose that different credentials are issued with different public (secret) keys which are chosen independently. The cooperation of some organization $O_{j'}$ with certain individuals does not help produce a forgery of the credential which is supposed to be issued by O_j. If it does, $O_{j'}$ and the individuals together must hold the secret key chosen by O_j, since the credential is based on a proof of knowledge. If an organization issues several different credentials, it must choose the corresponding secret keys independently and randomly in order to make the credentials secure.

5 Discussions

Shifting the credential system from an RSA setting to a discrete logarithm setting simplifies the process of validating pseudonyms by dropping cut-and -choose. The model proposed in this paper is closer to Damgård's model (see [Da88]) in the sense that the center will no longer be needed after the pseudonyms are validated, since each organization has its own secret key for issuing a credential without requiring the center to implement any computional task. Furthermore, individuals can validate their own secret keys which will be used when the signatures are necessary under the pseudonyms. But one notable special property of this credential system is that a version of the credential can be shown only once. If an individual wishes to show this credential to a second organization $O_{j''}$ after showing to $O_{j'}$, he must get another version from O_j. This property is suitable for a one-time credential (see [ChaP92]), such as a bank cheque. For other purposes, this property does not seem convenient. It is an open problem to construct a credential system based on discrete logarithms in which the credential, once issued, can be shown to many different organizations independently. Proving Conjecture 1 is another interesting open problem.

Acknowledgements

The author thanks Mike Burmester for discussions about this work, Bob Blakley for some suggested changes, and Dieter Gollman for presenting it at CPAC, as well as for simplification of the process of validation pseudonyms.

Appendix A

A general divertibility of the basic protocol is shown in Fingure 4.

References

[Bran94a]   S. Brands. Untraceable Off-line Cash in Wallet with Observers. In *Advances in Cryptology – Proceedings of CRYPTO 93*. Lecture Notes in Computer Science #773, Springer-Verlag, 1994, pp. 302–318.

[Bran94b]   S. Brands. Untraceable Off-line Cash Based on the Representation Problem. manuscript. To be published as a CWI Technical Report in January/February, 1994.

[CE86]   D. Chaum, J. H. Evertse. A Secure and Privacy Protecting Protocol for Transmitting Personal Information between Organizations. In *Advances in Cryptology - proceedings of CRYPTO 86*, Lecture Notes in Computer Science #263, pages 118-168. Springer-Verlag, 1986.

[ChaP92]   D. Chaum and T. P. Pedersen. Wallet Databases with Observers. In *Advances in Cryptology - proceedings of CRYPTO 92*, Lecture Notes in Computer Science #740, pages 89 – 105. Springer-Verlag, 1992.

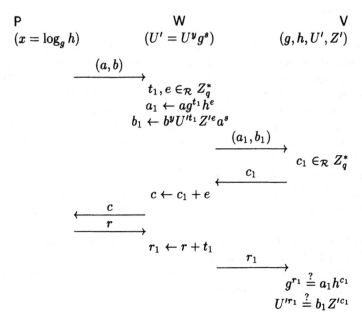

Fig. 4. General divertibility of the basic protocol

[CheDaP94] L. Chen, I. Damgård and T. P. Pedersen. Parallel divertibility of proof of knowledge. In *Advances in Cryptology - proceedings of EUROCRYPT 94*, Lecture Notes in Computer Science pages ? – ?. Springer-Verlag, 1995.

[Da88] I. B. Damgaard. Payment Systems and Credential Mechanisms with Provably Security Against Abuse by Individual. In *Advances in Cryptology - proceedings of CRYPTO 88*, Lecture Notes in Computer Science #403, pages 328 – 335. Springer-Verlag, 1990.

[DH76] W. Diffie and M. E. Hellman New Directions in Cryptography. In *IEEE Trans. Inform.*, IT-22(6):644–654, November, 1976.

[FS87] A. Fiat and A. Shamir. How to Prove Yourself: Practical Solutions to Identification and Signature Problems. In *Pre-proceedings of EURO-CRYPT 86*, pages 186 – 194. 1987.

[OO89] T. Okamoto, J. K. Ohta. Divertible zero-knowledge interactive proofs and commutative random self-reducibility
In *Advances in Cryptology - proceedings of EUROCRYPT' 89*, Lecture Notes in Computer Science #434, pages 134-149. Springer-Verlag, 1990.

[RSA] R. Rivest, A. Shamir, and L. Adleman A method for abtaining digital signatures and public-key cryptosystems In Commun. ACM Vol. 21, pp. 120-126, 1978.

[Sch90] C. P. Schnorr. Efficient identification and signatures for smart cards. In *Advances in Cryptology - proceedings of CRYPTO 89*, Lecture Notes in Computer Science, pages 239 – 252. Springer-Verlag, 1990.

[Sim84] G. J. Simmons. The Prisoner's Problem and the Subliminal Problems. In *Advances in Cryptology - proceedings of CRYPTO 83*, Plenum Press, pages 51–67. 1984.

A New Identification Algorithm

Kefei Chen
Mathematisches Institut
Arndt Strasse 2
D-35392 Giessen, Germany
e-mail: kefei.chen@math.uni-giessen.de

Abstract

Theory of codes with rank distance was introduced in 1985, which can be applied to crisscross error correction and also used to build some cryptographical schemes. In this paper, we propose a new identification algorithm based on rank distance codes. This algorithm is simple to describe, it has also advantages in both communications and memory bits.

1   Introduction

It is known that the problem of finding a codeword of given weight in a linear binary code is NP-complete [1]. This property can be used in cryptography. The idea of using error-correcting codes to build identification algorithm is due to Harari [5], but unfortunately his scheme was not practical. Since then, some schemes using coding problems have been presented [2, 4, 7].

In the present paper, we propose an identification algorithm based on rank distance codes. The main advantages of the new scheme are that it needs less memory and is simpler to implement than similar schemes in [6, 7].

2   Rank Metric and Codes

The idea of rank metric codes has been introduced in 1985 [3]. Let V be an n-dimensional vector space over a finite field $GF(q^m)$, where q is a power of a prime.

We fix a basis $\{u_1, u_2, \ldots, u_m\}$ of $GF(q^m)$. Then any element $x_j \in GF(q^m)$ can be uniquely represented in the form

$$x_j = a_{1j}u_1 + a_{2j}u_2 + \ldots + a_{mj}u_m.$$

Let $M_{m \times n}(q)$ be the set of all $m \times n$ matrices with elements from $GF(q)$, and define the map $\mathcal{A} : V \longrightarrow M_{m \times n}(q)$ as follows:

$$x = (x_1, x_2, \ldots, x_n) \longmapsto A(x) = \begin{pmatrix} a_{11} & a_{12} & \cdots & a_{1n} \\ a_{21} & a_{22} & \cdots & a_{2n} \\ \vdots & \vdots & & \vdots \\ a_{m1} & a_{m2} & \cdots & a_{mn} \end{pmatrix}.$$

The rank of the matrix $A(x)$ is uniquely determined by x. We need some ideas and results of rank metric codes, for the details cf. [3].

- The **rank** $r(x)$ of an element x of V is defined to be the rank of the matrix $A(x)$; The **rank distance** $d(x, y)$ of two elements $x, y \in V$ is defined by $d(x, y) = r(x - y)$.

- A subset C of V is called a **rank distance code** if we restrict ourselves to rank metric; C is called a linear (n, k)-code, if C is a k-dimensional subspace of V.

- We call $w_r(C) = \min\{r(x)|x \in C, x \neq 0\}$ the **minimum rank weight**, and $d_r(C) = \min\{d(x, y)|x, y \in C, x \neq y\}$ the **minimum rank distance** of C.

- Let C be a linear (n, k)-code, then $d := d_r(C) = w_r(C)$. C is also called a $[n, k, d]$-**code**, or a t-**error-correcting code**, where $t = [(d - 1)/2]$.

- Let $x = (x_1, x_2, \ldots, x_n)$ be a vector over $GF(q^m)$, λ be a non-zero element in $GF(q^m)$ and P be a regular $n \times n$ matrix over $GF(q)$, then $r(\lambda x P) = r(x)$.

3 The Identification Algorithm

Assume that a $[n, k, d]$-code over $GF(q^m)$ is given, and let H be the $(n - k) \times n$ parity check matrix of this code, which is used as a system parameter. Then, each user chooses a random vector $s \in V$ of rank r with $3r < d$, which serves as his secret key, it is also part of the system. The public identification is computed as

$$i := sH^t.$$

Now we describe an interactive protocol that enables any user (called prover) to identify himself to another one (called verifier). The identification algorithm is as follows, which should be repeated t times:

1. The prover chooses a random vector $x \in V$ and a random $n \times n$ regular matrix P over $GF(q)$, and sends the syndrome $c = xPH^t$ and $c' = xH^t$ to the verifier.

2. The verifier chooses a random element $\lambda \in GF(q^m)$ and asks the prover to send $w = x + \lambda s P^{-1}$.

PROVER A	VERIFIER B

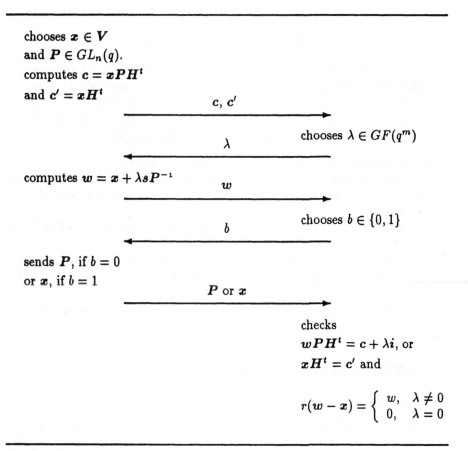

chooses $x \in V$
and $P \in GL_n(q)$.
computes $c = xPH^t$
and $c' = xH^t$

$$c, c'$$

chooses $\lambda \in GF(q^m)$

$$\lambda$$

computes $w = x + \lambda s P^{-1}$

$$w$$

chooses $b \in \{0, 1\}$

$$b$$

sends P, if $b = 0$
or x, if $b = 1$

$$P \text{ or } x$$

checks
$wPH^t = c + \lambda i$, or
$xH^t = c'$ and

$$r(w - x) = \begin{cases} w, & \lambda \neq 0 \\ 0, & \lambda = 0 \end{cases}$$

Figure 1: The identification scheme based on rank distance codes

3. After receiving w, the verifier asks the prover to reveal P or x. In the first case the verifier checks that $wPH^t = c + \lambda i$, and in the second case the verifier checks $xH^t = c'$ and $r(w - x) = r$ if $\lambda \neq 0$, $r(w - x) = 0$ if $\lambda = 0$.

We now show that the above is a zero-knowledge interactive proof system, i.e. it satisfies following properties:

Completeness: An honest prover who knows s will always pass this test.

Soundness: In each round the verifier can ask the prover $2q^m$ questions. A dishonest prover is able to answer correct $q^m + 1$ questions at most. (If he can answer correct $q^m + 2$ questions with $\{c, c', x, P\}$, then there are at least two distinct values λ_1 and λ_2, whose response vectors w_1, w_2 satisfy both condition:

$$w_1 PH^t = c + \lambda_1 i, \quad r(w_1 - x) = r$$
$$w_2 PH^t = c + \lambda_2 i, \quad r(w_2 - x) = r.$$

This implies that $(w_1 - w_2)PH^t = (\lambda_1 - \lambda_2)i$. Let $s' = \frac{1}{\lambda_1 - \lambda_2}(w_1 - w_2)P$, then $(s' - s)H^t = 0$, i.e. $s' - s$ is a codeword of C. Since $r(s' - s) \leq 3r < d$, it follows $s' = s$ and thus the secret key $s = \frac{1}{\lambda_1 - \lambda_2}(w_1 - w_2)P$ can be extracted from any $q^m + 2$ correct answers.) Consequently, the probability of success when the secret key s is not known is $(\frac{q^m+1}{2q^m}) \approx \frac{1}{2}$.

Zero-Knowledge: The simulator is constructed as follows:
At first A computes two vectors s_0 such that $s_0 H^t = i$ and s_1 such that $r(s_1) = r$ respectively. Then the following steps are repeated in each round.
— A chooses $x \in V$ and a regular matrix P, then sends $c = xPH^t$ to B;
— B chooses $\lambda \in GF(q^m)$;
— A chooses $j \in \{0,1\}$ and sends $w = x + \lambda s_j$ to B;
— B chooses $k \in \{0,1\}$. The simulation succeeds if $k = j$, otherwise the simulator repeats the above process.
Note that the above dialog will be arisen with the same probability if the verifier B talk to the true prover.

4 Security of the Scheme

The security of the scheme relies on the following items:

- The general decoding problem for linear codes is NP-complete [1]. The proposed algorithm based on the syndrome decoding problem for rank distance codes, which seems even harder than the general decoding problem of Hamming distance codes.

- The number of the vectors of rank r is

$$N_1 = \prod_{i=1}^{r} \frac{(q^n - q^{i-1})(q^m - q^{i-1})}{q^r - q^{i-1}} \approx q^{r(n+m)-r^2}.$$

- There are $N_2 = \prod_{i=1}^{n}(q^n - q^{i-1}) \approx q^{n(n-1)}$ regular matrices.

As a numerical example, consider the case where $n = 32$, $k = m = 16$, $r = 4$ and $q = 2$. There are three attacks on this scheme:

1. From $i = sH^t$ to find the solution of rank r. (Such a solution s is unique from $r(s) < d/3$, i.e. this scheme is collisionsfree.) Because it deals with rank distance codes, all of the known decoding algorithms are unable to be used. The equation $i = sH^t$ has $q^{mk} = 2^{256}$ solutions, these are too many for an exhaustive search.

2. Test all vectors of rank r. There are $N_1 \approx 2^{176}$ such vectors, therefore that is practically not possible.

3. By means of the knowledge of sP^{-1} to find the s. The number of the regular matrices ($N_2 \approx 2^{992}$) is so large that it will be infeasible to find the secret key.

5 Implementation of the Algorithm

The proposed algorithm is uses rank distance codes with small error correcting capacity. Because of the special properties of rank distance codes the new scheme has more advantages in comparison with other two similar schemes [6, 7].

- Without loss of generality we can assume that the system matrix H is given in the block form $H = (I, H')$, where I is the $(n - k) \times (n - k)$ identity matrix and H' is a $(n - k) \times k$ matrix over $GF(q^m)$. It requires 2^{12} memory bits.

- The secret key s can be stored in 512 bits, and the public key i can be stored in 256 bits.

		New scheme	Shamir's scheme	Stern's scheme
system parameter		2^{12} bit	2^{16} bit	2^{16} bit
secret key		512 bit	120 bit	512 bit
public key		256 bit	256 bit	256 bit
hash function		—	necessary	necessary
multiplications	A	2^{15}	2^{12}	2^{16}
(bit-wise)	B	2^{14}	2^{11}	2^{16}
communications		1792 bit/round	500 bit/round	950 bit/round
necessary round [a]		20	20	34

[a]The minimal round such that the probability for a successful cheat smaller as 10^{-6}.

Figure 2: Comparing RDC with PKP and SD

- The communication complexity of the scheme is (average) 1792 bits. Since the user must send vector c, c' (512 bits), w (512 bits) and either x (512 bits) or P (1024 bits) per round.

- The main computing task of the scheme is that the prover performs one matrix-vector multiplication over $GF(2^{16})$ and the verifier performs one matrix-vector multiplication over $GF(2^{16})$ every two interactions (on the average). It is equivalent to 2^{14} and 2^{13} bit-multiplications respectively.

- The new scheme does not need to use any cryptographical hash function. On the one hand hash function will increase the complexity of the system, on the other hand it is even a problem to find a secure cryptographic hash function.

References

[1] E. R. Berlekamp, R. J. McEliece and H. C. A. Van Tilborg, *On the Inherent Intractability of Certain Coding Problems*, IEEE Trans. Inform. Theory, IT-24(3) pp.384-386, 1978.

[2] K. Chen, *Improved Girault Identification Scheme*, IEE Electronics Letters, Vol.30, No.19, pp.1590-1591, 1994.

[3] E. M. Gabidulin, *Theory of Code with Maximum Rank Distance*, Problems of Information Transmission, Vol.21, pp.1-12, 1985.

[4] M. Girault, *A (non-practical) three-pass identification protocol using coding theory*, Proceedings of Auscrypt'90, LNCS Vol.453, pp.204-211, 1990.

[5] S. Harari, *A New Authentication Algorithm*, Coding Theory and Applications, LNCS Vol.388, pp.204-211, 1989.

[6] A. Shamir, *An Efficient identification Scheme Based on Permuted Kernels*, Proceedings of Crypto'89, Lecture Notes in Computer Science Vol.435, pp.606-609, 1990.

[7] J. Stern, *A New Identification Scheme Based on Syndrome Decoding*, Proceedings of Crypto'93, Lecture Notes in Computer Science Vol.773, pp.13-21, 1994.

Public-key Cryptography on Smart Cards

Andreas Fuchsberger, Dieter Gollmann, Paul Lothian*,
Kenneth G. Paterson**, Abraham Sidiropoulos

Information Security Group
Royal Holloway, University of London,
Egham, Surrey TW20 0EX, United Kingdom.

Abstract. Only recently, high performance smart card implementations of public key algorithms have reached the market, opening a new field of applications for such systems. We will survey the mathematical techniques behind this development, compare digital signature schemes in view of smart card implementations, discuss security management issues of smart card production, and present three applications to demonstrate the use of smart cards for security purposes.

1 Introduction

In many large scale distributed applications, a strongly centralized security management scheme will be prohibitively expensive and may be a major security risk in itself. Thus, we require schemes where most of the security relevant data are kept with individual participants. Even classified security is now drifting away from operating systems with multi-level security to solutions, like FORTEZZA, where users can access a distributed system from arbitrary nodes by keeping their keys on a PCMCIA card.

Public-key cryptography and smart cards are both valuable tools when designing decentralized distributed security systems. For example, Digital's Distributed System Security Architecture [8] uses public key cryptography for certificates and considers smart cards as a future replacement of password based identification and authentication. Public key cryptosystems on smart cards are thus a very attractive proposition for this kind of applications. However, performance has to meet application requirements before such devices can be seriously considered for practical use. A first major step in this direction was the CORSAIR smart card developed by Philips around 1990. Among other features, it offered a 512 bit RSA exponentiation taking less than 1.5 second [3]. Further progress has been made since and recently both Siemens and SGS Thomson announced smart cards which can perform a 512 bit RSA exponentiation in about 60 ms. At this stage, speed becomes less of an issue and the main considerations have to be memory space and support for implementing higher level protocols.

* This author's research supported by EPSRC Case Award No. 93315080 and Zergo Ltd.
** This author's research supported by a Lloyd's of London Tercentenary Foundation Research Fellowship.

In this paper, we will give a brief survey of the state of the art in smart cards with public key algorithms and refer to [26] for an excellent and detailed survey on the capabilities of currently available smart cards. We will explore the mathematical background and the restrictions governing fast implementations of public key algorithms on smart cards, discuss some applications, and point to management issues in a smart card lifecycle which have to be considered when using smart cards in secure applications.

2 Smart Cards

We start with a brief introduction to smart cards, quoting a definition from Eurocrypt'84 [39]:

> You all know what a smart card is: this standard dimensioned plastic card contains a micro electronics package including a memory and a microprocessor controlling read and write access to this memory.

Current smart cards have standardized interfaces, defining the location of the contacts [13, 14] and the transmission protocol [15], they usually have an 8 bit microprocessor and an arithmetic coprocessor to support cryptographic algorithms. At Eurocrypt'95, and not by accident again in France, a comprehensive overview of the current state of the art for such coprocessors was presented [26]. We will therefore restrict ourselves to a typical example, viz the SLE44C200 processor [37], to demonstrate the main memory characteristics of a smart card, i.e.

- 256 byte RAM,
- 9 Kbyte ROM,
- 2.5 Kbyte EEPROM.

The chips on the smart card are restricted in size and have to meet stringent reliability requirements with respect to temperature range, stress, and flexibility. This impacts on the fabrication technology and many current chips use a 1.2-μ CMOS process. In consequence, smart cards remain one of the few areas where memory size and memory requirements of algorithms are an important issue. Within this limited memory space, applications have to be programmed. As a security module, the smart card will store cryptographic keys in its EEPROM. The ROM mask may implement a basic operating system and 'higher level' instructions which execute cryptographic algorithms with the help of the arithmetic coprocessor. In turn, these instructions can be invoked to execute cryptographic protocols.

In the financial sector integrated circuit (IC) cards have now become sufficiently common to merit international standards on financial transaction systems using IC cards, viz ISO 9992 and ISO 10202.

3 Cryptographic Algorithms on Smart Cards

In this section, we outline somer popular public-key algorithms and compare their relative strengths and weaknesses from a smart card perspective. We conclude with a survey of methods proposed to increase their efficiency.

3.1 The RSA Encryption and Signature Schemes

These schemes [34] were the first practical public-key encryption algorithms. Suppose A wishes to send a message to recipient B. We assume that B has previously chosen a pair of large primes p, q, formed the product $n = pq$, and computed integers e, d such that $ed \equiv 1 \bmod \phi(n)$. B publishes n and e as his public key. His private key is d. We assume that messages are integers M with $0 \le M < n$. A, wishing to encrypt M computes ciphertext $C = M^e \bmod n$, which she transmits to B. Because of the relationship between e and d, we have

$$C^d = M^{de} \equiv M \bmod n \tag{1}$$

and B can recover M by computing $C^d \bmod n$ using the private key d.

RSA can also be used to obtain digital signatures. The signer, B, applies a publicly known hash-function h to the message to produce a short, typically 128-512 bit, hash-value of length compatible with modulus n. (Splitting the message into blocks and producing a separate signature for each block is both computationally expensive and potentially insecure.) Many hash-functions are based on block cipher algorithms, while others are 'dedicated', i.e. designed specifically for the purpose of hashing. They generally pose no major headaches for smart card implementation. B now presents (M, S) to A, where $S = h(M)^d$. To verify B's signature, A computes $h(M)$ and then checks that $S^e = h(M)$.

The security of RSA depends on the size of modulus n. Odlyzko [28] has recently forecast that 512 bit moduli will be vulnerable to factorisation in a couple of years and perhaps 768 bit moduli by the year 2004. These estimates are based on projections of computing power, algorithmic advances and continuing ability to organise disparate resources over the Internet. Thus 512 bit moduli are not suitable for the long-term protection of secrets. Barring major advances in algorithms, 1024 bit RSA will probably be secure for many years to come and seems likely to become commonly used soon.

The first hurdle when implementing an RSA-based system is the generation of primes and public and private keys. This can be done off-line, perhaps by some trusted third party, and although a computationally intensive process requiring careful implementation, does not represent the major problem. The main cost of executing RSA lies in the modular exponentiations $C = M^e \bmod n$ and $M = C^d \bmod n$. Square-and-multiply algorithms require, on average, $\frac{3n}{2}$ modular multiplications of n bit numbers. If, as is the case in signature verification, we can choose the parameter e to be only a few bits in size (common choices are $e = 3$ and $e = 2^{16} + 1$), the number of operations can be reduced. Recall however that we wish to carry out these multiplications on an 8 bit smart card with as little as 256 bytes of memory and that in many applications, the computation rather than verification of signatures needs to be carried out on the smart card.

3.2 The Digital Signature Algorithm (DSA)

The DSA, proposed by NIST in 1991, was derived from earlier signature schemes [5, 35] and has since become part of a US Federal Standard [7], mandatory for US government use. Unlike RSA, DSA does not have a public-key encryption capability. The DSA requires six basic parameters to form a signature:

- p, an L bit prime modulus, where L is a multiple of 64 and $512 \leq L \leq 1024$.
- q, a 160 bit prime divisor of $p - 1$.
- g, an integer of order q modulo p.
- x, a random or pseudorandomly generated integer, $0 < x < q$. Integer x forms the signer's private key.
- $y = g^x \bmod p$. Integers p, q, g and y form the signer's public key.
- k, a random or pseudorandomly generated integer, $0 < k < q$.

A fresh integer k is required for each message to be signed. The other parameters can be fixed. All of this information needs to be held on-card in an application where a smart card computes signatures. The signature for message M is a 320 bit pair (r, s), generated as follows:

- Compute SHA(M).
- Compute $r = (g^k \bmod p) \bmod q$.
- Compute $s = (k^{-1}(\text{SHA}(M) + xr)) \bmod q$, where k^{-1} is the multiplicative inverse of k modulo q.

For signature verification, the verifier calculates

$$w = s^{-1} \bmod q, \quad u_1 = \text{SHA}(M) \cdot w \bmod q, \quad u_2 = r \cdot w \bmod q$$

and compares r with $(g^{u_1} y^{u_2} \bmod p) \bmod q$. He accepts the signature if they are equal and rejects the signature otherwise.

The security of DSA rests on the difficulty of finding x from the public key $y = g^x \bmod p$. If an attacker can solve this discrete logarithm problem, then he can go on to form signatures for any messages he desires. Once again, the difficulty of the problem grows rapidly with the size of the modulus p. A recent implementation [21] showed the discrete logarithm problem to be easy for prime moduli of size 192 bits, and the conclusion of [21] was that the discrete logarithm problem is not much harder than factoring numbers of comparable size.

The computation of a signature requires the use of a hash-function, either SHA or a special-purpose function specified in the DSS, and the computations of r and s as specified above. To obtain r requires the modular exponentiation $g^k \bmod p$, and can be achieved using around 240 modular multiplications of L bit numbers using a square-and-multiply technique (recall that k is 160 bits). The calculation of s requires the computation of modular inverses. The DSS suggests the use of Euclid's algorithm for this purpose; Stein's adaptation requires only additions and subtractions of integers. However, many current smart cards have available a modular multiplier co-processor and use the identity $k^{-1} = k^{q-2} \bmod q$ to compute inverses through modular exponentiation [25]. Improvements in the efficiency of DSA have been obtained in [25], avoiding the computation of modular inverses altogether.

3.3 Zero-Knowledge Signature Schemes

A zero-knowledge protocol involves two parties, a prover and a verifier. Through execution of the protocol, the prover is able to prove knowledge of a piece of secret information to the verifier without the verifier (or indeed any third party) being able to discover anything about the nature of the secret, in some strict computational sense. These protocols can be used directly to give secure identification protocols and can be adapted to produce efficient signature schemes. In this section we concentrate on two zero-knowledge signature schemes.

The Fiat-Shamir Signature Scheme The Fiat-Shamir scheme [6] relies for its security on the difficulty of extracting square roots modulo n, where n is a product of primes p and q. This is equivalent to the difficulty of factoring n.

A central authority generates the public modulus $n = pq$, the primes p, q being known only to the central authority. The central authority then calculates a public vector $V = (v_1, \ldots, v_k)$ from I, the card dentity (typically including information such as cardholder name, card number, expiry date and the account number) using publicly known redundancy rules. Corresponding to V, the central authority generates a secret vector $S = (s_1, \ldots, s_k)$ with

$$(s_j, n) = 1 \quad \text{and} \quad v_j = s_j^{-2} \bmod n, \quad 1 \leq j \leq k.$$

Signature Generation To generate a signature on message M, the prover:

- generates a random vector $(r_1, \ldots, r_t)$, where $0 < r_i < (n-1)$ and computes $x_i = r_i^2 \bmod n$ for $1 \leq i \leq t$.
- calculates $h(M, x_1, \ldots, x_t)$, where h is a public hash-function and M is the message to be signed, and uses the first kt bits of this string as the entries e_{ij} of a $t \times k$ binary matrix E.
- computes the signature vector $Y = (y_1, \ldots, y_t)$ with $y_i = r_i \prod_{e_{ij}=1} s_j \bmod n$.

The signature on M is the pair (Y, E).

Signature Verification To verify a signature (Y, E) the verifier:

- computes public vector V from identity information I.
- computes $z_i = y_i^2 \prod_{e_{ij}=1} v_j \bmod n$, for $1 \leq i \leq t$.
- verifies that the first kt bits of $h(M, z_1, \ldots, z_t)$ are the entries e_{ij} of E.

Notice that a signature generated by the prover will always be verified, since

$$z_i = y_i^2 \prod_{e_{ij}=1} v_j = r_i^2 \prod_{e_{ij}=1} s_j^2 \prod_{e_{ij}=1} v_j = r_i^2 \prod_{e_{ij}=1} s_j^2 v_j = r_i^2 = x_i \bmod n.$$

The system can be broken by factoring the modulus n and extracting the secret vector S from the public vector V. Modulus n should therefore be at least 512 bits. Forging a signature for a message M' may also be possible if the forger can arrange the value of $h(M', x_1', \ldots, x_t')$ to be equal to the entries of a matrix E for which he has previously obtained a signature. It is suggested that kt be taken to be 64 or more to avoid this type of attack.

Signature generation and verification require on average $t(k/2 + 1)$ modular multiplications. For a 512 bit modulus, the private key size is $64k$ bytes while the public key (n, V) is $64(k + 1)$ bytes. The signature length is $64t + \frac{kt}{8}$ bytes (not including identity information). Thus it is possible to choose the values k and t to reduce the complexity signature generation and verification and the length of signatures at the expense of private key size. Some sample parameters are displayed in the table below.

k	t	private key size (bytes)	no. of prover multiplications	length of signature (bytes)
2	32	128	64	2056
8	8	512	40	520
32	2	2048	34	136

The Guillou–Quisquater Signature Scheme The Guillou–Quisquater signature scheme [9] is similar to the Fiat–Shamir scheme, but requires less storage and exchange of data at the expense of increased computation. A central authority generates a public modulus n, the product of two primes p and q, and then uses identity information I to generate a public value v again using publicly known redundancy rules. A secret value s corresponding to v is then generated; v and s satisfy $v = s^{-d} \bmod n$ where d is a public, typically 64 bit, integer.

Signature Generation To generate a signature on message M, the prover:

- generates a random number R and computes $x = R^d \bmod n$.
- computes $e = h(x, M)$ where h is a public hash-function.
- computes $y = Rs^e \bmod n$.

The signature on M is the pair (y, e).

Signature Verification To verify a signature (y, e) the verifier:

- computes v from the identity information I.
- calculates $x' = y^d v^e \bmod n$.
- calculates $e' = h(x', M)$.
- checks that $e' = e$.

The prover must execute $3 \log_2 d$ modular multiplications to compute a signature and the verifier must also execute $3 \log_2 d$ multiplications to verify a signature. With a special choice of d, these requirements can be reduced [10]. Recall that d is typically 64 bits in size. A signature (not including identity information) is twice the length of the public modulus n and will typically comprise 1024 bits.

3.4 Choosing a Signature Scheme: A Comparison of Algorithms

We compare the four signature schemes discussed above, paying particular attention to their implementation on smart cards. Some of the relevant issues are:

- the number of arithmetic operations needed to generate or verify a signature.
- the length of signatures, private and public keys.
- the amount of smart card RAM, ROM and EEPROM required to compute a signature (this will be heavily dependent on the use of co-processors and various optimisations).

In Table 1, we assume that 512 bit moduli are used for each scheme. We consider the three parameter sets $(k, t) = (2, 32), (8, 8), (32, 2)$ for the Fiat-Shamir scheme. We assume that the Guillou–Quisquater scheme uses a 64 bit exponent d. We consider two variations on a 512 bit RSA scheme: one with arbitrary verification exponent e and the other with $e = 3$. We assume no further optimisations. The figures for signature generation and verification are the number of 512 bit multiplications required for each scheme and do not include the computations required for verification of a user's credentials or for evaluating hash-functions.

	private key (bytes)	public key (bytes)	signature generation	signature verification	signature length (bytes)
Fiat–Shamir (2,32)	128	192	64	64	2056
Fiat–Shamir (8,8)	512	576	40	40	520
Fiat–Shamir (32,2)	2048	2112	34	34	136
Guillou–Quisquater	64	72	193	193	72
DSA	20	64	240	480	40
RSA, arbitrary e	64	128	768	768	64
RSA, $e = 3$	64	64	768	3	64

Table 1. Comparison of Computational Aspects of Signature Schemes

For Fiat-Shamir, Guillou-Quisquater and unrestricted RSA, there is a clear trade-off between computational costs and length of signatures (and consequent smart card-server communication load). DSA offers strictly better performance than RSA both in signature length and in multiplication count. From a smart card point of view, the advantage of Fiat-Shamir signatures lies in the fact that they minimise the computational resources needed: with the figures quoted for 512 bit moduli, roughly 5 times as much computation is needed for a Guillou-Quisquater signature, and more still for an RSA signature. The trade-off is an increase in signature length.

RSA with verifying exponent 3 is attractive when messages need to be signed only once but verified often; for then the cost of verification is very low and signatures are of a reasonable length. A possible application is in the generation of public-key certificates, which are signed by a certifying agency once and verified many times by other users.

Both RSA and Fiat-Shamir incorporate a public modulus which is a product of two primes and their security relies on the assumed intractability of factoring large numbers. However the modulus plays two different roles in the two schemes. In Fiat-Shamir the same modulus is shared systemwide and only the central authority knows its factorisation, which cannot be immediately obtained from the private key on a card. Key management is greatly simplified: there is no need for a public key or certificates and no danger if the private key on a card becomes compromised. Against this stands the fact, since there is only one system-wide modulus, factoring it may be worth a collaborative effort. Compare this with RSA, where each user has a unique modulus but compromise of the on-card private key only gives the ability to forge signatures for that one card.

One final issue is that of patenting. The Fiat-Shamir scheme is patented (U.S. Patent 4,748,668, May 31, 1988. European Patent 0 252 494, applied for Jan. 13, 1988). The status of Guillou-Quisquater patents is unclear. The situation for RSA and DSA is rather more complex. The US and European patents on RSA (and other types of public-key cryptography) are owned by Public Key Partners (PKP). They have been defended in the US but remain untested in Europe. PKP has also been granted a worldwide licence to use the DSA by NIST, but have stated that they will make the licence available on a royalty-free basis for personal, non-commercial and US Federal, state and local government use. The licensing situation is not clear for European governmental use at the moment.

There are many different trade–offs in Table 1 and the final choice of signature scheme will depend heavily on the proposed implementation. In particular the amount of ROM, RAM, EEPROM and processing power available will to a large extent determine the choice.

3.5   Fast Techniques for Asymmetric Algorithms

We have seen a common thread in the asymmetric algorithms presented above: they all require modular arithmetic computations with large (typically 512 bit) numbers. This is also true of many other signature schemes that we have not touched on here. Performing the computations required to form or verify a signature on a microprocessor with limited memory in a reasonable time is a formidable task and much research has been devoted to easing signature computation. We outline some of that work in this section.

Arithmetic Co-Processors The computational power of a smart card can obviously be increased by introducing onto the card an arithmetic co-processor, specialised for modular arithmetic operations. This single innovation is most responsible for the recent successful introduction of 512 bit public key algorithms on smart cards. The two main methods for speeding up integer arithmetics on these co-processors are customized data structures, and algorithm design.

In the first case, specific integer representations lead to faster algorithms. Signed-digit number have been used since the early days of computer arithmetics [33]. They allow to add integers in constant time, independent of their length, reduce the numbers of multiplications in a square-and-multiply exponentiation, on average from $\frac{3n}{2}$ to $\frac{4n}{3}$, and have been proposed for modular arithmetic [40]. There exist algorithms computing minimal signed-digit representations, see e.g. [17]. Similar advantages arise from carry-save representations, which, slightly modified as delayed-carry integers, are one of the ingredients of Brickell's RSA algorithm [1]. All redundant integer representations, by default, share a common disadvantage. They require more memory space than standard integer representations and memory still is one of the major restrictions imposed by smart cards.

Algorithm design can again attempt to speed up integer arithmetics. A judicious arrangement of carry-lookahead adders can reduce the average addition time and Booth's algorithm can be the basis for fast multiplication [36]. These ideas are suitable for smart card co-processors and form part of the background for the Siemens SLE44C200 RSA co-processor [37]. Precomputation can also contribute to faster exponentiation, but has obvious implications on memory requirements. However, most of the effort in algorithm design has gone into algorithms for modular reduction. Efficient reduction algorithms based on fast floating point arithmetic and rounding operations have been proposed as early as [27] and as recently as [31]. Again, memory overheads make these suggestions less attractive for current smart cards. More promising are algorithms which reduce the number of reductions by allowing intermediate results to 'overflow' and only occasionally subtract an appropriate multiple of the modulus [1, 40]. Finally, we mention Montgomery's algorithm [23], which achieves efficient reduction through the choice of the residue classes used to represent integers.

The methods mentioned so far apply to any modulus and any exponent. We can exploit specific circumstances to make further gains. In the most common example, the owner of a private RSA key, knowing the factors of the modulus, employs the Chinese Remainder Algorithm. Low weight exponents lead to faster square-and-multiply exponentiations by reducing the number of multiplications. Special moduli can make reduction particularly simple, see e.g. [18].

Server-Aided Computation Another option when computing signatures on a smart card is to enlist the computational power of an external server, located e.g. in the smart card reader. For RSA signatures, the card decomposes the private key d into a set of exponents which is sent to the server. The server performs a modular exponentiation of the message for each exponent and returns the results to the card. Finally, the card combines the appropriate exponentiations to form the signature. Similar protocols can be devised for other signature schemes.

These protocols are classed as dependent or independent according to whether the description of the private key is dependent on, or independent of, the private key itself. The first dependent protocol was proposed in [22] and represents the private RSA exponent as the scalar product of a secret binary vector and a public vector. A second protocol uses the Chinese Remainder Theorem to speed up signature generation. The time taken to generate a signature can be further reduced [20] if an addition sequence is used to speed up exponentiation. An independent protocol proposed in [32] decomposes the private RSA key in a completely different way using certain additive bases. It is more secure than the dependent protocols but suffers from the drawback of a high communication between the card and terminal. Performance and parameter selection for such schemes have been analysed in [2, 38].

There are two major problems with server-aided computations: firstly, they create a large amount of extra communication between card and server. Secondly, since we must pass a 'description' of the private key to the server, we may introduce security shortcomings. Various attacks on dependent protocols can be found in [2, 30]. With the recent progress in cryptographic co-processors on smart cards, the necessity for server-aided schemes seems to be on the decline.

4 Smart Card Lifecycles

The security of the mechanisms implemented on a smart card can easily be compromised by flaws in the management processes governing their creation and usage. The following section will therefore present the lifecycle of a smart card, illustrating the management problems which need to be addressed when using smart cards as part of an overall secure system.

The manufacture of a smart card consists of several processes [10]. Starting from the application requirements specification, individual specifications are prepared for the chip, card, mask, ROM software, and the application software. The ROM software is provided to the semiconductor supplier who manufactures the chips. The card fabricator embeds the chip in the plastic card and usually also loads the application software and normalisation data. Security and testing are fundamental aspects in the manufacture of a smart card and intrinsic to the total process. The newly manufactured cards then reach the hands of their respective holders for the productive part of their operational lives. Eventually they will be destroyed and replaced.

4.1 Manufacture

The manufacturing process starts with the design of the card operating system and the application software, following the principles applying to any software for use in security applications. This is in itself a non trivial task but at least the memory available in smart card chips is relatively small which limits the eventual size of the software. There have to be checks that the operating system meets its specification and also that no unintended features have been included.

The ROM mask of the operating system is then given to the chip manufacturer, who will return an implementation of the code for cross checking before manufacturing the batch of chips. This is in itself a useful integrity check but clearly one normally requires this code to be kept confidential and therefore its distribution should be carefully controlled. Furthermore, the manufacturer has to be accountable for all the chips made, some of which, due to yield failures, will need to be destroyed. Otherwise, an attacker may obtain raw chips to mount any form of counterfeit operation.

The application software will normally be designed and developed along a separate path and will be loaded into the PROM memory. Finally, the chip manufacturer would insert a secret manufacturing key into the EEPROM memory. The software in the ROM will have been designed to inhibit all functions without the correct presentation of this key and as such the chip is effectively locked. We must stress that the key generation process can be a weak link in the whole chain of development and manufacture.

The batch of chips is distributed to the fabricator whose task it is to embed the chips into plastic card. The role of the fabricator varies considerably between various customers and their services. As a very minimum the fabricator must test the complete IC card to ensure its operational state. In some cases the fabricator completely personalises the card to the requirements of the issuer. For simplicity we will assume this latter position. In order to undertake this software identification and personalisation process, the fabricator needs to 'unlock' the chip by entering the manufacturer's key. As the last step in the personalisation process, the fabricator will reset the manufacturer's key with a fabricator key before distribution to the card issuer. Once again, flawed cards will be destroyed and all cards have to be accounted for.

The card issuer on receipt of the personalised cards will unlock the card using the fabricator key and will set the PIN for the user and the transaction key that will be used as part of the final application. The issuer will also reset the fabricator's secret key to the card issuer's secret key. The card is now enabled for operation and is distributed to the user.

4.2   Use and Maintenance

A typical smart card can withstand a substantial amount of wear and tear during its lifetime, estimated to be up to ten years depending on the treatment of the card. In applications where a customer is required to enter a PIN to activate the card, the card will lock up if the customer fails to enter his PIN correctly for a predefined number of trials. When the customer returns the card to the issuer then the application can be reset by means of the issuer key.

4.3   Destruction and replacement

During the manufacturing chain or normal usage a number of defective or worn out cards develop. If the fault develops during the manufacturing process, it will be detected and the card will not be personalised and will remain useless. If the fault occurs during the operational lifetime of the card, the card will have to be returned to the issuer and several procedures will be followed to ensure the proper deactivation and/or destruction of the card. When the card reaches its expiry date, it will automatically cease operation and will have to be replaced.

Sometimes the card may have to be replaced for reasons other than malfunction or expiry of the card. If and when new cryptographic functions have to be added, the card may be replaced or re-programmed. Some of the newer cards, e.g. the Mondex card, offer such facilities as part of their specifications.

5 Applications

5.1 Access Control for Broadcast Television

Over the last ten years European states such as Belgium, The Netherlands and Germany have seen high rates of penetration of broadband cable for TV and radio signal distribution. In addition, medium to high power direct-to-home (DTH) satellites have increased the availably of broadcast TV channels right across Europe and have brought down the cost of complete reception equipment to below the price of home video-recoders. The very nature of DTH satellite technology and to a lesser extent broadband cable enables anyone with suitable equipment to receive these transmissions. Broadcasters therefore face the following problems:

- Satellite transmissions are available to everybody with suitable reception equipment right across Europe irrespective of national boundaries.
- Distribution rights may require a broadcaster to ensure that some programming is only available to a single state or a few states.
- Only legitimate subscribers should be able to receive the broadcasts.

The solution adopted by broadcasters is to encode their broadcasts and require subscribers to rent decoders for a monthly subscription fee, or buy a decoder that requires additional authorisation in exchange for a subscription fee. Currently three scrambling methods using additional authorisation and their variants are in wide spread use in Europe, viz Videocrypt, Eurocrypt, and Nagravision. Both Videocrypt and Eurocrypt use smart cards, while Nagravision is based on another form of token, similar to a smart card. Eurocrypt requires the D2-MAC broadcasting standard, whereas Videocrypt is compatible with three current colour broadcasting standards: PAL, SECAM and NTSC.

British Sky Broadcasting (BSkyB) was the first client to use Videocrypt in Europe. Sky Movies was the first channel to encrypt with Videocrypt on 5 February 1990. The Movie Channel and Sky Sports commenced encryption in April 1991, following the launch of the Astra 1B satellite. The British Broadcasting Corporation uses a variation of Videocrypt on their BBC Select service, modified for certain characteristics unique to terrestrial broadcasting e.g. ghosting. The rest of this section focuses on Videocrypt as an example of a pay-TV access control system based on smart cards [11]. Many of the details presented here have been extracted from information available on the Internet [19].

Videocrypt Videocrypt is a pay-tv scrambling system developed jointly by Thomson Consumer Electronics S.A. and News Datacom in the late 1980s (European Patent EP 0 428 252 A2). A smart card is a central part of the Videocrypt system. The card is pre-coded to determine a user's requirements and can subsequently be addressed by the broadcaster's Subscription Management Centre (SMC) utilising the decoder to amend the user's services [29]. The smart card can be used in a number of broadcasting modes, including:

- Clear Mode: Signals sent in the clear are recognised by the decoder and passed to the display without further processing.

- Free Access: Pictures are transmitted encrypted under a key available to the decoder and are decrypted directly without the intervention of a smart card.
- Controlled Access: Access to encrypted pictures is determined by the level of access authorised to the user's smart card.

Security can be addressed in a number of ways using a smart card. These include:

Chaining: An existing customer receives a new card which contains only part of a new encryption system, the remainder of the system is transmitted when the new card is inserted into the decoder and the subscriber complies with the instructions contained within the on-screen graphics.

Over-the-air addressing: Subscriber management centres can address individual smart cards through the decoder. This a vast improvement over other previous scrambling systems. The SMC can provide additional services, reduce service entitlements, send individual messages, blacklist and/or white list.

Cloning: A number of steps have been taken to stop smart cards being copied or cloned. The physical characteristics of the integrated circuit contained within the smart card make 'probing', the process of interrogating an IC by physical means, very difficult as the IC is likely to become damaged in the process. The specifications of the Videocrypt system recommend that the smart cards are replaced every six to twelve months, and each time this is done the algorithms contained in the smart cards could be changed. Any clone cards manufactured during this time would become obsolete.

Video Taping: Videocrypt offers a simple method for tracking down pirates who video high-value programming and then distribute it. A customer's unique identification number can be displayed on the decrypted screen for reference and future litigation. Identification codes can also be hidden in the picture and retrieved at a later stage.

Picture Encryption: Videocrypt encryption is a patented development of Active Line Rotation (Cut and Rotate principle). It encodes a TV picture by cutting each scan line of the picture in two pieces and transposing the resulting two line segments in the broadcast picture. There are 256 possible cut points and cut points near the picture border are eliminated. As each cut point in a single scan line differs from the next in the following scan line the signal has no viewing value to an unauthorised recipient.

In addition to the encrypted picture, a 32 byte control message is broadcast in one of the invisible scan lines sent during the Vertical Blanking Interval (VBI) in similar fashion to the teletext system. Approximately every 2.5 seconds one of these 32 byte messages is processed in the encoder by a secret hash algorithm which transforms the 32-byte message into a 60 bit value. These 60 bits are then used by a second algorithm to determine the cut points used for each scan line. This second algorithm can be seen as a Pseudo-Random Binary Sequence

Generator (PRBSG) that is reseeded every 2.5 seconds with a 32 byte value. On the receiver's side the decoder passes the 32 byte control messages to a smart card implementing the secret hash-function, allowing the decoder to generate the same cut point sequence as the encrypter and so to reconstruct the original image by re-exchanging line segments.

The 32 byte messages other than those used to reseed the PRBSG, are messages from the SMC for addressing individual smart cards for the purpose of subscription management. Smart cards may be activated, deactivated, subscription profiles changed or pay-per-view account information modified over-the-air. In addition the 32 byte messages contain a digital signature (currently 4 bytes) that allows the card to test whether these messages really originate from the encoder and have not been generated by someone analysing the card. This prevents chosen text attacks, where someone tries to probe the secret hash-functions within a smart card with carefully selected 32 byte messages. It also prevents the card accepting forged (re-)activation commands.

Sound Encryption: Videocrypt has the capability of encrypting sound sources to enhance the security of premium events. To date this security feature has not been utilised by broadcasters.

Security of Smart Card Based Pay-TV Systems Theoretically, smart card based pay-TV system are very secure, keeping all secrets on a tamper resistant smart card [24]. In addition, should these secrets become compromised, security can be regained simply by replacing the smart card.

In reality the situation is different. Since its introduction, Videocrypt has been under attack and in the last two years the security of the BSkyB smart cards can only be described as completely compromised. Clone cards have become available which implement the 'secret' hash algorithm but it is not known how the hash algorithm was leaked to the public. Programs are freely available on the Internet which, with the help of a smart card adapter, turn any PC into a Videocrypt smart card emulator. Recent reports even indicate that through poorly designed features of the chip, it was possible to entice the chip to reveal the contents of its secret and protected memory.

BSkyB could increase the frequency of replacing their smart cards, however with an estimated 4 million subscribers and an estimated cost of $1 per card they have so far chosen to remain with a replacement interval of 12 to 18 months, giving clone card manufactures plenty of time to build and distribute clone cards. BSkyB have reacted in a more economical way. The clone cards do not need to implement procedures for card activation, deactivation and pay-per-view functions, so their software is considerably simpler than the one in the real cards. This resulted in some differences between the reaction of the clone card and the original card on pathological 32-byte messages. These differences were used in electronic counter measures (ECMs) against clone cards several times in the past two-and-half years in order to deactivate clone cards. However it has been quite easy each time for clone card manufacturers to discover a solution and react by correcting the software in the clone cards.

5.2 Global System for Mobile Telecommunications (GSM)

The GSM cellular mobile telecommunications standard offered integrated security features in response to the widely reported security failures in earlier generations of mobile telecommunications networks. For example, frequency scanners were used to listen to mobile telephone conversations while some older telephones could be programmed to make calls using another user's identification number, hence billing them for the call. GSM provides authentication of a user to the network, confidentiality over the radio path and anonymity which protects the user's identity over the radio path. GSM constitutes the first large scale use of smart cards in mobile telecommunications as a convenient way of controlling subscriptions and preventing abuse of the network.

A subscriber interface module (SIM), commonly realised in the form of a smart card, is a key component of the GSM security system. No secret information is present in the mobile equipment and all algorithms apart from traffic encryption are carried out on the smart card. Most importantly, the mobile equipment does not take part in the authentication process and without a SIM, a phone cannot be used to make a call. From a user perspective, a SIM is convenient because a user may have one subscription but several pieces of mobile equipment, for example in the car and office, and can use the SIM in any public GSM phone. The SIM has the optional feature of a 4 digit user-changeable Personal Identification Number. If a user fails to identify himself after three attempts, the SIM becomes blocked and can be used no further.

Subscriber authentication starts with the SIM transmitting the user's Temporary Mobile Subscriber Identity (TMSI), comprising the address of the home network and unique identification of the user, to the visited network, which transmits it to the home network. Only the home network knows the relationship between the TMSI and a International Mobile Subscriber Identity (IMSI). The visited network receives a 128-bit random challenge R from the home network as well as the 32-bit value $S = A(R, K_i)$, which is computed using an authentication algorithm and a 128-bit IMSI-related secret key K_i. The visited network challenges the SIM with R and the SIM responds with $S' = A(R, K_i)$. Knowing R and S, the visited network can check $S = S'$. If this check holds, the SIM is authenticated to the visited network.

At the time GSM was defined, the performance of public key algorithms on smart cards was still inadequate and GSM uses symmetric cryptographic algorithms. The authentication algorithm, A38, is composed of two different functions: the authentication algorithm A3 and the cipher key generating algorithm A8. The latter generates a 64-bit cipher key K_c which is used to encrypt traffic over the radio path, thus providing confidentiality for the user. It is used with the proprietary A5 cipher algorithm, and resides in all pieces of GSM mobile equipment. The SIM stores, amongst other things, the A38 algorithm, K_i, K_c and the IMSI (and TMSI) .

The next generation of mobile telecommunications systems is currently being specified. Given the improvements in performance, public key systems are now a feasible alternative to the existing algorithms.

5.3 Electronic Cash

The introduction of smart cards in the financial sector has been mainly driven by the security failures of the existing magnetic stripe card technology. The information stored on the magnetic stripe card is easily read and forged. Smart cards offer the assurance that forgery will be an expensive and uncertain operation. They can provide cryptographic functions, carry more information and serve as a multiple payment applications platform. The two leading trends in the use of smart cards are credit card and electronic purse applications. We will describe one electronic purse system, Mondex, to illustrate the role of smart cards in future payment systems. Mondex will not carry a magnetic stripe on the card, whereas other emerging cards such as Visa's next generation credit card [41] still feature a magnetic stripe and use the chip or the stripe for greater versatility. Mondex, a joint venture between Natwest and Midland banks, is being marketed as a global scheme for an electronic purse system, i.e. an electronic alternative to cash for use in the retail and services business. Its features include:

- Mondex value is the equivalent of cash and is acceptable without need to check the identity of individuals offering Mondex value.
- Mondex value can be transferred from card to card.
- Mondex value can be downloaded by telephone transaction using special telephone devices.
- Voluntary locking of a Mondex card is possible using a simple key press. Unlocking requires entry of a PIN.
- Use of digital signatures and dedicated tamper resistant processor chip to handle cryptographic functions result in enhanced security.

Security features Mondex combines physical and procedural security features during the creation of value and cards and during normal operation. These features can be found in two basic elements in the Mondex system, the silicon chip embedded in the card and the software which controls the movement of value.

The chip embedded on the Mondex card is a specially-tailored security version of the Hitachi H8/310 smart card microprocessor, manufactured in a way that should prevent physical analysis and reverse engineering. Development and manufacture of the chip, embedding the chip on the card, and personalising the card in Mondex's Global Key Centre are all conducted in secure environments.

The personalised cards are customised and issued by the member banks. For each currency, a single 'originator' body formed by a number of member banks or a country's central bank, has the capability to create value in that currency. An originator's 'master value' card that can hold great amounts of value is the initial carrier of Mondex value and is kept under strict physical and electronic security. It can only transfer value to a set of 'bank value' cards and these cards will propagate and distribute the value through several card levels to customers and retailers. The value on each card is protected by cryptographic means.

A cryptographic Value Transfer Protocol validates and protects the transfer of value between cards. A transaction, e.g. between a customer's card and a shopkeeper's Mondex terminal, consists of three steps:

- Registration: The card information such as the value stored on the customer's chip, is supplied to the chip in terminal, which acts as the shopkeeper's store of value and performs validation checks.
- Authentication: Mutual authentication between the chip on the customer's card and the chip in the shopkeeper's terminal is performed.
- Value transfer: The shopkeeper's terminal requests payment and transmits a digital signature. The customer's card checks the signature and if the customer authorises the transaction, makes the payment with its own digital signature attached. The terminal checks the signature and sends a signed acknowledgement before updating its own stored value with the payment.

If the protocol is interrupted at any stage, recovery to prevent any duplication or unauthorised creation of value is possible.

Updates may be performed either by issuing a new generation of cards or by introducing new cryptographic mechansims without changing the cards.This is possible because the first generation of Mondex cards contains two cryptographic mechanisms, A and B say. Initially the card uses A but can switch to B when instructed to do so. If further new mechanisms are needed only a small number of BC cards needs to be issued, which will instruct any AB card to switch to B.

Fraud Detection and Containment The Mondex system provides several mechanisms to detect and contain the risk of fraud. Retailers' terminals and bank cashpoints capture transaction data providing Mondex with the ability to find suspicious patterns of behaviour. Every genuine transaction carries a unique sequence number and any inconsistency in the sequence numbers of transactions can indicate a problem with the card. In addition, each Mondex card can automatically assess activity, against a set of threshold parameters. If these thresholds are exceeded an internal flag is set to enable a member bank to disable the card. Transactions that try to exceed the limits set on the card, like value of a transaction, number of transactions or other preset criteria, also indicate that closer inspection is needed. Each Mondex card maintains three logs:

- A user transaction log that records data for the last ten transactions. These can be inspected using special equipment.
- A pending log which records the state of the current transaction allowing automatic error recovery in case of unsuccessful completion.
- An exception log, which records data of unsuccessful transactions. When this log is full the card is automatically disabled.

Naturally, all cards reported stolen, fraudulent or unauthorised can be deactivated from the banking system. Such cards can be isolated by retaining them at automatic teller machines (ATMs), locking them out, or placing them in 'hot lists'. A 'cascade' mechanism propagates hot lists through the card network and thus enables the locking of suspect cards by other cards, even if they have never used an on-line terminal.

6 Outlook

Technological advances will place more powerful microprocessors on smart cards. These improvements will certainly lead to further decreases in computation times but, as we have seen, speed has become less of an issue already with present smart cards. More importantly, memory will increase. Following a prediction by Hitachi [12], we can expect for a 0.3μ geometry

- 2 Kbyte RAM,
- 80 Kbyte ROM,
- 40 Kbyte EEPROM.

Eventually, the present 8-bit processors may be replaced by 16-bit processors. From a security perspective, the most important consequence would be the introduction of processors with a protected mode, which could form the basis of secure multi-application operating systems (MAOS). With such an operating system, applications from different organisations could be placed on the same card and the operating system has to prevent any interference between these applications and their security relevant data. Present applications, like Mondex, are still multi-function cards where all applications belong to a single organisation. The processors involved will still be relatively simple and thus may be amenable to formal proofs of correctness.

Sometimes, physical security is stated as another advantage of smart cards. While it is true that smart cards can be made reasonably tamper resistant, it would be unwise to rely too much on this feature. The discussion of the lifecycle has shown that there are many other ways of getting hold of sensitive information and the Videocrypt example has corroborated this view. It is more prudent to be prepared for regular changes in cards and algorithms to stay one step ahead of potential attackers.

References

1. E.F.Brickell, *A fast modular multiplication algorithm with application to two key cryptography*, Proc. of Crypto'82, pp. 51–60, 1982.
2. J.Burns and C.J.Mitchell, *Parameter Selection for Server-Aided RSA Computation Schemes*, IEEE Trans. on Computers, Vol.43, pp. 163–174, 1994.
3. D.de Waleffe and J.-J.Quisquater, *CORSAIR: A Smart Card for Public Key Crytposystems*, Proc. of Crypto'90, Springer LNCS 537, pp. 502–513, 1991.
4. W.Diffie and M. E.Hellman, *New directions in cryptography*, IEEE Trans. Inform. Theory, IT–22, pp. 644–654, 1976.
5. T.ElGamal, *A public key cryptosystem and a signature scheme based on discrete logarithms*, IEEE Trans. Inform. Theory, IT–31, pp. 469–472, 1985.
6. A.Fiat and A.Shamir, *How to prove yourself: Practical solutions to identification and signature problems*, Proc. of Crypto'86, Springer LNCS 263, pp. 186–194, 1987.
7. FIPS PUB XX, February 1, 1993, *Digital Signature Standard*, 1993.

8. M.Gasser and E.McDermott, *An Architecture for Practical Delegation in a Distributed System*, Proceedings of the IEEE Symposium on Security and Privacy, Oakland, pp. 20–30, 1990.

9. L.C.Guillou and J.-J.Quisquater, *A practical zero-knowledge protocol fitted to security microprocessor minimizing both transmission and memory*, Proc. of Eurocrypt'88, Springer LNCS 330, pp. 123–128, 1988.

10. L.C.Guillou, M.Ugon and J.-J.Quisquater, *The Smart Card: A Standardized Security Device Dedicated to Public Cryptography*, in G.J.Simmons, editor, Contemporary Cryptology, IEEE Press, 1991.

11. J.Hashkes and M.Cohen, *Managing Smart Cards for Pay Television, The Video-Crypt Approach*, Seminar on Conditional Access for Audiovisual Services, Rennes, France, 12-14 June 1990 (ACSA '90).

12. Hitachi Europe Ltd., em IC Card Devices — Towards 2000, presented at Smart Card 95, London, 1995.

13. ISO 7816-1:1987 *Identification cards – Integrated circuit(s) with contacts – Part 1: Physical characteristics*, Geneva: International Organization for Standardization, 1987.

14. ISO 7816-2:1988 *Identification cards – Integrated circuit(s) cards with contacts – Part 2: Dimensions and location of the contacts* Geneva: International Organization for Standardization, 1988.

15. ISO/IEC 7816-3:1989 *Identification cards – Integrated circuit(s) cards with contacts – Part 3: Electronic signals and transmission protocols*, Geneva: International Organization for Standardization, 1989.

16. ISO/IEC Draft International Standard 13818-1, *Information technology – Generic coding of moving pictures and associated audio information – Systems*, Geneva: International Organization for Standardization.

17. J.Jedwab and C.J.Mitchell, *Minimum weight modified signed-digit representations and fast exponentiation*, Electronics Letters, Vol.25, pp. 11171–2, 1989.

18. H.-J.Knobloch, *A Smart Card Implementation of the Fiat-Shamir Identification Scheme*, Proc. of Eurocrypt'88, Springer LNCS 330, pp. 87–95, 1988.

19. M.Kuhn, *Details.txt*,
ftp://cip.informatik.uni-erlangen.de/VideoCrypt/cardadapter.

20. C.-S.Laih, S.-M.Yen and L.Harn, *Two Efficient Server-Aided Secret Computation Protocols Based on the Addition Sequence*, Proc. of Asiacrypt'91, Springer LNCS 739, pp. 450–459, 1992.

21. B.A.LaMacchia and A.M.Odlyzko, *Computation of Discrete Logarithms in Prime Fields*, Designs, Codes and Cryptography, Vol.1, No.1, pp. 47–62, 1991.

22. T.Matsumoto, K.Kato and H.Imai, *Speeding up Secret Computations with insecure Auxiliary Devices*, Proc. of Crypto'88, Springer LNCS 403, pp. 497–506, 1990.

23. P.L.Montgomery, *Modular Multiplication Without Trial Division*, Mathematics of Computation, Vol.44, pp. 519–521, 1985.

24. G.Morgan, *Smart Cards for Subscription Television: VideoCrypt — a Secure Solution*, Proc. Smart Card '91, Agestream Ltd., Peterborough, UK, 1991.

25. D.Naccache, D.Raihi, D.Raphaeli and S.Vaudenay, *Can D.S.A. be Improved? — Complexity Trade-Offs with the Digital Signature Standard*, Proc. of Eurocrypt'94, Springer LNCS 950, pp. 77–85, 1995.

26. D.Naccache, *Arithmetic Co-processors: The State of the Art*, preprint, 1995.

27. M.J.Norris and G.J.Simmons, *Algorithms for high-speed modular arithmetics*, Congressus Numerantium, Vol.31, pp. 151–163, 1981.

28. A.M.Odlyzko, Talk given at Hewlett-Packard Symposium on Information Security, Royal Holloway, University of London, 19 December, 1994.

29. P.Peyret, G.Lisimaque and T.Y.Chua, *Smart Cards Provide Very High Security and Flexibility in Subscribers Management*, IEEE Transactions on Consumer Electronics, Vol.36, No.3, pp. 744-752, 1990.

30. B.Pfitzmann and M.Waidner, *Attacks on Protocols for Server-Aided RSA Computation Protocols*, Proc. of Eurocrypt'92, Springer LNCS 658, pp. 153-162, 1993.

31. K.C.Posch and R.Posch, *Modulo Reduction in Residue Number Systems*, IEEE Transactions on Parallel and Distributed Systems, Vol.6, pp. 449-454, 1995.

32. J.-J.Quisquater and M.De Soete, *Speeding up smart card RSA computations with insecure coprocessors*, D.Chaum, ed., Proc. of Smart Card 2000, Elsevier Science, Amsterdam, pp. 191-197, 1991.

33. G.W.Reitwieser, *Binary Arithmetics*, in Advances in Computers, F.L.Alt (ed), Vol.1, pp. 231-308, 1960.

34. R. L.Rivest and A.Shamir and L.Adleman, *A method for btaining digital signatures and public key cryptosystems*, Commun. ACM, ol.21, pp. 120-126, 1978.

35. C.P.Schnorr, *Efficient Identification and Signatures for Smart Cards*, Proc. of Crypto'89, Springer LNCS 435, pp. 239-252, 1990.

36. H.Sedlak, *The RSA cryptography processor*, Proc. of Eurocrypt'87, Springer LNCS 304, pp. 95-105, 1987.

37. Siemens, *ICs for Chip Cards – SLE44C200*, Data Sheet 09.94, 1994.

38. A.Shimbo and S.Kawamura, *Factorisation Attack on Certain Server-Aided Computation Protocols for the RSA Secret Transformation*, Electronics Letters, Vol.26, pp. 1387-1388, 1990.

39. A.Turbat, *Introductory Remarks*, Proc. of Eurocrypt'84, Springer LNCS 209, pp. 457-458, 1985.

40. N.Takagi and S.Yajima, *Modular Multiplication Hardware Algorithms with a Redundant Representation and Their Application to the RSA Cryptosystem*, IEEE Trans. on Computers, C-41, pp. 887-891, 1992.

41. Visa, *Chip Cards: More power to the Cardholder*, October 1994.

Integrating Smart Cards Into Authentication Systems

Gary Gaskell (DSTC)[1]
Mark Looi (QUT)[2]

Abstract

This paper presents alternative schemes for the integration of
smart card technology into the Kerberos authentication system.
A limitation of the initial interaction phase is identified and three
implementation options are proposed to overcome this weakness. A further three implementation options are described that
enhance the security of Kerberos authentication, however these
do not cryptgraphically overcome the identified limitation.

1   Introduction

The following paper outlines a number of options to integrate Smart Cards into authentication systems. In particular the Kerberos authentication system [KOHL] is considered.

The options have been designed to only require changes in the protocols in the initial
interactions with the Authentication Server (AS). This has the benefit of not creating
legacy problems for applications that are already secured using Kerberos.

A security issue with the Kerberos authentication system is the potential for an attacker to perform off-line attacks on the initial ticket from the Authentication Server.
This ticket is sealed using the password of the user. Using a password is a weak way
[GONG] to seal the initial ticket. An attack on the ticket can occur off-line by using a
password guessing approach. Once an attacker has an unsealed initial ticket, it is possible to masquerade as the user. The objective of this paper is to describe a series of options to integrate smart cards into Kerberos. All options but the first are targeted to
avoid this attack. It is acknowledged that Kerberos V5 has implemented "pre-authentication" in an effort to thwart this attack. The reader is referred to Section 4.1 for a discussion of this attack.

1.1   Overview

---

1. DSTC Pty Ltd operates the Co-operative Research Centre for Distributed Systems
Technology. Email gaskell@dstc.edu.au.
The work reported in this paper has been funded in part by the Cooperative Research
Centre program through the Department of the Prime Minister and Cabinet of Australia.
2. School of Data Communications, Queensland University of Technology. Email
mlooi@fit.qut.edu.au

Six proposals are included, and are listed as follows.

- Password release

   The card is only used to store a text password. A PIN is required before the card will release the password. This only requires changes in the client authentication program.

- Strong master key on card

   A cryptographic key is stored on the card. This key is chosen randomly and is considered to be the authentication information that Kerberos uses to identify the user.

- Session key decryption

   The Kerberos ticket from the Authentication Server is decrypted on the card. This allows the user's master key to remain securely on the card and only release a session key to the host.

- DES Challenge/Response

   A DES based challenge/response occurs between the Authentication Server and the card. A ticket to the Ticket Granting Server is only issued if this interaction is successful.

- RSA Challenge/Response

   An RSA based challenge/response occurs between the Authentication Server and the card. A ticket to the Ticket Granting Server is only issued if this interaction is successful.

- Zero Knowledge Proof

   Zero Knowledge Proof (ZKP) exchanges occur between the Authentication Server and the card. A ticket to the Ticket Granting Server is only issued if these interactions are successful. The ZKP to be used is the Guillou-Quisquater [GUIL] algorithm.

2   Objectives

The overall objective is to enhance the security of Kerberos-based authentication. Any changes performed should not change the way Kerberos and applications interact. The objective is to enhance the strength of the authentication stage of Kerberos, but to leave the remainder of Kerberos interactions intact. In this manner, legacy problems are avoided by not changing the way applications work with Kerberos. Integration of the outcomes of these projects into the Open Software Foundations's (OSF's) Distributed Computing Environment (DCE) should be possible.

The security of Kerberos authentication is greatly enhanced if the user's authentication information is more than just a password. The use of a smart card, requires a user logging in to not only recall a password (or passcode/PIN) but to also be in possession of a token. The six options outlined in this paper describe methods to include a token, such as a smart card.

Kerberos does not directly authenticate a user when that user applies for an initial ticket (TGT). A user's identity is only checked when the user presents the initial ticket and an "authenticator" to the Ticket Granting Server (TGS). The Authentication Server (AS) will issue a TGT for user A (the target user) to a user S if S requests it. That is, the user *A* is not authenticated at that stage. Another way to say this, is that the AS will issue a TGT for any particular user, to any other user that requests it. The caveat is that only the proper user should be able to use this TGT. The following stage of interaction with the TGS checks that an "authenticator" and the TGT is valid. The TGT is initially encrypted when it is sent from the AS and it must be decrypted. The TGT is encrypted using the target user's private key. The TGS checks that the TGT has been decrypted and also that the accompanying authenticator is valid. This is where the real authentication occurs in Kerberos.

The above paragraph describes the indirect (by inference) authentication of Kerberos. Kerberos in reality consists of three series of interactions. The first is with the AS, but as detailed above the TGT will be issued to anyone. The second series of interactions is with the TGS. The TGS checks the TGT and the authenticator, and this is where the actual act of authentication takes place. The third series of interactions is between the client and application server. An objective of the latter of these smart card integration options is to directly perform authentication with the AS. This is achieved at the expense of multiple interactions with the AS. However, as there are no changes to the TGS or server interactions no legacy problems for applications are created. (Applications only interact with the TGS of Kerberos and their own servers.)

A well known weakness of Kerberos is the use of the user's password to protect the TGT and session key. This can be attacked off-line by password guessing routines. It is claimed in [GONG] that even with password quality checks, the order of the attack is only 10^5 trials. This compares poorly with a brute force attack on DES which requires about 10^{17} trials. An attack on DES is considered as impractical for the purposes of this project, however the reader may wish to be aware of highly specialised equipment that could be built to attack DES encryptions [WEIN].

3   Technology Framework

3.1   Smart Card technology

Smart cards are plastic cards the size of a credit card. Embedded in the plastic of the card is a small silicon chip microprocessor. The microprocessor has been manufactured to make it virtually impossible to duplicate or read the contents of its registers. This is in complete contrast to magnetic stripe cards, which are simple to duplicate.

Smart cards are typically sold with an operating system already on the card. This operating system makes it possible to format and use various data areas on the microprocessor. It also provides a series of instructions and communications to allow applications on hosts to talk with and use the card. While this operating system provides a series of commands to perform read, write and encryption operations, it does not allow an application to actually have its own control logic on the card. Essentially these cards are limited to performance of commands from their operating system System Programming Interface (SPI).

3.2   The Kerberos Authentication Architecture

The Kerberos security architecture is essentially an authentication mechanism. It is intended for the client-server application programming paradigm. Kerberos operates with three major groups of interactions. These are the user interacting with the Authentication Server (AS), the client (user program) interacting with the Ticket Granting Server (TGS) and the client program interacting with the server (and using the Kerberos library APIs to secure an interaction). Further detail can be found in [KOHL].

The user obtains a credential from the AS server in the first interaction. The second interaction obtains a "ticket" for a service. The third interaction uses this "service ticket". It is possible in the first interaction for *any* user to obtain a credential for any other user. Kerberos authentication security is based on the principle that the wrong user should not be able to use this credential, as they do not have a password to decrypt it. The authors consider this a security weakness and propose smart card integration options in this paper to overcome this shortcoming.

4   Implementation Options

The following pages provide the detail of the various options for integrating smart card technology into the Kerberos authentication system. They are presented in a rising level of security. Any of the options could be implemented in isolation to any other option. Each option is independent of the others. The option that most suits an installation will depend on the risk as assessed and on the trust placed in the various encryption and protocol technology options.

The secret key based challenge/response option has not been explored in detail and so it is not included in the following discussions.

Legend to the diagrams:
- PIN   Personal Identification Number
- K_c User's master key - the password-based key
- TGS   Ticket Granting Server
- AS   Authentication Server
- Cuid   Client User ID
- $K_{c,tgs}$ Session key between the user and the TGS
- K_{tgs} Secret key of the TGS
- $T_{c,tgs}$ Ticket to the Ticket Granting Server for the user (it is encrypted by K_{tgs})
- rand   a random number
- Fn(rand) A function on a random number
- K_{pubC} The public key of the User (Client)
- K_{privC} The private key of the User (Client)
- K_{pubAS} The public key of the authentication server
- K_{privAS} The private key of the authentication server
- H(msg)   A hash function is performed on the value: msg.

The terminology *secret key* is used to refer to a symmetric cryptographic key, whereas, *public key* and *private key* refer to asymmetric cryptographic keys.

4.1 Smart card - password storage

A microprocessor card is used to store the user's password. The user supplies their PIN and ID to the Kerberos login program. The card receives the PIN and releases the user's password to the login program. This option requires minimal changes to the Kerberos system, namely alterations to the login program to extract the password text from the card rather than from the user's terminal process. Simple password storage provides no extra cryptographic security against password guessing attacks save that the text string that makes up the password can be complex, as the user does not need to remember the string.

Extra security is evident by the fact that the user now carries a "token" and that token is required to login to Kerberos. Kerberos could be changed so that it recognises both smart card and non-smart card users.

Advantages:
- A token is required to log in to Kerberos;
- All Kerberos interactions remain unchanged. The only change is to the login program (though a commercial implementation would insert smart card administration into the Kerberos administration server).

Disadvantages:
- This does not avoid password guessing attacks;
- It does not authenticate the user directly, but relies on the TGS to do it.

It is to be noted that when the AS receives the contents of message 2 (see Figure 1), it prepares a TGT and generates a session key for the user claimed in message 2. There is no possibility that the AS can authenticate the sender of message 2 and so it returns the TGT and a session key encrypted under the password-derived key for the claimed userid. The user (or attacker) receives the TGT and session key. Only the user should be able to decrypt these, however the ability to mount a password-guessing attack is a claimed weakness of Kerberos. Solutions have been proposed (pre-authentication [PA-TO]) that might address this, however these solutions are targeted at password guessing attacks and not this "delayed authentication". The pre-authentication solution also relies on "machine keys" on the clients' hosts, which makes system management more complicated. This approach may not be scalable and may not avoid this attack if the attacker can change the login program on a client machine. The last three options aim to defeat this claimed weakness, by requiring the user to directly authenticate to the AS, before a TGT will be issued.

Following step 5 of the protocol in Figure 1, the Kerberos client uses the password to derive the client's key to decrypt the TGT and the session key. If the TGT and encrypted session key were captured by an attacker, it is possible for the attacker to attempt to off-line password guessing attack.

This option is the simplest to implement. Potentially only the Kerberos login client needs to be modified. The random text string that forms the password can be handled in the normal way, when entering users into the Kerberos database. The AS uses the password from the database in the same way it would for any normal password. This is not possible in the following option.

The protocol steps below refer to figure 1 above.

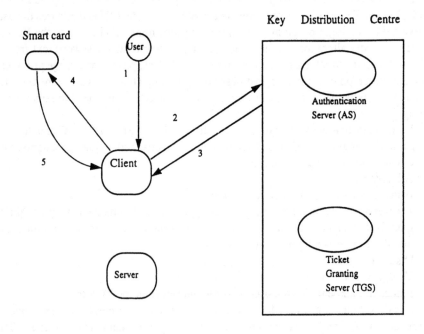

Figure 1 Initial Kerberos Protocols.

Step 1   Cuid, PIN
Step 2   Cuid, tgs
Step 3 $\{K_{c,tgs}\}K_c, \{\{T_{c,tgs}\}K_{tgs}\} K_c$
Step 4   PIN
Step 5   password

4.2   Smart card - master encryption key storage

A microprocessor card is used to store the user's secret key. The user enters their PIN and ID to the Kerberos login program. The card receives the PIN and releases the user's secret key to the login program. Using a smart card to store the user's secret key ensures a lost card poses no security threat. A user is not required to remember a cryptographic key, however the strong key allows the Kerberos tickets to be strongly protected. This prevents off-line password guessing attacks. This approach differs from the first option in that the key is a full strength DES key. That is, its strength is not limited by the fact it must be represented as a text string.

Extra security is provided by the fact that the user now carries a "token" and that token is required to login to Kerberos. Password guessing attacks on the tickets are also infeasible due to the use of a cryptographically strong user secret key. Kerberos could be changed so that it recognises both smart card and non-smart card users. This is not planned for this stage as it requires changes to the AS and the Kerberos databases.

Advantages:
- Off-line password guessing attacks become infeasible;
- A token is required to login to Kerberos;
- All Kerberos interactions remain unchanged. The only change is to the login program (though a commercial implementation would insert smart card administration into the Kerberos administration server).

Disadvantages:
- It does not authenticate the user directly, but relies on the TGS to do it;
- The user's secret key is released to a host computer;
- Key synchronisation between Kerberos and the smart may be complicated if changes are not made to the Kerberos server.

Similarly, as in the first option, the AS will issue a TGT for any user to any other user (or attacker) that requests it. The great benefit of implementing this option is that the password guessing attacks are effectively thwarted, because the full strength of the DES cryptosystem can be applied to protect the TGT and session key. This is directly due to the fact that the client's key can be randomly chosen from the entire spectrum of possible DES keys.

This option is more complicated to implement, as the Kerberos libraries must be modified to not convert the key (formally the password) from a string into a key, as it is already a key. This requires modifications to the login client, the AS server and the user database (if it is desired to distinguish between smart cards users and non-users).

The protocol steps below refer to figure 1. The steps are similar to the protocols in the password storage option, with the exception of step 5, where the user's master key is transmitted as opposed to the user's password.

Step 1   Cuid, PIN
Step 2   Cuid, tgs
Step 3 $\{K_{c,tgs}\}K_c, \{\{T_{c,tgs}\}K_{tgs}\}K_c$
Step 4   PIN
Step 5 K_c

4.3   Smart card - Session key decryption on card

A common security issue is the vulnerability of an encryption master key. A master key that sits on a host computer is only as safe as the security of the host computer. This option limits the vulnerability of the master key. The master key is stored on a smart card and never leaves the card to the local host. The master key is used internally to the smart card to decrypt the session key and TGS ticket. The session key and TGS ticket are then transmitted to the host computer. Therefore, the master key never has to be released to the host.

In this way, the vulnerability of the client is limited to duration of the life of the session key, which is typically one working day. This places a time limit on possible masquerading by an attacker, who obtained the session key due to poor host security.

Advantages:
- The user's master key is never exposed outside of the card (except in the trusted KDC);
- A token is required in the login process;

- Only the user's session key is exposed on the host, so that problems from trojan software are limited in time;
- Application interactions with the Kerberos security service remain unchanged. Disadvantages:
- The user is not authenticated directly, but relies on the TGS to do it;
- To program parts of the Kerberos login program (kinit) on a smart card will be more complicated than previous options. It is not clear that all cards will be able to decrypt the TGT.

The Kerberos login program passes the encrypted TGT and session key back to the smart card for decryption. As the common mode of encryption of DES is CBC, it is envisioned that the TGT may have to be passed in blocks to the card for decryption. If the card is not capable of DES-CBC, while maintaining states between block decryptions, then it appears that this card will not be useful. It is intended to assess the number of cards that are commonly available that could perform this service.

The amount of on-card decryption in this option is considered to be at a practical level between smart card processing and the level of security provided. The amount of on-card encryption processing is analysed in the Open Software Foundation Request for Comments 71.0 [GASK]. This OSF-RFC discusses the further options of decrypting service tickets on a smart card and also performing all the cryptographic processing on the card [KRAJ]. It is considered sufficiently secure for commercial purposes to maintain the secret only on the card, and expose the TGT and service tickets to the local host. It is to be remembered that the physical user is trusting the workstation to perform the business process and we conjecture that is sufficient for trusting the workstation with service tickets. The remaining options in this paper discuss options to enhance the security/trust in the log-in stage of Kerberos.

The protocol steps below refer to figure 1. The steps are similar to the protocols in the password storage option. In step 4 however, the encrypted ticket is sent to the smart card. In step 5 the session key (for the TGS) is returned from the smart card as opposed to the user entering their master password into the host.

Step 1   Cuid, PIN

Step 2   Cuid, tgs

Step 3 $\{K_{c,tgs}\}K_c, \{\{T_{c,tgs}\}K_{tgs}\}K_c$

Step 4 $\{K_{c,tgs}\}K_c, \{\{T_{c,tgs}\}K_{tgs}\}K_c, PIN$

Step 5 $K_{c,tgs}, \{T_{c,tgs}\}K_{tgs}$

4.4   Smart card - DES challenge/response

In standard Kerberos, any user can request a TGS ticket for another user from the AS. As the TGT is itself sealed, this is not normally a large problem but it is nevertheless a security weakness. This proposal aims to authenticate directly the requesting user in the initial interactions with the AS. This raises the security of the authentication system. A symmetric key challenge / response will be used in this option.

The protocol has not yet been selected. However, suitable symmetric cryptography challenge/response protocols exist and are published. A secret-key challenge/response approach may not be implemented, as an asymmetric system exhibits more advantages and therefore the exact benefit of a symmetric key cipher challenge is doubtful.

Advantages:
- Increased security in the authentication stage of Kerberos;
- The AS directly authenticates the user;
- Application interactions with the Kerberos security service remain unchanged.

Disadvantages:
- Parts of the protocol need to be executed on the card, which requires a special development kit.

The details of this option are not expanded upon, as it is expected that an asymmetric implementation is likely to be chosen over a symmetric implementation. It is therefore included more for completeness than for any other reason. Still it must be realised that some organisations prefer to trust symmetric key cryptography. Further reasons that may support this option include the patents issues with asymmetric cryptography and the issue that symmetric key cards are usually cheaper that asymmetric key cards.

4.5   Smart card with public-key cryptography

The aim of using a challenge / response interaction between the AS and the card is to authenticate the card to the AS. This provides the AS with the ability to decide not to issue a TGT to a false user. The combination of public-key cryptography and smart cards in the authentication stage allows the user and the AS to mutually authenticate each other in a strong manner.

This provides an appreciable increase in the amount of trust a user has in Kerberos. The use of public-key cryptography also increases the amount of trust an application server can have in a claimed identification. The solution is also possibly more scalable than standard Kerberos.

Advantages:
- Key management is simplified by the use of public-key technology. The trust model may also allow Kerberos to verify a chain of Certificate Authority Certificates;
- The AS directly authenticates the user;
- The user can mutually authenticate the AS (This may be required in some trust situations);
- Application interactions with the Kerberos security service remain unchanged.

Disadvantages:
- Reliance on a patented (US) cryptosystem (the RSA [RIVE] cryptosystem by default)
- Extra processing time in the AS for the public-key cryptography could be an issue. The RSA implementation on the card is assumed to be fast enough.

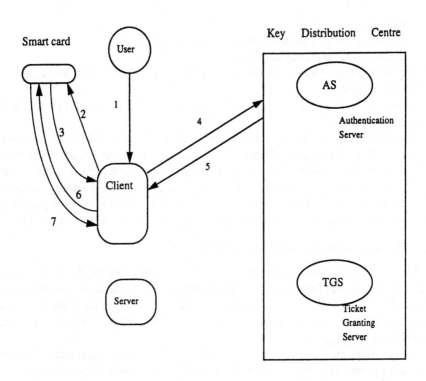

Figure 2 Kerberos with public-key authentication.

Step 1   Cuid, PIN

Step 2   PIN,H (Cuid, tgs, time)

Step 3 $\{H \text{ (Cuid, tgs, time)}\}K_{privC}$

Step 4   Cuid, tgs $\{H \text{ (Cuid, tgs, time)}\}K_{privC}$

Step 5 $\{\{K_{c,tgs}\}K_{pubC}, \{\{T_{c,tgs}\}K_{tgs}\}K_{pubC}\}, \text{time}\}K_{privAS}$

Step 6 $\{K_{c,tgs}\}K_{pubC}, \{T_{c,tgs}\}K_{pubC}$

Step 7 $K_{c,tgs}, \{T_{c,tgs}\}K_{tgs}$

4.6   Smart card with Zero-knowledge Proof

This option involves the use of the Guillou-Quisquater [GUIL] zero knowledge proof scheme. This scheme uses modular exponentiation and was designed to be implemented on a microprocessor card. It takes into account the limited processing abilities of the microprocessor, but is still apparently secure.

Advantages:
- The protocol was designed with a microprocessor card in mind;
- Apparently strong protocol, through the use of modular exponentiation;

Disadvantages:
- The algorithm is possibly protected by a patent;
- The response time is unknown and is expected to be slow due to the multiple interactions of this protocol.

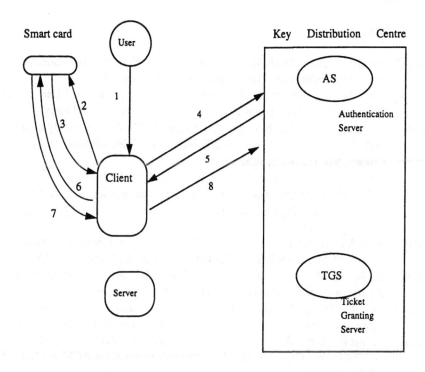

Figure 3 Kerberos with Zero-Knowledge Proof-based Authentication.
The protocol steps below refer to Figure 3 above.

Step 1   Cuid, PIN

Step 2   PIN

Step 3 $J, T=r^v \pmod{n}$, r is random $[1<r<(n-1)]$

Step 4   same message as three

Step 5   d (a random number $[0<v<(v-1)]$

Step 6   d

Step 7 $D=r.B^d \pmod{n}$

Step 8   same message as 7. The AS now checks that $T' = D^v.J^d \pmod{n} = T$

5 Conclusions

This paper has outlined six possible methods to integrate Smart Cards into the authentication stage of Kerberos. The methods have been presented in an anticipated order of implementation complexity which is also approximately the order of increasing assurance of security.

Recognised standards have been provided for wherever possible. It is the intention of the DSTC CRC to implement some of these options, in an effort to further understand the performance implications of each of these options.

6 References

[BELL] Bellovin S., and Merrit M., "Limitations of the Kerberos Authentication System", Proceedings of Winter USENIX Conference, 1991.

[GASK] Gaskell, G.I., Warner M, "Improved Security for Smart Card Use in DCE", Open Software Foundation Request for Comments 71.0, February 1995

[GONG] Gong L., Lomas M., Needham R., Saltzer J., "Protecting poorly chosen secrets from guessing attacks", IEEE Journal on Selected Areas in Communications, Vol 11, No. 5, June 1993.

[GUIL] Guillou L.C., Quisquater J.-J., "A Practical Zero-Knowledge Protocol Fitted to Security Microprocessor Minising Both Transmission and Memory", Advances in Cryptology-EUROCRYPT '88 Proceedings, Spring Verlag, Berlin 1988, pp123-128.

[KOHL] Kohl J.T., Neuman B.C., "RFC 1510: The Kerberos Network Authentication Service (V5)", IETF Request for Comments 1510., September 1993

[KRAJ] Krajewski M., Chipchak J.,, "Applicability of Smart Cards to Network User Authentication", Computing Systems, Vol. 7., No. 1, Winter 1994.

[PATO] Pato J., "Using Pre-Authentication to Avoid Password Guessing Attacks", Open Software Foundation Request for Comments 26.0, December 1992.

[RIVE] Rivest R, Shamir A., Adleman L., "A Method for Obtaining Digital Signature and Public-key Cryptosystems", Communicatiosn of the ACM, c.21, n.2, Feb 1978 pp120-126.

[WEIN] Weiner M, "Efficient DES Key Search", Rump session Crypto '93, University of California, Santa Barbara, Ausgust 20, 1993.

Smart-Card with Interferometric Quantum Cryptography Device *

Jaroslav Hruby

Group of Cryptology
of the Union of Czech Mathematicians and Physicists
P.O.B.21/OST, 170 34 PRAHA 7, Czech Republic

Abstract

We present an application of quantum cryptography with faint-pulse interferometry for smart-cards.

1 Introduction

Recently an interesting application of quantum cryptography (QC) for smart-cards using quantum transmission with polarized photons was presented by C. Crpeau and L. Salvail [1] at the EUROCRYPT'95 conference. Here we present this application with interferometric QC device because implementation of the necessary technology with polarized photons on a smart-card is less realistic.

The necessity to look for more secure smart-cards follows as the consequence of the fault case presented in the New York Times headline [2]:

"ONE LESS THING TO BELIEVE IN: FRAUD AT FAKE CASH MACHINE".

Generally there are two basic problems with existing identification systems using smart-cards:

1. the customer must type the PIN (Personal Identification Number) to an unknown teller machine which can be modified to memorize the PIN;

2. the customer must give the unknown teller machine the smart-card with information needed for the identification; in a dishonest teller machine it can be also memorized together with the PIN.

In this way such an identification system can fail.

Here we present a new identification system which in principle can be based on QC with the photon-pair interferometry [3] as well as faint-pulse interferometry [4] and which solves these two basic problems via the following way:

1. PIN will be typed directly to the card for activation of the chip and phase modulator, and no PIN information will be exchanged with the teller machine.

---

*This work was supported by the GA CR under grant 202/95/0002

2. The information needed for the identification of the card inside the teller machine will be protected against eavesdropper copy via QC methods (for more details of QC see articles in ref. [5]).

It is known that in the ordinary teller machine without quantum channels all carriers of information are after all physical objects. The information for the identification of the card is enclosed by modification of one or few their physical properties. The laws of the classical theory allow the dishonest teller machine to measure and copy the cryptographic key information precisely. This is not the case when the nature of the channels is such that the quantum theory is needed for the description. Any totally passive or active eavesdropper (E) can be viewed in a quantum channel.

The present development of optical fibres and related technologies present the possibility for the construction of the QC device for our application. The distance for quantum transmission is very short here. The problems appearing in application of QC in optical communications are negligible.

2 The quantum cryptographic smart-card

A scheme of the QC smart-card with interferometric quantum entanglement is shown in Fig. 1.

The continuous line ($-$) presents the quantum channel (optical fibre) and the break line ($---$) is the public channel. The double line ($=$) presents the electric connection from the source and the triple line ($\equiv$) the signal pulse connection between modules which is 'secure' together with modules against external measurements of electromagnetic fields from signal pulses.

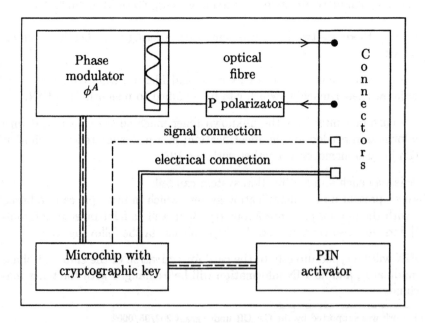

Figure 1. Schematic representation of the QC smart-card

This card consists of:

1. a PIN activator, which can have the form of the ordinary card light-source calculator with sensor keys; the user puts the PIN on the card to identify itself in a secure distance from the teller machine (i.e. unknown verification quantum cryptographic device) and a secure outer area; the card will be blocked when the sequence of three incorrect PIN's will be given; when the activated card is not used in short time period ($\sim 20s$), the activation is closed;

2. a microchip with the implemented cryptographic key $\{0,1\}^n$, which is long enough for the given QC protocol; the microchip is activated when correct PIN is put on the card;

3. a phase modulator (PM), which transforms the cryptographic key $\{0,1\}^n$ to the values of the optical phase under the following encoding rules:

   (a) "0" is randomly encoded by phase shift 0° or 90° ;

   (b) "1" is randomly encoded by phase shift 180° or 270° .

   Practically it can be done with the lithium niobate PM [6] or via another of high-speed integrated phase modulators (switching speeds ~ 100 MHz) which is thin enough for the card implementation; there is also the possibility of using two PM — one for the first code basis 0° or 180°, denoted I, and the other for the second code basis 90° or 270°, denoted II, and switching between them to represent randomly encode "0" or "1" from the cryptographic key, coming as the signal from the microchip. PM with optical fibre needs polarized light and the same polarization must be preserved in the whole QC apparatus. The PM presented here works as polarizator. It gives the same polarization as all P polarizators in the apparatus. PM plus polarizator can be constructed in one integrated element;

4. a part of quantum channel which is constructed from the short-range optical monovide fibre with high interference visibility [3]; the quantum channel from an optical link conserving polarization is better for PM with optical fibre. There are optical connectors for connecting the quantum channel to the teller machine and electric connectors for the energy which is necessary to microchip and phase modulator function. There is also a part of public channel to obtain information from the QC teller machine and its connector. In the optical connectors the polarization can be lost. It is the main problem for the polarimetric scheme of QC smart-card. Here in the interferometric scheme there is no electro-optic polarization switching as in [1] and only one polarization is conserved for all QC protocol.

Present technologies give the possibility to construct the card without connections via reflection in phase modulator and electromagnetic radiation energy for the smart-card function. In this form the QC smart-card includes only phase modulator and a short part of the optical quantum channel from the whole QC interferometric device. All other parts of the QC interferometric device are in the teller machine.

In this way the QC smart-card controls the phase shift in one arm of the QC interferometric device which consists of two extended Mach-Zender (MZ) interferometers [7]. The QC smart-card protects itself against a fake QC interferometric device that sends in bright pulses using part of the secret key from the microchip for random attenuating.

The attenuated ordinary pulses will be undetectable by the QC interferometric device, but bright pulses will be detected. A fake QC teller machine via measurements of bright pulses produces mistakes in the testing secret on bits which determines the QC smart-card by attenuating.

The MZ interferometer consists of two beam-splitters, short and long optical arm, and an adjustable phase shift ϕ in the long arm.

As is usual we shall call the two parties which want to correspond and to identify together via 'conjugate coding', as A (it will be *Alice* as the QC smart-card) and as B (it will be *Bob* as the QC teller machine).

3 The quantum cryptographic teller machine

The interferometric QC teller machine can be based on two principles: on the principle of the photon-pair interferometry [3] and on the principle of the faint-pulse interferometry [4]. From the practical point of view the second one is better because only two photon-counting detectors are needed (in the first case we need twice more). So we concentrate our attention only on the QC teller machine based on the faint-pulse interferometry which has a scheme as shown in Fig. 2.

The faint-pulse sharing QC teller machine B consists of the following:

1. two identical unbalanced MZ interferometers which are connected in series with optical fibre where the QC smart-card A is part of the first one. In Fig. 2 beam-splitters are denoted as $/$ or $\backslash$ in MZ interferometers. Dispersion effects in the out-of-balance interferometers can be avoided by reduction of bandwidth and operation close to the dispersion minimum - 1300 nm - in conventional fibres. Of course all optical paths inside B can be done with ordinary optical channel with P polarizators preserving constant polarization. Using optical fibres in quantum channels in B, P polarizator is substituted by a polarization controller.

2. pulsed light source; a pulsed semi-conductor laser source with the attenuator can be used (the source is attenuated to contain in average less than 10% of photons per pulse in transmission fibre). There is the monovide optical fibre with the same physical properties as in A. The P polarizators conserve constant polarization in QC device. There are optical and electrical connectors for connecting A with B (here the connection without conditions on the preserving the polarization is technically easier than in the case of polarized version of the QC card). There is the phase modulator switching ϕ^B which works in the long arm of the second interferometer on the same principle as in A; it means A controls ϕ^A and B controls ϕ^B.

3. two photon-counting detectors with nanoseconds efficiency. Better detector performance is achieved via using the gate width of the pulse much less than the pulse repetition rate;

4. a server computer for memorizing measurements and providing QC protocol. All cryptographic keys of honest users A are preserved and are transformed on the private sequences ϕ^A via the same QC protocol as tested A. The verification of the same private sequence $\phi^A = \phi^B$ is done. When $\phi^A \in \phi^A$, B accepts.

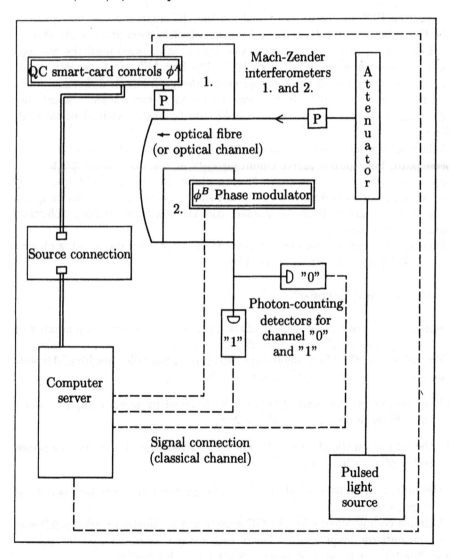

Figure 2. Scheme of the QC teller machine based on the faint-pulse key sharing system

If dishonest A knows almost nothing about ϕ^B and is trying to impersonate the honest A, it will be detected with probability $p > 1 - 2^{-\alpha n}$ where $\alpha > 0$ and n is sufficiently large security parameter which is given by dimension of the cryptographic key.

4 Simple quantum transmission between QC smart-card and teller machine

After putting PIN and identifying itself on the QC smart-card A, user puts A in the QC teller machine B, and the following identification process will start:

A light source emitting short pulses of light is attenuated until the average photon number per pulse is less than 10%. The first MZ interferometer works by separating the pulse in time and produces the path length difference much longer than the pulse width. In this way are prepared two pulses — short and long — which propagate to the second MZ interferometer with identical path difference.

When the leading pulse takes the longer path and the trailing pulse takes the shorter path, both pulses arrive simultaneously at the final beam-splitter and interference between them is seen. Events where the pulses take the long (or short) path in both interferometers will not show interference but can be gated out as they will arrive later (or earlier) compared with the timing reference originated in the source.

Having gated out the non-interference effects the probability of seeing photon in "1" or "0" output detectors is given by [8]

$$P_{1/0} = \frac{1}{4} m d l \left[1 \pm \cos(\phi^A - \phi^B) \right],$$

where ϕ^A, ϕ^B are phases on A, B phase modulators, m – the mean number of photons per pulse, d – the detection efficiency, and l – the lumped transmission.

The output on the photon-counting detectors in the teller machine, depending on the choosing ϕ^A and ϕ^B, will be the following:

0) the output on the channel "0" will be received by server computer when the difference phase $(\phi^A - \phi^B) = 2n\pi$;

1) the output on the channel "1" will be received when the difference phase $(\phi^A - \phi^B) = (2n + 1)\pi$.

Absolute security, as is usual in QC, can be guaranteed by random switching between phase bases.

A simple coding scheme for the QC smart-card application works as follows:

A transmits the cryptographic key in such a way that randomly selects $\phi^A = 0, \pi$ coding bit $b =$"0" and $\phi^A = \pi/2, 3\pi/2$ coding bit $b =$"1" .

B chooses how to measure the incoming photon either by reading it in the basis $I \equiv (0°, 180°)$ or in the basis $II \equiv (90°, 270°)$.

Quantum Oblivious Mutual Identification protocol between A and B is the same as in the ref. [1] because there the polarization bases $\{\text{+}, \times\}$ corresponds to our phase bases $\{I, II\}$.

From the identification protocol it follows that QC smart-card and QC teller machine who have the secret strings ϕ^A and ϕ^B respectively want to test whether $\phi^A = \phi^B$ without revealing their values. For this purpose A and B are willing to use the quantum and public channels. The transmission of information $c \in_R \{0, 1\}$ phased in the basis $\phi \in_R \{I, II\}$ hides all information about ϕ to anybody who has almost no a priori information on the values of c and ϕ and the principle of protocol follows from [1]:

Let A and B share $\phi = \phi^A = \phi^B \in_R \{I, II\}^n$.

In order for A to prove to B that A knows ϕ, A could transmit a string $c \in \{0, 1\}^n$ phased in the basis ϕ. Therefore, it is enough for A to choose a random codeword c from a code C_n which she sends to B. B then measures it in ϕ to obtain the decoded string c'. If the quantum channel is noiseless, then $c = c'$ and B accepts. In the realistic case there is a parameter ε as error rate of the quantum transmission. Suppose A a B share $\phi = \phi^A = \phi^B$. This implies that for any $\varepsilon_0 > 0$ and except with vanishingly small probability, B will decode c'_{real} as c which is at Hamming distance less than $\Delta(c'_{real}, c') \leq (\varepsilon + \varepsilon_0)n$ given n sufficiently large.

If there is a large number of codewords in C_n, it could be the case that measuring c in E bases ϕ^* hides almost all information about ϕ and B does not accept. From the other side, if A does not know ϕ, it could be very unlikely that A succeeds in sending c' close to codeword c as long as codewords are not to close to each other.

By using the modification in ref. [1] the complete Quantum Oblivious Mutual Identification A and B was proved. It can be implemented to our QC smart-card application. There is no need for photons traveling from B to A. All may be implemented with the photons going in a single direction from A to B.

5 Conclusions

In this paper we show the possibility of implementing the necessary technology on a QC smart-card by using faint-pulse key sharing system and using the protocol for the Quantum Oblivious Mutual Identification. An eavesdropper will be detected in our system because his active or passive measurements produce mistakes.

Manufacturers should be encouraged to develop cheap photon-counting detectors for optical time-domain and thin high-speed integrated phase modulators. We show a positive solution on one of open problems presented in ref. [1].

Acknowledgments

The author thanks J. Janecko for his interest and helpful discussions.

References

[1] C. Crpeau and L. Salvail, "Quantum Oblivious Mutual Identification", Advances in Cryptology - EUROCRYPT'95, Lecture Notes in Computer Science 921, Springer, 1995, pp. 133–146.

[2] One Less Thing to Believe In: Fraud at Fake Cash Machine, New York Times 13 May 1993, pp. A1 & B9.

[3] J. G. Rarity and P. R. Tapster, "Fourth-order interference effects at large distance", Phys. Rev. A, **45** (1992) 2054;
J. G. Rarity, J. Burnett, P. R. Tapster and R. Paschotta, "High-Visibility Two-Photon Interference in a Single-Mode Fibre Interferometer", Europhys. Lett. **22** (1993) 95.

[4] P. D. Townsend, J. G. Rarity and P. R. Tapster, "Single Photon Interference in 10 km Long Optical Fibre Interferometer", Electron. Lett. **29** (1993), 634;
P. D. Townsend, J. G. Rarity and P. R. Tapster, "Enhanced Single Photon Fringe Visibility in a 10 km-long Prototype Quantum Cryptography Channel", Electron. Lett. **29** (1993), 1291;

[5] Journal of Modern Optics, Special Issue: Quantum Communication, vol. 41, n. 12, December 1994.

[6] Leaflet of UT Photonic (UNIPHASE), Inc., 1289 Blue Hills Avenue Bloomfield, CT 06002, About $LiNbO_3$ 1320 nm APE^{TM} Phase Modulator

[7] C. H. Bennett, "Quantum Cryptography Using Any Two Nonorthogonal States", Phys. Rev. Lett. **68** (1992) 3121

[8] J. G. Rarity, P. C. M. Owens and P. M. Tapster, Quantum random-number generation and key sharing, pp. 2435–2444 in ref. [5].

Cryptographic APIs

Dieter Gollmann

Information Security Group
Royal Holloway
University of London

Abstract. The demand for security services requiring cryptography is now sufficiently widespread to make appropriate security APIs increasingly valuable tools in the design of secure systems. We will examine some of the APIs currently under development, like GSS-API, and argue that the designers of such APIs have to be explicit about the key management issues relevant to their interfaces.

1 Introduction

Layered design is a well established method for developing complex systems. Systems are viewed at different levels of abstraction, progressing from a general description of the requirements they should meet to succesively refined specifications and implementations. Interfaces between the layers hide details below from the layer above to facilitate, for example, the implementation of an application on several platforms or the exploration of different design alternatives at the lower levels. In software engineering, application program interfaces (APIs) allow an application programmer to call routines which are provided at a lower systems level.

This approach is, of course, also attractive for the design of systems that provide security services and a number of such APIs have now been proposed, most notably the Internet GSS-API [1]. Security APIs which refer to a module that mainly provides cryptographic functions are called cryptographic APIs. Again, the designers of security APIs subscribe to a layered approach. As a typical example, the layered model in Fig.1 shows the classification of security services envisaged at one time by X/Open [6]. It should be noted that X/Open's more recent preliminary specification of a GCS-API [7] does not refer to this model when explaing the goals of the API. Adhering to security APIs, and to the underlying design methodolgy, should help to generate applications which are interoperable or which are portable between different platforms. However, there are some additional issues which are quite specific to security APIs, viz

- exportability, and
- allowing application writers to use security services without having to be security experts themselves.

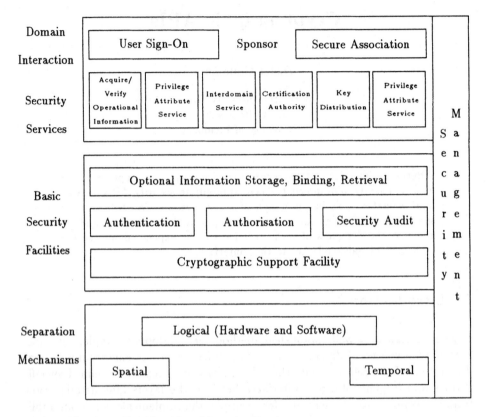

Fig. 1. X/Open – Classification of Security Services

The first issue is political. In many countries, and most importantly in the US, export (and import) restrictions on cryptographic products apply. An application program which does not directly call cryptographic algorithms potentially could avoid such problems.

The second issue is of intrinsic importance for the design of security applications. As mentioned above, the API hides implementational details from the application programmer. On one hand, this refers to the implementation of the cryptographic algorithms. In some security APIs the programmer does not even have to specify the algorithm, which is then chosen by the service invoked. *Algorithm independence* may be a desirable feature in itself, for example with respect to exportability, and the degree of cryptographic expertise required (*cryptographic awareness* [7]) is a reasonable yardstick for comparing security APIs. This aspect is well covered in existing security APIs.

However, the effective provision of security serices does not only depend on the proper implementation of the relevant security mechanisms, but also on proper key management (security management). Key management is usually not part of processing a service request, which leads to situations, like in Fig.1, where security management becomes a separate issue, orthogonal to the layers of the model. To achieve the desired separation between application writers and security experts, we also have to decompose key management and be explicit about the parts of this task that are within the responsibility of the application writer.

2 Security APIs and Key Management

We have stressed that the definition of any security API should explain its relation to key management. Key management has a number of facets, and one should at least consider the following four areas,

- key generation,
- key storage,
- key transport,
- key usage.

The last item refers to keys which should, for example, only used for generating digital signatures or only used for verifying signatures. Such applications are familiar in the financial sector but APIs emanating from traditional operating systems security tend to neglect this issue.

There are, of course, two locations of a security interface which make its relation to key management particularly simple.

1. Key management is left entirely to the layers below the interface. The upper layer does not perform any key management operations.
2. Key management is left entirely to the layers above the interface. The lower layer assumes that it is given appropriate keys for the task it is asked to perform.

Let us examine the advantages and disadvantages of this first two basic alternatives. At the top layer of a security model, we usually find applications which

- have to meet certain security requirements, but
- are not expected to contribute to the provision of security services.

It is preferable to assume that application programmers are not security experts and, therefore, should be relied on as little as possible to participate in security management. Therefore, it is reasonable to define an interface of the first kind for this top layer. The layers below such a top level interface provide much more than just an implementation of cryptographic algorithms. The GSS-API [1] comes close to this definition. However, there the application programmer is still required to perform elementary security management operations like establishing a security context and ensuring that calls are issued in the appropriate security context. GSS-API is discussed in more detail in Section 4.

A very simple *cryptographic module*, which only executes cryptographic operations for given keys and data, could be accessed through an interface of the second kind. The interfaces surveyed in Section 6 could serve as typical examples. Here, the cryptographic module has to rely on the upper layers to prevent the misuse of keys and to ensure the authenticity of security relevant data.

There is obviously a considerable distance between these two basic interfaces and well-designed intermediate interfaces would be very useful. GCS-API tries to fill this gap.

3 Security Boundaries

Clear security boundaries are essential in the design of secure systems. Security APIs help to structure those systems and, moreover, should allow to separate security responsibilities between the writers of application programs and security services respectively. We will use again a simple cryptographic module (CM) as our example to demonstrate the potential security boundaries a cryptographic API could constitute. Four possibilities are of particular interest,

- confidentiality of keys,
- integrity and confidentiality of specific data fields,
- integrity of messages,
- access to the CM.

Confidentiality of keys is of paramount importance for the security of cryptographic schemes. It is therefore reasonable to assume that keys will never pass in clear through the cryptographic interface. If some master key or initialisation key has to be entered in clear, then we assume that this is done through a separate interface. Keys generated by the CM should not depend in any predictable way on values, which are available outside the CM. Any data that is used in key generation and could compromise the key, like a (pseudo) random seed, has to be generated within the CM. If keys are generated by some other module, appropriate care has to be taken so that only encrypted keys leave that module.

In most cryptographic applications we do not want to process a single amorphous block of data but a structured message containing fields like addresses, time stamps, or nonces. There exist cryptographic interfaces which include calls reading a secure clock [3]. If the CM should protect time stamps and nonces, then they have to remain within the CM. If they are passed to the layer above the cryptographic interface, then we have to rely on this higher layer to protect their integrity. In such a situation, time stamps and nonces could as well be generated by a module other than the CM without loss of security. Of course, it may still be convenient to combine supporting services, like a clock, with the CM to reduce the number of modules within the system.

Next, consider how a larger message may be processed. It is obviously up to the layer which assembles a message to do so correctly. If this message cannot be passed directly as an argument to the CM, then it may help to be able to keep some state information in the CM to control its processing. Further issues are intermediate results which should not pass out of the CM, and composite calls whose sequence of execution should not be interrupted, like loading a key before it is used in some cryptographic operation. A cryptographic interface which is able to deal with state information will enhance security. However, it may be difficult to find a generic interface which deals with all technical variations of implementing state information.

The overall security system has to ensure that keys are only used by authorized users/applications for authorized purposes. Some aspects, like identification and authentication, are more appropriately located at layers above the CM, as they may require the execution of protocols, access to databases, etc. These operations are not at the same level of granularity as typical CM functions like 'sign' or 'encrypt'. However, identification and authentication are necessary to control access to the CM itself. The NIST cryptographic API [5] contains service calls for user authorization, but the CM does not use this authorization information internally and leaves the actual access control to the upper layers. Furthermore, the user database referred to by the authorization mechanisms may be located outside the CM. In such a situation, identification, authentication, and access control could be performed by a module other than the CM without loss of security.

There remains one aspect of access control which should be enforced by the CM. If usage of keys should be restricted to specific services, then it should be possible for the CM to enforce such 'access rules'. Typical examples are IBM's control vectors or tagged keys. The cryptographic interface is only marginally affected by key control. When setting a key, a call also has to define the type of that key. Secondly, return signals may also have to indicate type violations. The method of control can be specific to the CM and need not be standardized.

In the following sections, we will discuss existing security APIs and, in particular, examine their relation to key management.

4 GSS-API

To facilitate source-level portability of Internet security services, a Generic Security Services API (GSS-API) has been proposed [1]. The basic goals of the GSS-API design are mechanism and protocol independence. Typical Internet applications would be secure electronic mail or secure remote login services. Typical GSS-API callers are thus communication protocols themselves. The basic security elements of GSS-API are credentials, tokens, security contexts, and status codes.

Credentials contain the security relevant data required by peer entities to establish security contexts with each other. There is no standardized credential structure. It is left to the lower layers to define their content but the same structure may be used by different mechanisms. Credentials have to be handled properly by the underlying system to maintain the security of the application. Credentials are not passed as arguments of GSS-API calls but by reference to credential handles.

Tokens are the data elements passed between GSS-API callers. A recommended structure for tokens is given in Section 4 of [1].

Security contexts capture the information related to the management of the security services. Security contexts are established between peers using credentials. Status flags are set to indicate which features are desired, e.g. a *mutual_req_flag* to indicate that mutual authentication is requested. Setting up security contexts is also supported by status codes, e.g. a GSS_CONTINUE_NEEDED status which indicates that initializing a security context has not been completed. Major status codes are mechanism independent while minor status codes may be specific to the underlying security mechanism. This state information gives the GSS-API caller an opportunity to perform limited causality checks with respect to security relevant calls. It is the task of the caller to make sure that calls are issued in the correct security context.

GSS distinguishes between calls for initializing a security context and peer entity authentication on one side, and calls for data origin authentication, data integrity, or data confidentiality on the other side. Overall, there are

- four credential management calls,
- nine security context calls,
- seventeen support calls,
- four message calls: GetMIC, VerifyMIC, Wrap, and Unwrap.

GSS-API is a high level interface with respect to the service calls. With respect to security management, GSS-API is at a lower, albeit still rather high level. It relies on the lower layers to define and protect most of key management data. However, some security management tasks have to be performed by the application programmer and it is left very much to the caller to get these management issues right.

GSS-API is an algorithm independent interface requiring no cryptographic expertise but some programming discipline when setting up a security context.

Due to its origins, it is a reasonable interface for 'stand-alone' security services but not a standardized basis guaranteeing interoperability of different security services. Among the security APIs we discuss, GSS-API works at the highest level of abstraction, but there is a danger that it is 'brought down' in a rather haphazard way by introducing a growing number of security management features. There has been a considerable increase in the number of management related calls between Versions 1 and 2 of GSS-API.

5 GCS-API

X/Open's Generic Cryptographic Service API (GCS-API) [7] is an algorithm and application independent cryptographic interface to the Cryptographic Support Facility (CSF) of Fig.1. It supports the production of integrity check values, encipherment, hash functions, and random number generation. It also addresses the generation and derivation of keys, export and import of keys, and storage and retrieval of keys. Authentication of CSF callers is regarded to be potentially within the scope of the API but is not yet covered.

GCS-API achieves algorithm independence by putting algorithm and key specific information into a data structure called 'cryptographic context' which can be referenced by a context handle. There exists calls for retrieving and releasing cryptographic contexts. Cryptographic service calls then refer to such a context to specify the algorithm and key to be used in their execution.

The interface contains calls which do not handle cryptographic keys in the clear (cryptographic selecting calls) and those which do (cryptographic enforcing calls). These two types of calls constitute very different security boundaries and only trusted users should be allowed to use cryptographic enforcing calls. Protection of cryptographic keys is now shared between the CSF and the mechanisms at the application level which control access to the API calls.

GCS-API puts considerable emphasis on key retrieval and uses an elaborate key life cycle model with a corresponding hierarchy of key encrypting keys. Key management calls are located at a somewhat higher level of abstraction than the message calls in the API. For reasons of export control, GCS-API also contains calls for installing and deinstaaling algorithms in the CSF.

GCS-API is still being developed. It is located at a level below GSS-API, while still being algorithm independent. Its rôle as a secury boundary deserves further clarification and such a discussion may gain from viewing GCS-API as the collection of a number of different interfaces.

6 Cryptographic APIs

In this Section we collect a number of cryptographic APIs, which are all at a lower level than GCS-API. All these interfaces require the application writers to have some security expertise and contribute to security management.

6.1 IBM Cryptographic Application Programming Interface

The Cryptographic Application Programming Interface [2] is part of IBM's Common Cryptographic Architecture. It is strongly geared towards financial applications and it is the low level inteface to a common cryptographic module for a variety of IBM systems. It provides (cryptographic) data operation, key management, and PIN management services. This interface does not offer public key cryptography and IBM has proposed a new interface which extends the cryptographic services in this direction.

Data keys outside the CM are encrypted under key encrypting keys. However, for reasons of backward compatibility, the interface includes *encode* and *decode* service calls, performing ECB encryption and decryption with keys supplied in the clear. Control vectors allow for cryptographic key separation. For certain checksum computations, internal chaining vectors are maintained. In addition, there exist PIN management calls to support ATM applications. The IBM interface provides service calls supporting key managment, and internal controls in the CM are part of overall key management. Some management calls require specific authorization, e.g. through a PIN or a mechanical key.

6.2 MOSAIC Cryptologic Interface

The MOSAIC Cryptologic Interface is a C library for the TESSERA Crypto Card [3]. Names have changed and this project is now known as Fortezza. The interface refers to a PCMCIA card containing a suite of cryptographic algorithms. The algorithms include symmetric encryption, and public key cryptography for digital signatures but not encryption.

Key management reflects the particular application of the interface. Each card belongs to an individual user. The user's private keys are loaded by a *site security officer*, before switching the card into user mode. The card can control access to itself through a PIN mechanism. The provisions to specify key usage are limited and under the control of the user.

The interface supports key management through calls for key generation, random number generation, etc. The calls for generating time stamps or random numbers write their result into a buffer. This buffer has to be protected to prevent these values from being manipulated by entities outside the card.

When assessing the general applicability of the MOSAIC Cryptologic Interface, one should always keep in mind the specific assumptions about the intended application of the corresponding cryptographic module. It permits individual users to access a distributed system from any node in a secure manner by performing security relevant operations on their own PCMCIA card.

6.3 NIST Standard for Cryptographic Service Calls

NIST has produced a draft standard for Cryptographic Service Calls [5]. These calls are issued to a cryptographic module (CM) as described in [4], which implements cryptographic mechanisms and also generates keys. For public key cryptography, it is assumed that each CM belongs to an individual user and stores this user's public and private key. Public keys of other users are contained in certificates which may be cached by the CM.

A *VerifyUser* call supports the authentication of callers. However, it is then left to the application to invoke services in a consistent manner. The CM assumes that a particular user is logged in and service calls do not check that they may be executed. There seems to be no way for the CM to detect when it is misinformed by the application.

The *GenRandNum* call outputs the random number generated. As explained in Section 3, we then rely on the layers above the CM to use these random numbers properly. The NIST interface also allows the CM to maintain state information through the *SetCount* and *ReadCount* calls, which are included to meet the requirements of some key management standards. Limited controls on key usage can be based on the distinction between key encrypting keys, data encrypting keys, and keys for computing Data Authentication Codes.

The API contains a number of calls related to security management. Secret (symmetric) keys are stored in a database and the CM has calls for retrieving keys from the database and loading keys into the database by reference to a *user_id* and a *key_id*. The CM thus supports some key managment tasks but it is again left mainly to the application to implement good security management.

7 Conclusion

We have emphasized that the evaluation of any security API has to include an analysis of its relation to key management. In this respect, there exist two generic security interfaces,

- an interface *on top* of key management, and
- an interface *at the bottom* of key management.

Comparing different APIs is made more difficult by the fact that calls like 'sign' will appear in APIs at all levels of abstraction. Superficially, all these calls mean the same, i.e. sign some message. However, in some circumstances 'sign' means 'get the approriate key and sign' whereas other interfaces require the caller to specify the key directly. In the first case, responsibility for using the correct signature key rests with the service routine, in the second case with the application.

If the key management proper is left to layers above the CM, then it is not too difficult to define a cryptographic interface. The interfaces in Section 6 can serve as good examples. The main difference between these interfaces lies in the scope of cryptographic algorithms and of the security management functions they support. It may well be that these choices are more contentious than any service call to a cryptographic function.

It may be held against such low level interfaces that they require application writers to be security experts themselves and that they leave the CM open to misuse. There are good reasons to demand that a CM should be protected against misuse and that these protection mechanisms should be placed in the CM itself. The feasability of such a cryptographic interface has yet to be examined.

There is also a slight semantic problem with low level interfaces. Although it may be technically correct to refer to entitites at the layer above the interface as 'applications', they are very specific security applications rather than applications as perceived by the user. Some sources predict considerable economic benefits if a standard cryptographic interface is adopted due to the wide range of applications that will make use of this interface. However, it is doubtful whether there really is a large number of competent application writers for low level interfaces or, as a matter of fact, whether there really is a demand for a large variety of 'intermediate' security applications.

Application writers at a high level interface need not be experts on the cryptographic algorithms they invoke. They should also be relieved of key management duties. An arrangement like in the Zergo Security Architecture may illustrate how this can be achieved. This architecture has been developed for an environment where there exists a recognized high level rôle for a security manager. The separation between applications and key managment is resolved vertically by defining key management as another application. There are three interfaces, all at the same level,

- a Management API, providing functions for the operation, monitoring and control of the cryptographic facility,
- an Applications API, presenting a standard interface for calling Security Services like Encrypt, Decrypt, Digital Signature, Verify Digital Signature,
- a Support API providing an interface for cryptographic support functions like Key Management or Access Control.

Applications establish keys (and algorithms) either through *Connect* and *Release* calls or by specifying their source and destination, and the security manager has set up a database with the appropriate keys before the application starts.

In our previous discussions we have noted that some security APIs treat message calls and key management calls at different levels of abstraction. This further indicates that it could be advisable to separate these two aspects into different APIs in general.

Finally, note that most of the APIs covered in this paper are still under development. Comments made here may no longer apply to more recent versions than those referenced here.

References

1. J. Linn, *Generic Security Service Application Program Interface*, Version 2, Internet RFC 1508, 2 June 1995
2. IBM, *Cryptographic Application Programming Interface*, 2nd Edition, October 1990
3. The MOSAIC Program Office *Cryptologic Interface Programmers Guide*, Revision P1.2, 1 February 1994
4. NIST, *Security Requirements for Cryptographic Modules*, FIPS PUB 140-1, January 1994
5. NIST, *Cryptographic Service Calls*, Draft FIPS PUB XXX, 23 May 1994
6. X/Open, *Distributed Security Framework*, X/Open Snapshot, Draft 3, 18 August 1994
7. X/Open, *Generic Cryptographic Service API (GCS-API)*, Draft, 14 February 1995

Foiling Active Network Impersonation Attacks Made In Collusion With An Insider

Selwyn Russell[1][2] *

[1] CRC for Distributed Systems Technology
[2] Information Security Research Centre
Queensland University of Technology
2 George Street, Brisbane 4000, Australia
email: selwyn@fit.qut.edu.au

Abstract. This paper examines the problems of transmission in a network with links subject to attack by a wire tapper who seeks to create bogus messages to deceive a recipient by impersonation of another. Many cryptographic solutions have been proposed which assume the attacker is acting in isolation without knowledge of secret keys. These solutions fail if the attacker obtains private key information. Most computer crimes involve a person inside the enterprise, so these proposals do not cover the most perilous scenarios. If the wire tapper is in collusion with an insider at the impersonated enterprise, and thereby learns the private key, conventional defences will fail. In this paper, a paradigm is described whereby, even if the wire tapper is in collusion with one insider at each of the innocent network parties and each insider is able to insert, delete, modify messages, and divulge secret keys, any attempted deception will be detected.

1 Introduction

This paper considers the problem of preventing impersonation attacks by a network wire tapper who is in collusion with an insider.

Statistics indicate that an employee of an organization is the most likely to attempt computer crime. There have been many cases of an insider originating an unauthorized message and sending to another. An offender is more difficult to apprehend when the attack originates from somewhere in the communications network, as can be sadly confirmed by administrators of university computers connected to the Internet.

A Chief Information Officer with the even most stringent of security measures implemented throughout his area of responsibility faces a security problem source he cannot control – the communications network. This is a greater problem for those who use international communications.

* The work reported in this paper has been funded in part by the Cooperative Research Centres Program through the Department of the Prime Minister and Cabinet of Australia

We are concerned with electronic commerce, where messages are transferred across the network without a real time interactive dialogue. A self contained message is sent, and the receiver acts upon it at some later stage. Meanwhile, the sender has become engaged in other tasks. This is the electronic mail model, in contrast with the TELNET style of two way real time communications.

A hazard of using electronic network communications is an attack by an active "wire tapper" [8]. Such an attacker may modify, insert, or delete, messages inside a network. By pretending to be someone else, the attacker uses messages to deceive a victim into doing something which will benefit the attacker. Cryptographic defences against impersonation attacks where the attacker does not know secret information belonging to the impersonated party have been devised and are well known [6] [7]. These defences involve encryption of some item using key information known to the alleged sender. They essentially only verify that the actual communicator knows a key believed to be known by the alleged communicator. The logic is: X knows a key K. Y has demonstrated knowledge of key K. Therefore Y is X. The weakness with these defences is that they assume that

- X is a unique individual, not an enterprise of many persons, some of whom may be hostile to the enterprise;
- no attacker knows that secret key;
- no attacker is able to make use of the secret key, perhaps without knowing its value, as in a sealed security module.

These authentication methods cannot prove that the key is not being used by an impersonator who is using it illegally, and cannot distinguish between use by the owner and use by a thief. Their usefulness to the receiver depends upon the effectiveness of the sender's key security management, which is usually beyond the sphere of influence of the receiver. They provide no defence against an attacker who knows, or is able to use, the secret key.

We seek a solution so that if a wire tapper had access to the private key, and therefore is able to defeat authentication of origin tests, we can still prevent a successful impersonation attack. In this paper, we will develop a solution, starting with the addition of an explicit acknowledgement, and adding further features to provide the required security.

A remark about terminology: in this paper, we have retained the A – B message path, but the reader should note that we are not limiting the discussions to single individuals, the Alice – Bob scenario, but the A and B stand for enterprises comprising many persons.

1.1 Definition of the Problem

A and B are enterprises connected by an insecure communications network which is outside the control of either enterprise. The network may have an active wire tapper some times at work inside it.

Each enterprise is using an asymmetric cryptosystem to produce digital signatures of electronic documents and to encrypt session keys for communications.

Messages are signed and enveloped, as specified for example, in RSA PKCS Number 1 [5], "Public-Key Cryptography Standards, Number 1, RSA Encryption Standard"[3].

Each enterprise has a Data Processing Centre (DPC) which is isolated from the insecure network by a communications group. The communications group has the following functions:

- Accept outgoing messages from the DPC, process them for communications and send them into the network.
- Digitally sign outgoing messages using the enterprise's private key (and add certificates as needed) so that a receiver may conduct conventional authentication of origin checks on the outgoing messages.
- Log all outgoing messages.
- Receive incoming mesages from the network.
- Log all incoming messages.
- Perform authentication of origin tests on signed incoming messages.
- Using the enterprise private key, unenvelope incoming messages which have been signed and enveloped by the sender.
- Pass to the DPC those incoming messages which pass authentication of origin tests.

(Note that an active wire tapper in collusion with an insider in the communications group can also sign (and envelope) outgoing messages and read any incoming message.)

An enterprise does not know the security arrangements in force inside another enterprise, and is unable to determine the security of the private key.

Each enterprise consists of many individuals, some of whom are authorized to use the private key for digital signatures or for other encryption purposes. An enterprise has no way of knowing details of these individuals in another enterprise, such as how trustworthy they are, or if any are assisting an attacker.

When enterprise B receives a communication with a digital signature identical to one which would be generated with the private key of enterprise A, and the communication content claims it has been sent from A, the problem faced by enterprise B is "Has this communication really been sent by enterprise A?". As shown earlier, it could have been sent by the collective efforts of an insider and an active wire tapper.

A firewall would not protect the enterprise in this case because we are assuming the wire tapper has access to all relevant internal secrets via the insider.

---

[3] Succinctly: "signing" is the encryption of the digest of the message with the sender's private key; "enveloping" is the encryption of the contents of the message with a random symmetric key (DES in this case) which is then encrypted with the receiver's public key.

2 General Requirements of a Solution

We require an overall system involving A and B for verification of transmission of communications between A and B such that B is able to verify if a communication has been sent from A, even if there is an insider in the communications group of A who is in collusion with an active wire tapper in the network. We will consider the worst case, where the insider is able to add, delete, and modify messages, whether incoming or outgoing, is able to use the private key, is able to pass information to the active wire tapper, and that the active wire tapper is able to use the private key.

We assume enterprises are similar, so that if enterprise A can have an insider, then enterprise B could also have an insider. We will consider the worst case, where these insiders are both in league with the active wire tapper, and the solution must foil such attacks.

For B, the gist of the acceptance criterion will be "If A confirms that the transmission from A has occurred, accept the message as being from A."

3 Basic Strategy of a Solution

There are a small number of components to the basic strategy.

1. Verification Feedback request.

   A wire tapper and insider in collusion can defeat conventional authentication of origin protection methods. The problem is that these methods attempt to provide all of the authentication information with the message, and such information can be created by the wire tapper and the insider. Since it is not possible to have anything in the incoming message which will convince B that it could not have been created as an act of fraud, we must resort to other means. We propose "verification feedback", whereby feedback is sought and obtained from A as to the genuineness of the transmission.

   Our strategy for a solution is derived on the fact that a wire tapper's message did not originate from the enterprise which is being impersonated. This fact cannot be changed by the impersonators. Even though B has doubts regarding an incoming transmission, A will know exactly all authorized transmissions which have been sent. The basic strategy is for B to enquire of A regarding the received message, and for A to inform B. These actions must take place in an environment where there may be an insider at A and at B involved in network communications. This simple strategy is complicated considerably by the fact that the active wire tapper can insert, delete, and modify any message in the network, and by the potential actions within the enterprise by the insider.

2. Secure Log of Outgoing Messages.

   Another essential component of the solution is a log of outgoing messages which is accurate and complete, and beyond the interference of an insider. This log is consulted to determine if the transmission of a particular message is contained in it. To prevent tampering by an insider in the communications group, the log needs special treatment.

We require that a record of an outgoing message will appear in the log if and only if it was sent from that location. The log needs to be secure from an insider at A who might be able to use A's private key and add a false entry to the log as part of an impersonation attack on another enterprise, i.e. anyone at A who might be able to use the private key must not have write access to the log.

A problem of access to the log arises. Enterprise A probably does not want the log to be available for read access by all who may at any time receive an authorized message from A. In some cases, this would involve opening up access to millions of other enterprises world wide, some of whom will be competitors. We seek a solution whereby log information is retained as accessible by only members of the sending enterprise.

3. Restructure of the Communications Group.

The insider at A must be involved with outgoing communications for an impersonation to have a good chance of success. To prevent the insider at A from intercepting the feedback verification request from B before it reaches the log manager (and replying with a fraudulent affirmative reply), we need a new architecture of the communications group.

4. All enterprises need similar communications groups.

If we take the view that all enterprises should conform to the same basic architecture, then enterprise B should also have a similar structure to that at enterprise B, or one which is functionally equivalent.

5. Extended acceptance criterion.

The network and internal structure have been changed and we need an extended acceptance criteria, "Accept the message from the alleged originator as being from that enterprise if and only if verification feedback requests are affirmative."

4 Basic Architecture of a Solution

The components of the architecture are based on the above strategy.

1. Verification Feedback request.
2. Secure Log of Outgoing Messages.

To prevent tampering by an insider in the communications group, the log in our first architecture is a separate machine with a separate administration, as shown in figure 1.

We require that a record of an outgoing message will appear in the log if and only if it was sent from that location. The log needs to be secure from an insider at A who might be able to use A's private key and add a false entry to the log as part of an impersonation attack on another enterprise, i.e. anyone at A who might be able to use the private key must not have write access to the log. This is done fairly easily by separating the communication

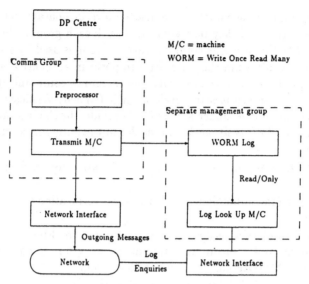

Fig. 1. Use of a log for verification of transmissions

system from the log system, i.e. those with access to the communication system have no write access to the log system. This structure is depicted in figure 1.

Messages are received by the communications group from the enterprise Data Processing Centre and are processed by and sent through the transmission/communication computer system which automatically writes to the log system via a secure channel. In this architecture, the final stage of processing by the communications group comprises a physically secure, simple no-login, non-programmable, dedicated single task system which accepts ready-to-transmit messages and outputs them to a Network Interface Controller or modem, whilst simultaneously writing a log record via another port. The log record travels via a secure cable to the log machine, the main task of which is to maintain a write once read many data base, e.g. on a WORM drive. The log machine is a relatively simple, non-programmable system running dedicated tasks. The only write to the data base is via the secure port from the dedicated transmission machine. Communication with the log machine is via a secure channel, e.g. by suitable encryption. Part of the log record could be an encapsulation of the essential information regarding the transmission enveloped with the public key of the intended recipient. This small block of core information (e.g. message identity, message digest, time, date) could be supplied to the intended recipient upon request.

3. Restructure of the Communications Group.

To prevent the insider at A from intercepting incoming communications, we organize the communications group into two independent groups, one for incoming messages, and one for outgoing messages.

To achieve complete separation, we also require a separate network and a separate private/public key pair associated with each group. The public key associated with the outgoing communications group is used by a receiver of a message supposedly from this enterprise to perform authentication of origin tests. The public key associated with the incoming communications group is used by other enterprises to envelope their (signed) verification feedback requests. Verification feedback services are provided by the incoming communications group. In this situation, an insider in the outgoing communications group of A

- cannot physically intercept incoming verification feedback requests from B;
- cannot read incoming signed and enveloped verification feedback requests from B;
- cannot formulate a verification feedback reply to B because the private key unique to A's incoming communications group is not accessible by anyone in A's outgoing communications group.

4. All enterprises need disjoint communications groups.

If we take the view that all enterprises should conform to the above architecture, then enterprise B should also have two disjoint communications groups. If we entertain the case of an insider at B, then we must have two groups at B or the above operation will fail because the insider at B will delete any negative acknowledgement from A and will substitute a positive acknowledgement, thus deceiving B into accepting and processing the fraudulent message.

5. Extended acceptance criterion.

Because of the possibility of a colluder in either of the communications groups of B who would indicate a positive reply had been received from A, the acceptance criterion for a message to be passed to the main data processing group must be extended to "Accept the message from the incoming communications group only if both of the local communications groups state that a positive verification feedback reply has been received from A."

6. Two verification feedback requests, one from each communications group.

There is the request from the incoming group at B, as before. This request must be passed to the outgoing communications group at B for transmission to A. The outgoing group at B also sends a verification feedback request for the same received message. This may be as a result of receiving the request from the incoming group for transmission to A, or it may result from the addition of details of the received message being added to the incoming log.

7. Multiple time out alarms, to guard against deletion of requests and replies.

Each communications group needs a local log with a time out, as guard against requests being deleted and never returned. We are assuming there may be one insider in the enterprise, so one of the alarms could be suppressed. However the other will still activate. Even if both were suppressed, the extended acceptance criterion means that the DPC will not process a received message until both communications groups signify its authentication of origin. The default is thus to not accept anything.

5   Implementation Issues

5.1   Sending a Message

Messages sent will be either an initial normal transmission of a cycle or a reply to a request for verification feedback.

When sending an original message, the procedure is as with conventional non-feedback proposals. Although a network entity has two network addresses, one for outgoing messages and one for incoming, as far as the transmission equipment is concerned, there is only one enterprise destination address for sending messages. Each address is associated with a separate public key. For incoming messages, as with conventional proposals, only one address and one public key are ever used, and they are for conventional authentication of originating network address (although this fails in the collusion scenario). For outgoing messages, as with conventional proposals, only one network address and one public key are used for an enterprise. Unlike conventional proposals, the address and key for incoming are different from those for incoming.

5.2   Receiving a Message

Incoming messages will be either an initial message or a request for verification of an earlier normal transmission.

The differences from conventional procedures occur when a message has been received and verification feedback is required.

– If a normal message is received, the incoming group prepares a verification request and passes to the outgoing group. If a negative reply is returned or if no reply is received within a time out period, the received message is rejected. (As suggested earlier, a verification request may also be sent by the receiver's outgoing group.)
– If a request for verification is received, the puzzle is solved and the log consulted. The results of both are signed and enveloped into a reply and passed to the outgoing group for transmission.

6   Operation of an Implementation

This section describes the operation a sample implementation based on the strategic principles given above, to be read in conjunction with figure 2.

1. A message is sent to B's incoming communications group and is allegedly from A.
2. The message is received by B's incoming communications group.

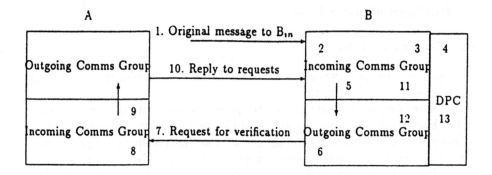

Fig. 2. Summary of verification feedback steps

3. The message is passed from B's incoming communications group to the main Data Processing Centre.

4. The message is temporarily stored at the main Data Processing Centre until verification is completed.

5. B's incoming communications group creates an enquiry for A to verify the transmission. The query is written into a local write once read many log in an encrypted form decryptable only by B's incoming communications group, and a watchdog timer is started. (This log is not the secure one which is written automatically when a message is transmitted but it may be a similar type of system.) If the timer expires before a reply from A is received, further action is required. The query is passed to B's outgoing communications group for transmission.

6. B's outgoing communications group creates an enquiry for A to verify the transmission. The query is written into a local write once read many log in an encrypted form decryptable only by B's outgoing communications group, and a watchdog timer is started. (As with B's incoming communications group, this may be a system similar to that used for automatically logging outgoing messages). If the timer expires before a reply from A is received, further action is required.

7. The two enquiries are sent to A's incoming communications group.

8. A's incoming communications group receives the two enquiries, decrypts them, accesses the log, verifies the message transmission, and composes a signed and enveloped reply to B's incoming communications group and composes a signed and enveloped reply to B's outgoing communications group.

9. The replies are passed to A's outgoing communications group for transmission to B's incoming communications group.

10. A's outgoing communications group sends the replies to B's incoming communications group.

11. The incoming communications group at B receives and unwraps its reply, and retrieves from its local log the details of the request. Action is taken depending on whether the reply is positive or negative.

12. The incoming communications group passes the reply for B's outgoing communications group to the outgoing communications group, which unwraps its reply, and retrieves from the details from its local log. Action is taken depending on whether the reply is positive or negative.

13. Finally, the main Data Processing Centre is notified whether or not to abandon the received transmission.

The time flow is shown in figure 3. In this diagram, A_{in} refers to the incoming communications group at A, A_{out} refers to the outgoing communications group at A, B_{in} refers to the incoming communications group at B, B_{out} refers to the outgoing communications group at B.

6.1 Resistance to Attacks

Suppose a message is received by B and is allegedly from A. We will assume there is an insider in A's outgoing group who is in collusion with an active wire tapper. Together they can create a message as if it came from A. Further we will assume that there is only one insider in an organization. Therefore A's incoming group has no insiders, but there may be one in B's incoming group.

To enquire of A, B's incoming communications group must pass the query to their local outgoing communications group to send, but an insider in B's outgoing communications group could delete the query. As a defence, the query is written into a local write once read many log in an encrypted form decryptable only by B's incoming communications group, and a watchdog timer is started. If the timer expires before a reply is received, further action is taken and the attackers fail.

Alternatively, perhaps the insider is in B's incoming communications group, who would later tell the main Data Processing Centre that a positive confirmation was received from A. To guard against this, B's outgoing group must also report a positive reply. An easy way to do this is for B's outgoing communications group to append to B's incoming communications group's query its own query. To guard against deletion of the message, a local log entry containing the query in a form decryptable only by B's outgoing communications group is made. To accept confirmation, B's outgoing communications group will expect to receive a positive reply signed and enveloped from A's incoming group.

In either case, the query sent from B's outgoing communications group to A's incoming communications group thus has a request for confirmation of the designated transmission from B's incoming communications group, and a request from B's outgoing communications group. For convenience, we have included these together, but an implementor may prefer to have a query from the incoming communications group and a separate query from the outgoing communications group.

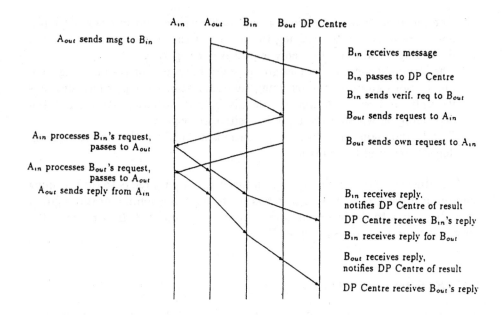

A$_{out}$ sends msg to B$_{in}$

A$_{in}$ A$_{out}$ B$_{in}$ B$_{out}$ DP Centre

B$_{in}$ receives message

B$_{in}$ passes to DP Centre

B$_{in}$ sends verif. req to B$_{out}$

B$_{out}$ sends request to A$_{in}$

A$_{in}$ processes B$_{in}$'s request, passes to A$_{out}$

B$_{out}$ sends own request to A$_{in}$

A$_{in}$ processes B$_{out}$'s request, passes to A$_{out}$

A$_{out}$ sends reply from A$_{in}$

B$_{in}$ receives reply. notifies DP Centre of result

DP Centre receives B$_{in}$'s reply

B$_{in}$ receives reply for B$_{out}$

B$_{out}$ receives reply, notifies DP Centre of result

DP Centre receives B$_{out}$'s reply

Fig. 3. Time flow of verification feedback steps

If the wire tapper deletes the queries, the time outs will alarm and the message will not be accepted by the DPC. If the wire tapper is in collusion with an insider at B, one of the queries could be modified or falsified, but not both, because ther is no more than one insider at B. If all is well with the transmission, A's incoming communications group receives the queries, decrypts them, accesses the log, verifies the message transmission, and composes a signed and enveloped reply to B's incoming communications group and composes a signed and enveloped reply to B's outgoing group. The replies are then passed to A's outgoing communications group for transmission to B's incoming communications group.

If the reply by A is negative, i.e. the message is an impersonation and there could be an insider in A's outgoing communications group, then the insider in A's outgoing communications group will know the reply must be negative, and will attempt to delete the reply. If this is done successfully, B's incoming communications group's timer will eventually expire and sound an alert. Similarly with the timer set by B's outgoing communications group. If there is an insider in either of these communications groups, its alarm will be suppressed, but the other will trigger security action. The insider and the wire tapper are unable to create a fraudulent reply because there is no access to the private key of the incoming communications group at A.

Each communications group at B receives and unwraps its reply, and retrieves from its local log the details of the request it sent. The received reply is compared with the requested information.

Finally, the main Data Processing Centre is notified whether or not to abandon the received transmission. The default probably should be to abandon the message after a pre-specified time has elapsed if a positive confirmation has not been received from both the local incoming communications group and the local outgoing communications group.

If there is an insider in A's outgoing communications group in collusion with an active wire tapper, an impersonation of A may be attempted on B. However, with the paradigm above, even with another insider in B's incoming communications group or in B's communications outgoing group, the deception will fail.

7 Summary

We have presented a method which overcomes the limitations of conventional authentication when an insider is in collusion with an active wire tapper.

References

1. Caelli W., Longley D., & Shain M., *Information Security Handbook*, MacMillan Publishers Ltd., England, 1991.
2. International Standards Organization, "ISO/DIS 7498-2 – Information Processing Systems – Open Systems Interconnection Reference Model – Part 2: Security Architecture", ISO, Geneva, July 1987.
3. Rivest R.L., Shamir A., & Adelman A., "A Method for Obtaining Digital Signatures and Public Key Cryptosystems", *Communications ACM*, Vol 21 Feb 1978, pp 120-126.
4. RSA Data Security, Inc., " Public-Key Cryptography Standards" RSA Data Security, Inc., Redwood City, CA, 3 June 1991.
5. RSA Data Security, Inc., " Public-Key Cryptography Standards, Number 1, RSA Encryption Standard" RSA Data Security, Inc., Redwood City, CA, 3 June 1991.
6. Simmons, G., "An impersonation-proof identity verification scheme", *Advances in Cryptography - Proceedings of Crypto 87*, Springer-Verlag, 1988.
7. Woo T.Y.C., Lam S.S., "Authentication for Distributed Systems", *IEEE Computer*, January 1992.
8. Wyner A.D., "The Wire Tap Channel", *Bell System Technical Journal*, Vol 54 No 8, Oct 1975.

The CASS Shell

G. Mohay, H. Morarji, Q. Le-Viet, L.Munday
School of Computing Science
email: g.mohay@qut.edu.au
and W.Caelli
School of Data Communications

Queensland University of Technology
Brisbane, Australia

Abstract The goal of the Computer Architecture for Secure Systems (CASS) project [1] is to develop an architecture and tools to ensure the security and integrity of software in distributed systems. CASS makes use of various cryptographic techniques at the operating system kernel level to authenticate software integrity. The *CASS shell*, the work described in this paper, is on the other hand a secure shell implemented on top of UNIX[1] System V Release 4.2 (UNIX SVR4.2) to achieve the same purpose but in an operating system independent manner. The CASS shell carries out cryptographic authentication of executable files based on the MD5 Message-Digest algorithm [2] and presents a closed computing environment in which system utilities are safeguarded against unauthorised alteration and users are prevented from executing unsafe commands. In order to provide cryptographic authentication and other cryptographic functions such as public-key based signatures, *in hardware*, the work has also involved the incorporation of an encryption hardware sub-system into SVR4.2 operating on an Intel 80x86 hardware platform. The paper describes the structure and features of the CASS shell and the development and performance of both the hardware and software implementations of the cryptographic functions it uses.

1. Introduction

1.1 Goals

The goal of the Computer Architecture for Secure Systems (CASS) project is to develop an architecture and tools to ensure the security and integrity of software executed by a system [1].   To achieve this goal, software is executed within a CASS environment only if it passes an authentication check to confirm its integrity.   This check uses a signed message-digest of the program text produced at the time the program was developed for the application environment. The CASS project, of which

---

[1] UNIX is a registered trademark of UNIX System Laboratories Inc.

this work on the CASS shell is but a part, includes a modified Modula-2 compiler and linker which carry out static checking to determine whether the program meets specified security criteria, such as avoiding certain 'unsafe' language features, not importing low level library modules and meeting other possibly application specific constraints. If a program meets all these various security criteria, a certificate consisting of a signed message digest of the program based on the MD5 Message-Digest algorithm [2] is generated. This previous work on CASS located the authentication check of the signed message digest in the operating system kernel and consequently required modifications to the (UNIX SVR4.2) platform kernel. The focus of the present work in contrast has been to implement the CASS authentication check without modification of the kernel and thereby to provide a portable version of CASS. In consequence, we have developed the CASS shell, a restricted environment which implements MD5 based authentication of executable files and which also controls users' activities to safeguard application software from unauthorised alteration and to protect the system from unsafe commands. This paper describes the structure and features of the CASS shell: a restricted UNIX shell [3] with an integrated mechanism to validate the integrity of executable files. It also describes the incorporation into SVR4.2 of cryptographic software used by CASS and the CASS shell. This software was implemented initially as a pseudo-device driver which implements the cryptographic functions entirely in software and then as a hardware device driver which accesses the hardware cryptographic features of the Eracom PC-ASM encryption board [4]. The latter implements DES (Data Encryption Standard [5a]), ANSI X9.9 MAC [5b] and RSA (Rivest-Shamir-Adleman [6]) functions thus providing *hardware based* cryptographic authentication and public-key signature support for CASS. The development and performance of this driver software is discussed.

1.2 The Restricted Environment

Restricted environments have been applied in various ways from reducing the accessing scope of users and eliminating the number of commands that can be executed, to encapsulating the working environment. The main purpose of restricted environments is to enhance the security of the system. In a UNIX environment, this may be achieved by three different methods with various levels of flexibility and security [7].

The first method is to replace the login shell with a special program. Normally, special programs only provide a limited number of facilities with different protocols from an ordinary shell; they usually serve specific purposes. Most restricted environments using special programs use a different set of commands or commands with different syntax to eliminate unauthorised users. For these reasons, special programs have limited scope of application.

The second method is to replace the login shell with a restricted shell which accepts only a subset of commands. Commands and features that are considered insecure are refused and users cannot create and organise their own file systems with the result that the working environment is reduced to the home directory only. For application systems, these restrictions are unreasonable. On the other hand, these restricted shells have loopholes that may be used to break the closed environments by allowing users to invoke a different non-restricted shell or to use commands such as *LS* to investigate the structures of the supposedly inaccessible global file system. The restrictions imposed by such a restricted shell may not apply to shell scripts, so users can use shell scripts to escape from the restricted environment.

The third method uses the *chroot* (change root) UNIX command to restrict the accessing scope of a user. The *chroot* command changes the users' view of the root of the system. It creates a virtual root and applies that notion on all users' processes to hide details of the file system. This approach provides the strongest protection to the system [7] and requires special privilege from a super-user to set-up a restricted environment. However, this is impractical for a system with a large number of users because it lacks flexibility and is difficult to manage.

The above methods are either inflexible or not very secure, making them unsuitable for the CASS project. The approach adopted in CASS is therefore somewhat different to all of these but is based on the second method above. The rationale for and nature of the CASS shell are described in the following sections: section 2 describes the shell design and components and section 3 describes the implementation of the shell. Section 4 then describes the development and performance of the cryptographic drivers incorporated into our SVR4.2 platform. Section 5 concludes by outlining further developments intended for the CASS Shell.

1.3 MD5 Message-Digest Algorithm

One way to check a file for unauthorised tampering is to derive a message-digest using a known formula and then to compare that value against the expected value. The MD5 Message-Digest algorithm is currently used as the basis for generating 'message digests' of executable files in the CASS project. The MD5 algorithm was developed by R. Rivest and has been published as Internet RFC #1321 with source code for an implementation [2]. It is an algorithm which takes as input a message (an executable file) of arbitrary length and produces a 128-bit digest. The RSA algorithm is then used as the basis for the following signature mechanism. The MD5 signature of a file in CASS is generated by using a private key to sign its MD5 digest. The integrity checking mechanism consists of recalculating the MD5 digest and comparing it with the (public key) decrypted MD5 signature. All executable and shell script files are submitted to a certifier and have MD5 signatures appended before they are made available. This mechanism protects the system against users producing their own MD5 signatures for files that contain unsafe features. An advantage in using MD5 Message-Digest is that its code is compact and very fast on 32-bit machines. Moreover, the MD4 algorithm, an earlier version of the MD5 algorithm, is accepted as strong enough for most applications [8]. So the MD5 algorithm, which is theoretically proven more secure than the MD4 [8], has been adopted for use in CASS to produce encrypted signatures for executable files. The actual algorithm used in CASS is based on a different implementation, developed by C. Plum for different use, in the public domain [9].

2. Structure of the CASS Shell

2.1 Design

The CASS shell is designed and implemented so that it:
- provides a restricted UNIX environment, in which the execution of certain commands are not allowed and the accessing of information outside the designated environment will be restricted,
- is as flexible and as transparent as possible to users in providing basic UNIX facilities, and
- has an integrated mechanism to verify the MD5 signature and therefore the integrity and authenticity of executable files.

The CASS shell provides the restricted shell safeguards described above while providing a more flexible working environment. It acts like a filter to reject unsafe commands from users and yet provides development facilities similar to those of the Bourne shell. It is not possible to break out of the restricted environment so that a secure, closed environment is enforced.

In this closed environment imposed by the CASS shell, users can create and manipulate files and subdirectories only within their home directories as all commands including *cd*, *mkdir* etc are interpreted and if necessary rejected by the shell. The *cd*, *mkdir* and *rmdir* commands can be used to traverse and maintain the file structure only within the bounds of the file store sub-tree of the closed environment. Environment variable PATH can be defined to determine the execution path within the provided environment. As with other restricted shells, the CASS shell reduces the number of commands available for users; commands that are considered unsafe will be rejected. To be flexible, the number of disallowed commands can be varied or different sets of disallowed commands may apply to different users or groups of users.

To meet the specific requirements of CASS, the new shell integrates the MD5 checking algorithm into the loading and executing procedure: all files have to pass the validation process before they can be executed otherwise they will be rejected. The scope of the checking can be applied with two options: check all executable files or check users' files only. To protect the closed environment, the CASS shell will prevent intentional or accidental accessing of data and files outside the provided working environment. It may reject the command or convert it to an acceptable form.

In summary, the new shell associates features and components of previously described methods in providing a restricted environment but with increased flexibility and provides additional integrity assurance.

2.2 Components of the CASS Shell

Working Environment

For each user, the home directory is considered the root of the system file structure. A user can manipulate files and data within this directory and its sub-directories   only.

The CASS shell rejects all retrievals of files and data stored outside this environment and hides the real organisation of the file system. The CASS shell accepts a normal *.profile* which is stored in the parent directory of the user's home directory. The profile contains similar information as do other shells for setting up the working environment such as search path, environment variables, etc. In order to prevent attempts to break the closed environment while the system is executing the *.profile* during the login process, the *.profile* traps and ignores interrupt signals from the keyboard. The *.profile* and other files used likewise to initialize the environment are stored in the parent directory and only authorised users can access or modify them. For flexibility, the CASS shell allows users to have their own profiles in their home directories. The shell will process these profiles in a similar way to a normal shell.

Command Interpreter

The command interpreter of the CASS shell, as for other shells, performs its tasks through a sequence of steps. The next step is executed only if the previous one terminates normally. Firstly, it accepts the user's command and does the preliminary syntax checking. The next step is to validate any I/O redirection and piping between commands. For I/O redirection, standard files are closed and user-defined files are opened to be used in I/O redirection or piping. Then, commands and arguments are marshalled into the required format. The CASS shell performs this task somewhat differently. It converts all input paths to a format that hides the actual file structure (ie. /users/.../user1/f1 to /f1) and that format is used to communicate between users and the CASS shell. The last step is to submit valid commands to the command execution mechanism. If an error occurs during the execution of any step, an appropriate error message will be displayed and the execution is terminated. The shell will continue to input new commands.

Command Execution

The CASS shell carries out command execution in different ways according to the following classification of command types or categories:

- disallowed commands which contain unsafe features; these commands are predefined and will be rejected if they are invoked
- system commands including built-in and external commands; these commands will be executed with or without MD5 checking depending on a predefined switch
- user's commands which are divided into two sub-categories:

  - shell scripts which are interpreted by the shell and scanned for unsafe commands and constructs, and
  - executable files, which will be executed only if validated by the MD5 checking.

The classification of commands is represented in internal lists within the shell.

These lists are stored in external files in the parent directory and can be modified by authorised users only. During the login process, the CASS shell loads these lists of command into memory and they are fixed for a login session. The execution of a command will follow the rules for that category of command.

3. Implementation

3.1 System Profile

As a user logs in, the normal shell is invoked first and will execute the system profile to perform environment settings. The first command in the system profile is to trap interrupt signals to prevent a possible exit from the restricted environment during the initialisation. The last two commands in the system profile are: invoking the CASS shell and *exit*. The order of these commands is to prevent a break from the CASS shell returning to an unrestricted shell. When a user exits the CASS shell, the actual shell will also automatically terminate. All variables and environment settings in this profile will be stored and used by the CASS shell. Users can access and modify settings and variables defined in their home directories only.

3.2 Shell Execution Order

After execution of the system profile in the parent directory, the normal shell initiates the CASS shell. The first step is to look for and execute the user's *.profile*, if it exists. Then, the CASS shell creates and loads the lists of system commands or restricted commands from external files in the parent directory. The CASS shell then completes the initialisation and allows the command interpreter to accept user's commands. The command interpreter and CASS shell terminate on receiving the *exit* command and this will also terminate the login session.

3.3 Command Interpreter

The command interpreter is implemented as a function which accepts single characters at a time and groups them into commands and arguments. These commands, then, are checked to find syntactical errors such as conflicts between I/O redirection symbols, missing file names for I/O redirection or piping, etc. This implementation follows the rule of the Bourne shell for output redirection files: if the output file name already exists, it will override the old content without warning (C shell will reject the command in this case). If the command is syntactically correct, the shell will open files for redirection and will check for possible errors in opening files for I/O redirection.

The next step is to check for possible accessing outside the closed environment. The command interpreter will convert input path names to path names relative to the actual file structure. Unusual but legal path names such as ../../../../abc can also be converted to an actual path name and then be checked for validity (inside the closed environment). These conversions are transparent to users and a command can be passed to the next step only if it passes all validations.

3.4 Command Execution

Classification Mechanism

A data structure is needed to build internal lists of system commands, those which are allowed and/or those which are disallowed. Static data structures are not desirable since the number of commands may change unpredictably. The selected data structure must also have a fast searching algorithm to speed up system response time. Extensible hashing is ideal but since the distribution of commands is unknown, no effective hash formula can be used. The AVL tree algorithm is another candidate with search performance of $O\ log(n)$ but it occupies more memory space and its implementation is fairly complicated. The selected data structure is the *Skip-list* with algorithms as described in [10], which also has a search performance of $O\ log(n)$ independent of the distribution of data.

The design decision was made to store the list of allowed commands only. A command which does not exist in the list is considered as invalid. This decision while conservative is for precisely that reason more secure and has the advantage that we may store the path names for valid commands in the list. With this approach, the system performance is also improved; to execute a command, the shell can load the command directly without following one path after the other in the defined PATH variable to locate the command. However, this is achieved at the cost of using more memory space.

To carry out command classifications, there is a list of built-in commands which is static and hard-coded because these commands such as *cd*, *pwd*, variable assignments, etc. will be executed by the CASS shell itself as opposed to being executed by a utility. These built-in commands are special in the sense that they could potentially be used to break the restricted environment or investigate the actual structure of the file system, were it not for their execution by the CASS shell.

In classifying a command, the first step is to search the list of built-in commands. If the command is not in the built-in list, the list of system commands will then be searched. If it is not found in the internal lists, the search continues by following the PATH variable to locate the file within the user directory structure. An error message will be displayed for the non-existence of the command and control is returned to the command interpreter. Otherwise the command is loaded to be executed.

MD5 Checking

In systems in which the restricted environment applies to all users, the MD5 checking for system commands may be omitted for performance reasons since command files are considered to be immune to unauthorised alteration. The omission of MD5 checking for system commands can be defined for each login session. To check system command files, the MD5 checking function needs special privilege to access these files as they are owned by the system. The SUID (set user ID) facility is used to set the owner of the MD5 checking function to the system so it can access system files. All users' executable files have to be checked, script files do not as they are interpreted and checked for safety by the shell.

Execution Mechanism

The execution of commands is done by the *fork ()* and *exec ()* system calls. If a command is run on the foreground the parent process will wait until the child process is terminated before returning to the command interpreter. If it is run on the background, the parent process will create another process to monitor the result of the command and return to the command interpreter.

After syntax validation and classification, commands are executed differently depending on their categories. Built-in commands are executed "in place" by the CASS shell. Systems commands or user executable files, after passing the MD5 checking function (in the case of system commands, this will only be checked if the option for doing so had been selected previously) will be executed with full path names.

3.5 Testing and Performance

The command interpreter behaves similarly to the command interpreter from the Bourne shell and is relatively transparent to users for most basic UNIX commands. Commands submitted to the CASS shell produce similar results as when submitted to the Bourne shell. Rejection of disallowed commands and acceptance of valid commands take place as defined earlier.

As was expected, performance of the CASS shell when MD5 generation and checking are switched off is comparable to other 'normal' shells, both performing similar functions. Our preliminary figures of processing performance using a simple shell script executed 100 times showed an overhead of a factor of two when MD5 generation and checking is switched on. This is arguably an acceptable level of overhead, probably not even noticeable for most users, and well worthwhile given the security benefits provided.

The MD5 Message-Digest software has been separately tested on a variety of file sizes ranging from 0 bytes to quite large files; its performance is discussed in the next section.

4. UNIX SVR4.2 Cryptographic Device Driver

4.1 Cryptographic Functions

The CASS shell and the larger CASS project of which it is a part require not only the MD5 implementation described earlier, but also some additional cryptographic functions. In this section we describe the performance and incorporation into UNIX SVR4.2 of the cryptographic driver software which provides that additional functionality. Our early implementation led to the development of a pseudo-device driver which implements these functions in software, and this was followed by implementation of a hardware device driver to utilise the Eracom PC-ASM encryption board.

The PC-ASM encryption board is a 16-bit card that can be used with any IBM-compatible PC AT and provides a variety of encryption services based on the Data Encryption Standard (DES) [5] and Rivest-Shamir-Adleman (RSA) [6] algorithms. It provides the usual modes of DES operation - Electronic Code Book (ECB), Cipher Block Chaining (CBC), and Byte Cipher Feedback (BCF) and supports ANSI X9.9 Message Authentication Code (MAC) generation [7]. It provides the usual RSA encryption services of key generation and modular exponentiation. DES keys can originate from the application using the board, or from the board itself. The keys can be single or double length. The board has a non-volatile, tamper-resistant memory which is used to store keys and data. Each board also has a unique hardware key that can be folded with key data to produce a final key unknown even to the legitimate user. The board has a secure clock, and can produce cryptographically strong random byte streams. In addition, the board is capable of background processing when not otherwise in use.

4.2 UNIX SVR4.2 Pseudo-Device Driver

A pseudo-device driver [11] for DES encryption services was initially developed as a STREAMS driver for Unix SVR4.2. This provides ECB encryption/decryption and a 16-bit MAC (Message Authentication Code) generation using the DES ECB algorithm. This driver is no longer being used as the hardware device driver is now available, however it allows for some useful comparisons to be made between the performance of the hardware and software implementations.

4.3 Porting the SCO PC-ASM Device Driver to Unix SVR4.2

SCO Software

The SCO Unix 4.0 device driver for the PC-ASM board is a character-type driver [14] and consists of a total of 43 *.c* and *.h* source files, and one *.asm* (assembly) file. Most of these files form part of two libraries (the high and low level message libraries respectively), while the remainder make up the driver entry point functions [12] and various OS-dependent functions. These are all linked together to form a driver object file, which is then linked with the kernel.

While using the libraries unmodified would have saved time initially, it was felt they would be unlikely to work correctly under SVR4.2. Another important consideration was the need to be able to optimize the device driver for SVR4.2. For these reasons it was decided to rebuild the libraries.

Porting problems

While much of the code was able to be used unmodified, differences between SCO UNIX and UNIX SVR4.2, and the respective code syntax and structure made the modifications non-trivial.

The main problems were: :
- SCO uses a 32-bit Microsoft compiler *cl* and assembler *masm*, while SVR4.2 uses the standard Unix tools *cc* and *as*

- Much of the code uses Microsoft syntax and extensions; this includes frequent use of the // comment and use of various modifiers unique to Microsoft; the *cc* compiler does not recognise these
- Two of the source files contain inline assembly in the Intel format; SVR4.2's *as* assembler reads the AT&T format, which is very different
- One of the source files is written entirely in Intel-format assembly language
- Some of the kernel function names differ
- SCO allows interrupt handlers to be set up at run-time, while SVR4.2 must set them up during a kernel rebuild
- The SCO code uses various kernel functions not present under SVR4.2
- In general the code has low portability, and the interfaces between code layers could have been better designed to facilitate portability.

Solutions to Porting Problems

Since SVR4.2 uses the *cc* compiler, it was necessary to remove all Microsoft extensions. This involved conversion of all // comments to the conventional /* ... */ form and the replacement of Microsoft-specific modifiers with standard modifiers together with some other minor syntactic changes relating to the use of modifiers.

Files consisting partly or entirely of assembly language had to be converted from Intel format since there is no Intel assembler available under SVR4.2. All SVR4.2 assembly language was implemented as inline code.

Since the names of several of the kernel functions [15] differ between SCO and SVR4.2, it was necessary to determine the purpose of these SCO functions and replace them with the appropriate SVR4.2 functions calls. A more difficult problem was the fact that the SCO source calls several kernel functions that are not present under SVR4.2 :

- memset
- memcpy
- memcmp
- add_intr_handler

The first three are generic memory manipulation functions, ie they use void pointers. Thus they have to be written in assembly. Since these functions are present in the SVR4.2 standard C library, three simple programs were written, each calling one of the memory functions. The resulting executables were then run through *debug*, and the assembly code copied. The code was then placed as inline assembly in three memory functions with the required names. Finally, the new kernel routines were tested and their results verified against the corresponding standard library functions.

The other SCO kernel function *add_intr_handler* allows interrupt handlers to be set at run-time. Because of this SCO device drivers do not need an interrupt driver entry point [13]. SVR4.2 can only set interrupt handlers during a kernel rebuild. Thus it was necessary to comment out the *add_intr_handler* function and provide an interrupt entry point that called the interrupt handler function defined in the SCO source.

Performance

The performance of the pseudo-device driver and the Eracom PC-ASM board in combination with the SCO, UNIX SVR4.2 and DOS drivers is presented below. These figures were obtained by timing 500 encryption operations on 0 bytes of data and 32 kbytes of data respectively with an 8 byte key and calculating T:

$$T \ (kb/s) = 32/(t_{32} - t_0)$$

where t_{32} and t_0 are the times for one encryption operation on 32 kbytes and 0 bytes of data respectively. The time t_0 (typically 1.4 ms) essentially measures the overhead of the encryption calls, and is subtracted to produce a throughput independent of operating system overhead. For smaller data sizes (less than 64 bytes) the overhead becomes non-trivial and must be taken into account to determine true throughput. (These figures were produced using zero wait states on the PC-ASM board; all figures relate to code written in assembly language except for the figures in the last column and the MAC figure for SVR4.2 (PC-ASM) which relate to code written in C.)

Throughput (kbytes/sec)	SVR4.2 PC-ASM	SCO 4.1 PC-ASM	DOS PC-ASM	SVR 4.2 Pseudo
CBC	1080	1020	1033	-
ECB	1083	1017	1032	3.81
BCF	269	264	265	-
MAC	1153	1346	1550	6.21

The relative performance of the software MD5 generation on the one hand and the hardware MAC operation on the other have also been measured and are presented below. What is noteworthy is that MD5 is consistently faster than the hardware MAC operation which is probably the joint result of a number of factors which we are investigating further: the relative inherent complexity of the two algorithms, possible contention on the i/o bus and the several layers of software, including both library software and the driver itself, involved in invoking the hardware MAC.

Data Size (bytes)	MAC (milliseconds)	MD5
0	1.88	0.37
10	4.32	0.49
100	4.43	0.54
1000	5.26	0.95
10000	14.49	6.22
100000	116.51	59.10

5. Conclusions

The CASS shell provides a secure environment which incorporates all the security advantages of a restricted shell but in addition also provides a degree of development flexibility lacking in typical restricted shells. In addition, it also provides authentication of all software which is executed by the restricted user. The environment, therefore, successfully safeguards the system from interference by a restricted user and simultaneously provides such a user with protection from virus or Trojan horse software.

The present environment does not safeguard users executing the normal shells nor are they in any sense restricted. Further development of the CASS shell will extend the scope of the work by investigating the effects and feasibility of enforcing the restricted shell for all users except super-user thereby addressing that very issue. While the current CASS shell provides the capability of preventing the execution of disallowed commands and uncertified executable files, it does not prevent executable files from themselves executing other programs (e.g., *exec* kernel calls). Prevention of system abuse via *exec* calls can currently only be prevented through the certification mechanism. In future versions of the software, we will develop a secure C library in which the *exec* family of library calls carries out checks identical to those in the current CASS shell before making the kernel call. The certification mechanism is at this stage achieved offline by simply attaching an appropriate signed MAC to the executable file resulting from a compile and link operation. The CASS project, of which this work on the CASS shell is but a part, includes a modified Modula-2 compiler and linker which provides the option of disallowing 'unsafe' language features such as pointer casts and other low level features imported from the SYSTEM module. We will be incorporating use of this 'trusted' compiler and linker into the CASS shell environment with a view to providing an online certification protocol which is currently lacking.

Acknowledgements

This research was partly supported by an Australian Telecommunications and Electronics Research Board grant and a Queensland University of Technology Research Encouragement Award. This support is gratefully acknowledged.

References

[1] Mohay, G., Caelli, W., Gough, K.J., Holford, J., Low, G. *CASS - Computer Architecture for Secure Systems*, ACSC 16, Australian Computer Science Conference, Feb 3-5, 1993. Griffith University, Brisbane, Australia.
[2] Rivest, R. *The MD5 Message-Digest Algorithm*, Technical Report, Internet, April 1992. RFC #1321.
[3] Curry, D.A. *UNIX System Security* Addison-Wesley, Reading, MA, 1992.
[4] ERACOM Pty. Ltd., Burleigh Heads, Queensland 4220, Australia.
[5a] Smid, M.E. and Branstad, D.K., *The Data Encryption Standard: Past and Future*, Proceedings of the IEEE, vol. 76, no. 5, May 1988, pp 550-559.
[5b] ANSI X9.9 (Revised) *American National Standard for Financial Institution Message Authentication (Wholesale)* American Bankers Institution, 1986

[6] Rivest,R.L., Shamir, A. and Adleman, L., *A Method for Obtaining Digital Signatures and Public-Key Cryptosystems*, Communications of the ACM, Vol. 21, No. 2, Feb 1978, pp 120-126.

[7] Farrow, R. *UNIX System Security* Addison-Wesley, Reading, MA, 1991

[8] Garfinkel, S. and Spafford, G. *Practical UNIX Security* O'Reilly & Associates, Inc., Sebastopol, CA, 1991

[9] Plum, C. *Truly Random Numbers* Dr. Dobb's Journal, November 1994, p.

[10] Pugh, W. *Skip-lists: A Probabalistic Alternative to Balanced Trees* Communications of the ACM, Vol 33, No. 6, June 1990, p. 668-676, 1990

[11] Downey, S, *DES Pseudo Device STREAMS Driver* Technical Report, Queensland University of Technology, November 1993.

[12] ERACOM, *Encryption Services Application Program Interface*, ERACOM Pty Ltd, Burleigh Heads, Queensland, Australia, 903-33-00 Rev B3 edition, May 1994.

[13] ERACOM, *RSA Encryption Services Application Program Interface*, ERACOM Pty Ltd, Burleigh Heads, Queensland, Australia, 909-33-00 edition, February 1994.

[14] Pajari, G, *Writing Unix Device Drivers*, Addison-Wesley, Third Ed., 1992.

[15] Unix Press, *Device Driver Interface/Driver-Kernel Interface Reference Manual (Intel Processors)*, Prentice-Hall, Englewood Cliffs, New Jersey, September 1992.

[16] Linn, J, Internet RFC 1508, Geer Zolot Associates, Sept 1993.

Author Index

Springer-Verlag
and the Environment

We at Springer-Verlag firmly believe that an international science publisher has a special obligation to the environment, and our corporate policies consistently reflect this conviction.

We also expect our business partners – paper mills, printers, packaging manufacturers, etc. – to commit themselves to using environmentally friendly materials and production processes.

The paper in this book is made from low- or no-chlorine pulp and is acid free, in conformance with international standards for paper permanency.

Lecture Notes in Computer Science

For information about Vols. 1–954

please contact your bookseller or Springer-Verlag